A-LEVEL GEOGRAPHY

COURSE COMPANION

Letts Aids

Clifford Lines MSc (Econ)
Formerly Assistant Director,
East Sussex College of Higher Education

Laurie Bolwell MA, DPhil, FColl P
Head of Department of In-service Development
Brighton Polytechnic

Charles Letts & Co Ltd
London, Edinburgh & New York

First published 1982
by Charles Letts & Co Ltd
Diary House, Borough Road, London SE1 1DW

Reprinted 1983 (twice)
Second edition 1984
Reprinted 1984
Third edition 1986
Fourth edition 1988
Revised 1991

Illustrations: Illustra Design Limited/Peter McClure/Tek-Art

British Library Cataloguing in Publication Data
Lines, C.J.
A-level geography : course companion. —
4th ed. — (Letts study aids).
1. Geography — Examinations, questions, etc.
I. Title   II. Bolwell, L.H.
910'.76   G131

ISBN 0 85097 822 X

Printed in Great Britain by BPCC Wheatons Ltd, Exeter

Acknowledgements

The Publishers wish to thank the following for permission
to use illustrations in this book.

Aerofilms: **fig. 2.2**, **fig. 9.4**; Nelson (Chandler, T.J., *Modern Meteorology and
Climatology*, 1972): **fig. 3.2**; Oxford and Cambridge Schools Examining
Board: **fig. 3.3**; *Geography* (Dury, G.H., 'Rivers in Geographical Teaching',
vol. XLVII, 1963): **fig. 5.4**, **fig. 6.2**; UTP (Hilton, K., *Process and Pattern in
Physical Geography*, 1979): **fig. 6.1**; Hodder and Stoughton (Abbott, A.J.,
Bradshaw, M.J., and Gelsthorpe, A.P., *The Earth's Changing Surface*, 1978):
fig. 6.3, **fig. 9.5**; Heinemann (Dury, G.H., *Environmental Systems*, 1981): **fig.
7.1**, **fig. 7.2**, **fig. 7.3**; CUP (Small, R.J., *The Study of Landforms*, 1978): **fig.
8.2**; Longman (Ollier, C.D., *Weathering*, 1975): **fig. 12.6**; Harper (Haggett,
P., *Geography – A Modern Synthesis*, 1972): **fig. 13.1**; Prentice Hall
(Kormondy, E.J., *Concepts of Ecology*, 1969): **fig. 13.2**; Allen and Unwin
(Simmons, I.G., *Biogeographical Processes*, 1982): **fig. 13.5**; Oliver and Boyd
(Daniel and Hopkinson, *The Geography of Settlement*, 1979): **fig. 17.1**, **fig.
17.2**; Liverpool University Press, *Town Planning Review* (Lawton, R., vol. 43,
1982): **fig. 19.1**; Gower (Johnson, B.L.C., *India: Resources and
Development*): **fig. 19.3**; Wiley (Van Valkenburg, S., and Held, C.C.,
Regional Geography of Europe, 1952): **fig. 20.4**; *Geographical Review*
(Taaffe, E.J., Morrill, R.G., and Gould, P.R., 'Transport Expansion in
Underdeveloped Countries: a Comparative Analysis', vol. 53, 1963): **fig. 22.3**;
McGraw-Hill (Brock, J.O.M., and Webb, J.H., *Geography of Mankind*):
fig. 23.3, **fig. 23.4**; OUP (Grove, A. T., *Africa*, 1978): **fig. 26.2**; Scottish
Examination Board: **fig. 27.3**, Bell (Henshall, J. D., and Momson, R. P.,
A Geography of Brazilian Development, 1974): **fig. 28.1**.

Preface to the fourth edition

The first edition of this book appeared in 1982 and was an instant success because it was designed to meet the expressed needs of A-level students. Consultations with groups of students had revealed a need for guidance on how to plan studies, which books to read, how to structure answers to questions and the techniques required to achieve an optimum performance at the examination. As a consequence this book was written as a study companion which is always available for support and advice in specific areas of knowledge, skills and techniques, as well as on the problems of planning a study programme, writing essays and preparing for the final examination.

Like other bodies of knowledge, Geography is evolving constantly as new concepts are introduced and old ones discarded. Changes have also taken place in the design of examination questions, with a greater emphasis on data and stimulus-response questions and a corresponding decrease in questions requiring an essay-type answer. Concern about the environment has brought about changes in some syllabuses with local, regional and global issues forming the basis for a number of questions. At the same time there has been a shift in emphasis away from Regional Geography – the study in depth of particular areas which form geographical regions. In this edition we have introduced a section on Regional and Environmental Issues which will give you a greater awareness of some of the problems which can result from human activity in the natural environment. It will also show you how some of these issues are examined at A level with examples from recent question papers.

In making these changes we have taken care to retain the structure and basic content which made the earlier editions of this book so popular. Whether you are studying for A level, AS level or the Scottish Higher Grade examination we are convinced that this *Companion* will give you more confidence in your ability to succeed, more enthusiasm to read widely about the subject and a deeper understanding of what the examiners expect from you in the examination.

C J Lines
L H Bolwell
July 1990

Contents

Part I

Introduction

HOW TO USE THIS BOOK

This book has been written for A, AS level and Scottish Higher Grade examination candidates and our main objective has been to provide a book which will act as a guide and companion to geography students throughout the time they are studying for these examinations.

The student who has recently been successful at GCSE can be bewildered and sometimes overwhelmed by the shift in emphasis which A or AS level demands. For the majority of students this is the first opportunity they will have to accept responsibility for their own pattern of learning and rely on their own resources to organize a study programme which is most suited to their needs. The freedom which is associated with A level studies requires a personal discipline which will enable students to work independently. Most A level candidates have a great deal of time available for personal study and there is considerable evidence that many students cannot organize this study time effectively and do not prepare adequately for the examination.

This book has been designed to give you guidance on some of the essential material you will require during the course. It also suggests sources which should be consulted and contains advice on the techniques which are required, to pass the final examination.

One book alone cannot give you all the facts and techniques you will require during the course. Your teacher will probably recommend a small group of specialist texts which will be supplemented by many other books and articles in journals and magazines such as *The Geographical Magazine* and *New Scientist*. (Television programmes, e.g. *The World About Us* and *Horizon* may also provide interesting and relevant information.) This book is not intended as a substitute for specialist text books and journals. It has been written to complement them by examining some of the most significant topics in the syllabuses and by demonstrating the techniques and skills which are essential when answering questions – both assignments set during the course or A or AS level examination questions.

The reports issued by the various Boards emphasize that many candidates have not been prepared adequately for the examination, and, through inexperience or lack of confidence, achieve poor results or fail. In this book we have drawn on our experience as examiners and the comments of the Chief Examiners to describe the techniques and skills which will help you to make the most of the opportunities which the examination situation provides.

On the pages which follow this introduction an analysis has been made in tabular form of how the topics which appear in this book relate to the syllabuses set by the Examination Boards. This information is intended to be a useful guide which may need checking since some of the syllabus details change from time to time.

Instead of attempting a superficial coverage of topics which appear in these syllabuses, we have selected thirty topics and examined them in depth. For each topic the following details have been provided where appropriate; the essential information required about the topic, the concepts which underpin the subject matter, the various theories and viewpoints which exist and the relationships of the topics to others you are likely to study.

In a further section, typical questions have been selected for analysis and answer plans suggested. Study of this section will help you to understand how a wide variety of questions should be approached and answered.

A final section lists suitable books for further reading. During the course you may wish to add to this list as new material becomes available.

If this book is used as it was designed, as a guide and companion throughout the A and AS level or Scottish Higher Grade course, we believe you will increase your ability to study intelligently without close supervision and demonstrate this self-reliance at the examination.

THE DIFFERENCES BETWEEN GCSE AND A, AS LEVEL GEOGRAPHY

When you were studying for GCSE you may have thought that A level was rather similar but more difficult. This is an easy trap to fall into since at first sight there are some similarities in the syllabuses. For example, many aspects of physical and human geography, map work and regional studies are common to both GCSE and A level syllabuses. Moreover, the central concern, that of the interaction of man with his physical environment is evident at both levels.

The similarity is, however, superficial and your A level studies will have a greater significance if you fully appreciate what is expected of you and the approach which is required at this level. Just as there are differences between the GCSE and A level syllabuses, differences also occur when the examination questions are analysed. One of the functions of this book is to consider a variety of questions, to identify what is required by the examiner and to suggest suitable structures for the answers.

What the examiner will expect

Most GCSE candidates take seven or more subjects in the examination. A level subjects are selected more carefully since most candidates find that three subjects are as many as they can cope with. Consequently, the A level examiner will expect more of you in terms of breadth and depth of knowledge as well as in your ability to apply this knowledge and related concepts to specific situations or problems.

The successful A level answer paper is one which convinces the examiner that you have a good grasp of the geographical facts, and an understanding of the processes involved and the related problems which exist. It should also demonstrate your ability to understand and use geographical models, statistical data and a comprehensive range of skills and techniques.

Quantitative differences

When GCSE and A level examinations are compared there are some obvious quantitative differences. Most examining boards set two papers at GCSE, each designed to last for 2 hours. At A level there are usually three papers, each one being timed for 2½ hours. Not only are the examinations longer but also more time is provided to answer each question. Whereas a GCSE paper may require four questions to be answered in 2 hours, A level papers may require only three questions to be answered in 2½ hours, giving 50 minutes for each question.

The new AS level examination consists of one or two papers lasting 2½ to 3 hours.

These quantitative differences are not as formidable as they may appear at first sight. The normal two year time-span between GCSE and A level gives you the opportunity to study geography in depth as well as breadth, and there should be no difficulty in writing answers in the time provided. If there is any difficulty it is likely to be in deciding what should be included and what excluded in the time available. The selection and presentation of material in a manner which will demonstrate your ability to be succinct as well as erudite is of considerable importance.

The significance of facts

Some GCSE questions test your knowledge of geographical facts and little else. You may have spent many hours of revision before the exams learning sketch maps, climatic statistics, industries associated with particular cities, crop requirements and many other parcels of facts. These facts are required to answer questions like this one.

(i)   On the outline map of the USA which is provided show the main areas of corn (maize) cultivation.
(ii) State the conditions under which corn is grown in the region.
(iii) Describe the major changes in agricultural land use in the Corn Belt.

By contrast an A level question on the same subject reads

Outline and explain the changes which have taken place in farming in the Corn Belt over the last forty years.

This question requires the candidate to make intelligent use of the knowledge at his or her disposal rather than to test a number of memorized facts. Moreover it concentrates on the concept of change over a period of time, a concept rarely considered at GCSE, where the emphasis has been on understanding skills, values and issues as well as on knowledge. You will still need a wide range of facts to answer many A level questions but the facts are not the final product. They are required to support ideas and arguments and to substantiate hypotheses.

A more analytical approach

A level demands a much more analytical approach than does GCSE. This becomes apparent when sets of questions at both levels are examined. The repeated use of certain verbs is common in both GCSE and A level questions but there is a difference between those used at each level. In GCSE the most common directions are describe, write an account of, explain, locate and describe, define, account for and what is meant by. At A level explain and account for recur but they are joined by others such as assess, discuss, justify, analyse, consider, compare, comment on, critically examine and to what extent would you agree with?

The A and AS level approach expects the candidate to have acquired sets of values; to appreciate the dynamic character of the environment in time and space, to weigh up the available evidence and make informed judgements about problems and issues, to understand the geographical processes, concepts and general principles by which spatial patterns may be explained and not to be content with straight forward descriptions.

A more mature approach

The increased sophistication required at A and AS level is evident from the wording of the questions and analysis of a large number of questions reveals certain recurring themes.

(a) Inter-relationships Some questions examine the inter-relationships which exist within the physical environment or between the physical and human aspects of the environment. The candidate is expected to link one aspect with another even though they may have been taught as separate units. For example,

How far is the transport network more closely related to the distribution of population than to relief and drainage? Give examples from one or more developed countries to illustrate your answer.

(b) Problem analysis Instead of asking for descriptions or factual statements some questions focus on problems which have geographical implications. For example:

Examine the problems of siting new airports in developed countries.

To answer this type of question it is necessary to have a detailed knowledge of specific siting problems which have arisen and then to be able to summarize them in general terms. Relevant factual evidence is essential and the examiner will give high marks to the candidate who can show an ability to categorize the factors involved.

(c) Use of models and concepts Models, which are generalizations of some significant features or relationships in geography, are frequently used in A level studies. In the examination a diagrammatic form of a model may be given which must then be explained and sometimes related to a particular situation. Alternatively a particular model is referred to, as in this question.

Describe Weber's model of industrial location. Show how far the model can be used to explain the location of **either** the iron and steel industry of one country, **or** motor car manufacturing in one country.

An understanding of concepts is also important at A level as this question illustrates.

With reference to specific examples discuss the concept of the urban hierarchy.

Concepts should be referred to in answers whenever possible, including answers to map questions. The Southern Universities' Joint Board adds this statement at the end of the compulsory map question.

N.B. Credit will be given for the incorporation in your answer of concepts, terminology and techniques acquired from your study of physical and human geography.

(d) Advanced skills and techniques Many questions are set to test your ability to use geographical skills and techniques. Sometimes the questions give you the opportunity to write about practical experiences describing skills and techniques which have been gained during field work.

A level questions reflect the trend in geography towards a more scientific and statistical approach which requires objectivity and rigour in collecting, measuring and interpreting data in physical and human geography. This question typifies the A level emphasis on quantitative methods.

Discuss the methods you would use and the problems involved in identifying and mapping **either** (a) a Central Business District, **or** (b) rural land use.

(e) Data and stimulus response There are an increasing number of questions in which information is given, usually in map or diagram form or as a statistical table and candidates are required to interpret the data and answer questions directly or indirectly based on the information provided. For example:

The table overleaf gives information on non-agricultural employment in northern New England (Maine, Vermont and New Hampshire) and southern New England (Massachusetts, Rhode Island and Connecticut) in 1965, and the changes which occurred between 1965 and 1980.

Non-agricultural employment per cent				
Northern New England			Southern New England	
	1965	1965–80	1965	1965–80
A	37.6	+19.1	36.2	+1.4
B	10.5	+37.9	9.5	+3.1
C	51.9	+94.2	54.3	+57.9
All	100.0	+59.8	100.0	+32.2

A manufacturing industry
B construction, communications, gas, electricity and water industries
C trade, government, finance and services

(Cambridge, June 1986)

(f) A systems approach A number of syllabuses including Oxford and Cambridge and WJEC have, to some extent, adopted a systems approach. The emphasis in this approach is on the inter-relationships between variables such as vegetation, land and capital. These variables interact with one another and are influenced by such features as energy flows. Changes in one variable set in train changes throughout the system. These changes may be brought about by human activities such as cutting down rain forests or damming a river. As geographers we are concerned with the processes by which changes occur in a system and in the complex inter-relationships which exist between the variables. References to systems, including explanatory diagrams, occur where appropriate throughout this book. There follows an example of an A level question which is based on a knowledge of interaction within a system:

Discuss the ways in which human activity in agriculture and forestry may affect the hydrological characteristics of a drainage basin.

(g) Emphasis on process Whereas GCSE is concerned with knowledge of physical and man-made phenomena such as corrie lakes and the location of nuclear power stations, A and AS level are more concerned with causes and the processes of development and change. This is clearly evident in many questions on landforms. At GCSE the questions are concerned mainly with identification and the basic reasons for the formation of the landforms, whereas A level questions require a detailed account of the process at work. For example:

Analyse the processes that have led to the formation of **two** of the following: loess; drumlin; corrie (cirque); inselberg; meander.

Analysis Tables

Analysis of A level examination syllabuses 1991-92

unit no		AEB	Cambridge	Cambridge	JMB	JMB	London	London	NISEC	Oxford	Oxford and Cambridge	SEB Higher Grade	WJEC	unit no
	Year	1991/2	1991	1992	1991	1991	1991/2	1991/2	1991	1991	1991		1991	
	Syllabus	626	9050	9050	B	C	210	219		9845	9630			
	Number of papers	3	3/4	2/3	2/3	2/3	3	2	3	2	6	2	3	
	Compulsory map question	●Alt	●Alt											
	Field work or course work: local study: practical	20%	28%†	25% Alt	20% Alt	20% Alt	*	35%	15%	28%	$16\frac{2}{3}\%$ Alt	25%	15%	
	Special paper (optional)	●	●		●		●	●		●	●		●	
	PRACTICAL													
1	Statistical methods	●	●Alt	●Alt	●	●	●			●	●Alt	●	●	1
2	Ordnance Survey maps	●	●Alt	●Alt	●	●	●					●	●	2
3	Weather maps	●	●Alt		●	●	●		●	●			●	3
	PHYSICAL													
4	Weathering	●	●	●	●	●	●			●	●	●	●	4
5	Slopes	●	●	●	●	●	●			●	●	●	●	5
6	Rivers and river valleys	●	●	●	●	●	●	●	●	●	●	●	●	6
7	Drainage patterns	●	●	●	●	●	●	●	●	●Alt	●	●	●Alt	7
8	Glaciation	●	●	●Alt	●	●	●			●	●	●	●Alt	8
9	Coastal landforms	●	●	●Alt	●	●	●	●	●	●Alt	●Alt	●	●Alt	9
10	Desert landforms	●	●	●Alt	●	●	●			●Alt	●Alt			10
11	Weather and climate	●	●	●	●	●	●		●	●	●	●	●	11
12	Soils	●	●	●	●	●	●	●Alt	●	●	●Alt	●	●	12
13	Plant communities: ecosystems	●	●	●	●	●	●	●	●	●	●Alt	●	●	13
	HUMAN													
14	Growth and distribution of population	●	●Alt	●	●	●	●	●Alt	●	●	●Alt	●	●Alt	14
15	Movement of population	●	●Alt	●	●	●	●	●Alt	●	●	●Alt	●	●Alt	15
16	Rural settlement patterns	●	●Alt	●	●	●	●		●	●	●	●		16
17	Central place theory	●	●Alt	●	●	●	●		●	●	●	●	●	17
18	Urban land use	●	●Alt	●	●	●	●	●	●	●	●	●	●	18
19	Cities and their problems	●	●Alt	●	●		●	●	●	●	●	●	●	19
20	Agricultural land use	●	●Alt	●Alt	●	●	●	●	●Alt	●	●Alt	●	●	20
21	Location of industry	●	●Alt	●	●	●	●	●	●Alt	●	●Alt		●	21
22	Transport and transport networks	●	●Alt		●	●	●	●Alt		●Alt	●Alt			22
23	Less-developed countries	●	●Alt	●Alt		●	●	●Alt	●	●Alt	●	●	●Alt	23
	REGIONAL AND ENVIRONMENTAL ISSUES													
24	United Kingdom: regional problems		●Alt		●	●Alt	●	●Alt	●		●	●	●Alt	24
25	France: population and regional development		●Alt		●	●Alt	●			●	●Alt	●Alt		25
26	West Africa: population issues		●Alt			●Alt	●						●Alt	●Alt	26
27	India: development issues					●Alt	●						●Alt		27
28	Brazil: regional strategies		●Alt			●Alt				●		●Alt	●Alt	28
29	Exploitation and conservation	●	●Alt	●			●		●Alt	●Alt		●	●Alt	29
30	Pollution	●	●Alt	●			●	●	●Alt	●Alt	●	●	●Alt	30

Alt–Alternative   †–Either in addition to or instead of the paper on Geographical techniques
*–Optional Geography project. Separate grade not shown on certificate, certificate endorsed 'with Geography Project'.

Analysis of AS level examination syllabuses 1991-92

	AEB	COSSEC		London		NISEC	Oxford	WJEC
Year	1991/2	1991	1991	1991/2		1991	1991	1991
Syllabus	988	8460	8461	210	219		8746	
Number of papers	1	1/2	1/2	2	1	1	2	2
Compulsory map question								
Field work or course work: local study: practical	30%	25% Alt	25% Alt	20%	36%*	30%	20%	20%

unit no		AEB	COSSEC 8460	COSSEC 8461	London 210	London 219	NISEC	Oxford	WJEC	unit no
	PRACTICAL									
1	Statistical methods	•			•		•	•		1
2	Ordnance Survey maps				•		•	•		2
3	Weather maps		•		•		•			3
	PHYSICAL									
4	Weathering		•						• Alt	4
5	Slopes		•						• Alt	5
6	Rivers and river valleys	•	•		•	•	•		• Alt	6
7	Drainage patterns	•	•		•	•	•		• Alt	7
8	Glaciation		• Alt						• Alt	8
9	Coastal landforms	• Alt	• Alt			•	•		• Alt	9
10	Desert landforms		• Alt							10
11	Weather and climate	•	•		•	•	•		• Alt	11
12	Soils	•	•			•	•		• Alt	12
13	Plant communities: ecosystems	•	•			•	•		• Alt	13
	HUMAN									
14	Growth and distribution of population	•			•	•	•	•		14
15	Movement of population	• Alt			•	•	•	•		15
16	Rural settlement patterns	•			•	•		•		16
17	Central place theory	•			•	•		•	• Alt	17
18	Urban land use	•			•	•	•	•	• Alt	18
19	Cities and their problems	• Alt			•	•	•	•	• Alt	19
20	Agricultural land use	• Alt		• Alt	•	•		•	• Alt	20
21	Location of industry	•		• Alt	•	•		•	• Alt	21
22	Transport and transport networks									22
23	Less-developed countries	•			•	•	•	•	• Alt	23
	REGIONAL AND ENVIRONMENTAL ISSUES									
24	United Kingdom: regional problems			• Alt	•	•				24
25	France: population and regional development			• Alt				• Alt		25
26	West Africa: population issues	• Alt		• Alt		•	• Alt			26
27	India: development issues	• Alt		• Alt		•		• Alt		27
28	Brazil: regional strategies			• Alt	•	•		• Alt		28
29	Exploitation and conservation	•				•	• Alt		• Alt	29
30	Pollution	•	•	•	•	•	• Alt		• Alt	30

Alt – Alternative
* – Including 16% for decision making exercise

Examination boards

AEB	The Associated Examining Board Stag Hill House, Guildford, Surrey GU2 5XJ
Cambridge	University of Cambridge Local Examinations Syndicate Syndicate Buildings, 1 Hills Road, Cambridge CB1 2EU
COSSEC *(AS only)*	As for Cambridge, or Oxford and Cambridge
JMB	Joint Matriculation Board Devas Street, Manchester M15 6EU
London	University of London Schools Examinations Board Stewart House, 32 Russell Square, London WC1B 5DN
NISEC	Northern Ireland Schools Examinations Council Beechill House, 42 Beechill Road, Belfast BT8 4RS
Oxford	Oxford Delegacy of Local Examinations Ewert Place, Summertown, Oxford OX2 7BZ
O and C	Oxford and Cambridge Schools Examinations Board 10 Trumpington Street, Cambridge and Elsfield Way, Oxford
SEB	Scottish Examinations Board Ironmills Road, Dalkeith, Midlothian EH22 1BR
WJEC	Welsh Joint Education Committee 245 Western Avenue, Cardiff CF5 2YX

Study skills and examination techniques

At least 80 per cent of your time as a student will be spent on private study so it is very important for you to acquire those skills which will enable you to study effectively. Many hours can be wasted reading books from which you learn very little, or drawing elaborate maps and diagrams which are soon forgotten.

Study will involve you in collecting information, analysing it, clarifying your thinking, assimilating knowledge and expressing yourself clearly. No one is born with these skills, nor are they obtained accidentally, they must be acquired by conscious effort and practice. Here are some suggestions which will help you to develop these skills and make the most of your study time.

(a) Establish targets Research has shown that a learning period of about forty-five minutes produces the best relationship between understanding and remembering. Set yourself study tasks which can be achieved in this period of time and then take a break for fifteen minutes or longer before attempting another period of work. Plan reasonable targets which you can achieve in each study session, e.g., to read twenty pages and make notes.

(b) Focus on essentials There are large numbers of books and articles which deal with topics in the A or AS level syllabuses. Some of this material is inappropriate or duplicates what is written better elsewhere. Try to focus on sections of books, avoid extraneous material and select what you read intelligently.

(c) Select key words and phrases When you read a section of a book, practise selecting words or phrases which will help you to remember what the section is about. These words can be written down for further reference and used as personal notes (see (d) below). Here is a section from Estall, R. C. and Buchanan, R. O., *Industrial Activity and Economic Geography*, Hutchinson, 1973, in which the key words and phrases are shown in italics.

'The *first* of these *external economies* derives from the principle mentioned above, that a *firm* may *buy in* more of its requirements. By this means individual firms can *"contract out"* of making (probably on a smaller and comparative expensive scale) certain *necessary parts*. These they purchase from *another firm* which *specializes* in the production of those parts. Such a specialized firm with its *larger scale of production,* can use *specialized machinery* and reduce its *overhead costs*. It may also gain in other ways, for example from economies in *bulk purchasing* of materials.

The firms that are buying in the parts can therefore expect to *acquire* them *more cheaply* than by making them themselves, and the *whole* of their *capital, accommodation* and *labour* will be available for the processes they perform themselves. This process of drawing upon a large number of *sub-contractors* is well exemplified by many important industries and *affects their location.*'

(d) Note taking Far too many students write notes as they write essays, in linear sentences. About 90 per cent of what is written is wasted material and will never be remembered. It is the key words, concepts and phrases which need to be remembered and with practice you can abandon linear notes and learn more effectively by recording only the key words. This skill takes some time to acquire and can best be learned in stages by first writing down long phrases but not sentences and then, after a time, reducing your notes to just the key words and phrases. This form of note-taking is suitable for notes made while reading or during a lecture. Remember to record the author and title. Sometimes the page number is also useful for future reference.

(e) Make a topic summary When you need to plan an essay or summarize a topic, the most effective method is by making a topic summary. First, print in the centre of a sheet of paper the core theme or topic title. Then draw lines from this centre, making as many lines as the number of sub-themes you can distinguish. Print the titles of the sub-themes at the end of the lines. Along each line print key words or phrases which are associated with the sub-theme. You can always add new lines as you go along and use arrows to show relationships between different sub-themes on the diagram. This topic summary, or web, as it is sometimes called, will help you when planning an essay (the sub-themes could become sections or paragraphs), when revising a particular topic or as a summary of a chapter in a book. Printing the words will produce a diagram which can be referred to with ease at a later date.

Here is an example of a topic summary which was prepared by a student before answering the following question.

For any **one** country of your choice, describe and account for the major patterns of internal migration since 1950. The student chose Brazil for his answer.

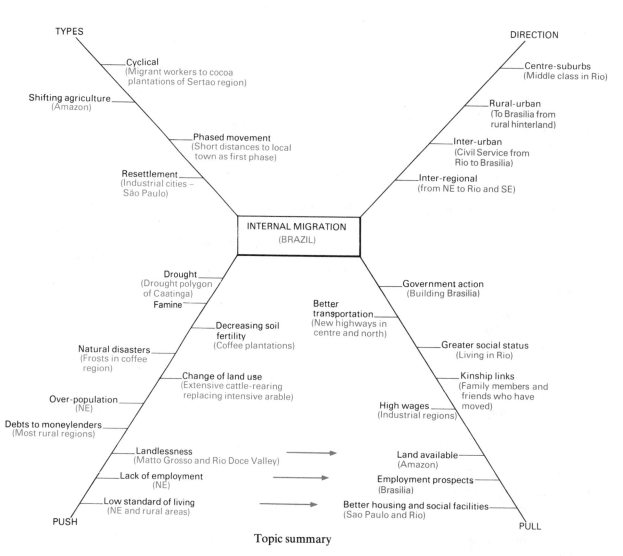

TYPES

Cyclical
(Migrant workers to cocoa plantations of Sertao region)

Shifting agriculture
(Amazon)

Phased movement
(Short distances to local town as first phase)

Resettlement
(Industrial cities – São Paulo)

DIRECTION

Centre-suburbs
(Middle class in Rio)

Rural-urban
(To Brasilia from rural hinterland)

Inter-urban
(Civil Service from Rio to Brasilia)

Inter-regional
(from NE to Rio and SE)

INTERNAL MIGRATION
(BRAZIL)

Drought
(Drought polygon of Caatinga)

Famine

Decreasing soil fertility
(Coffee plantations)

Natural disasters
(Frosts in coffee region)

Change of land use
(Extensive cattle-rearing replacing intensive arable)

Over-population
(NE)

Debts to moneylenders
(Most rural regions)

Landlessness
(Matto Grosso and Rio Doce Valley)

Lack of employment
(NE)

Low standard of living
(NE and rural areas)

PUSH

Government action
(Building Brasilia)

Better transportation
(New highways in centre and north)

Greater social status
(Living in Rio)

Kinship links
(Family members and friends who have moved)

High wages
(Industrial regions)

Land available
(Amazon)

Employment prospects
(Brasilia)

Better housing and social facilities
(Sao Paulo and Rio)

PULL

Topic summary

(f) Practising techniques Many aspects of practical geography and statistical data cannot be learned effectively by seeing examples or reading about them in books. To understand the techniques involved it is essential to practise such things as using formulae, drawing sketch maps, compiling charts and plotting graphs. You will find suitable exercises in some text books and in past examination papers. Always try to reproduce diagrams and sketch maps neatly since marks are not awarded for untidy work.

(g) Further reading The earlier in the course you can equip yourself with the study skills described above, the more effective will be your performance as a student. Further information about study skills can be found in Buzan, T., *Use Your Head,* BBC, 1975 and Rowntree, D., *Learn How to Study,* Macdonald and Janes, 1970.

THE EXAMINATION SYSTEM

There seems to be a deliberate attempt to prevent candidates from knowing how the examination system works. One Board, however, The Associated Examining Board, has produced a booklet which states in its introduction,

The administration of public examinations tends to be shrouded in myth and mystery. The purpose of this brief account is to take away the mystery by describing the organization and working methods of The Associated Examining Board.

Although each Board has its own working methods, the patterns are broadly similar and some of the main features are summarized below.

(a) Setting the examination The examination papers are set many months before the actual date of the exam by the Chief Examiner, sometimes assisted by a team of setters. The wording of the questions is checked by a Moderator who must ensure that the questions are clear, unambiguous, of the correct level of difficulty and based on material in the syllabus which the

Board has published. Marking schemes for each question are also prepared when the question paper is set and scrutinized by the Moderator who can suggest changes when necessary. Further checking is carried out by a sub-committee before the paper is printed.

(b) Marking the examination The geography papers are marked by a panel of Assistant Examiners who are normally A level geography teachers themselves. Each examiner marks several hundred papers in the three weeks which follow the examination and no examiner is allowed to mark the work of candidates he or she may have taught. Examiners only know the candidates by the names and numbers on the examination papers; they have no knowledge of the school or part of the country in which they live. Examiners rarely mark all the papers of one candidate, Paper 1 will go to one examiner, Paper 2 to another and so on.

After receiving the examination papers and marking schemes, and before the marking begins, all the examiners attend a standardization meeting with the Chief Examiner. At this meeting each question is discussed and points raised about how the marks are to be awarded in specific examples. In some cases photocopies of scripts are marked by examiners at the meeting to check that all are using the same standard before the marking programme begins.

Assistant Examiners send batches of marked scripts to the Chief Examiner for checking and further random sampling of scripts is carried out to ensure that all Assistants are marking to the correct standard. When all the marks are received statistical distributions are worked out, but only rarely is it necessary to adjust an examiner's marks at this stage.

(c) Checking and awarding grades All scripts are checked in considerable detail by office staff when they have been marked to ensure marks have been added correctly and that the correct marks are recorded on computer sheets. If a candidate has good marks on two papers and poor marks on the third, the poor marks are checked before being accepted.

A panel of Awarders assisted by the Chief Examiner studies the examination papers and marking schemes to decide the standard of answer required for each grade. The Awarders read many scripts and decide as a group the mark range for each grade of the examination. Grades are not awarded on a statistical basis, nor must there be a certain percentage of candidates in each grade. What determines each grade is the content and standard of work which varies.

After every examination some candidates and teachers are disappointed with their results. If a request is made the work of such candidates will be checked by the Board.

Special provision can be made at the examination for handicapped students. Special consideration can be given to students taken ill during the examination if they have completed the paper. The Boards deal as fairly as possible with each student and make every effort to mark and grade accurately. There is an appeal system and schools can appeal against an award if a candidate fails, or gets a lower grade than expected and there is evidence that the candidate should have done better.

EXAMINATION TECHNIQUES

(a) A revision schedule It is essential to plan a revision schedule for the weeks leading up to the examination. Bearing this in mind the timetable which follows has been prepared assuming that the examination will take place during the first week of June. A similar time-scale would apply to other months of the year.

Mid-March Draw up a revision schedule, planning weekly programmes for late March, April and May and increasing the work-load for the Easter holiday period. Review weaknesses which may have shown up at the mock examination and allocate additional time in the schedule to weak areas. Plan the schedule so that there is at least a week available before the examination to refresh your memory of the most important points.

Late March, April and May Follow your revision schedule allocating an hour or more each day with an increased revision work-load during the Easter vacation.

The week before the exam Spend this week glancing back over your notes at some of the most important areas you have revised and the diagrams and maps you need to remember. As it is very important to enter the examination refreshed and with a clear mind, do not attempt any revision during the twenty four hours before the examination.

(b) Answering questions Here is the procedure you should adopt when you are permitted to read the examination paper.

Selecting the questions

(a) Read the rubric at the top of the paper.
(b) Read all the questions carefully.
(c) Mark all the questions you might be able to answer.

(d) Reduce the number of questions you have marked to the number required by the rubric.

(e) Read again the questions you will attempt, underline key words and ensure that you appreciate exactly what is involved in answering each one.

(f) Choose the order in which you will answer them, leaving the weakest until last.

(g) Allocate the remaining time available for each question, allowing ten minutes at the end of the exam for checking and revision. (Do not allow yourself to over-write for more than five minutes on any one question. Space can always be left for additional material to be added later, if time permits.)

(h) Check once more that you are following the rubric and have chosen wisely.

Answering each question

(a) Read the question carefully, noting the key words you have underlined.

(b) Check whether the question falls into sections which can be dealt with separately. Decide whether maps and diagrams will be included in your answer.

(c) Draft a brief answer plan, listing the main sections of your answer and noting down key words, concepts and examples. Judge approximately how much time you should allow for each section.

(d) Read the question once more to ensure that your plan answers all the points required by the question.

Common faults If this procedure seems tortuous and repetitive to you, here is a summary of criticisms made by the Chief Examiners about the A level answer papers they have read.

(a) Time is wasted writing plans which are then ignored.

(b) Candidates fail to answer the question as it was asked.

(c) Answers are unstructured or do not keep to the structure of the question.

(d) There is a failure to concentrate on the main theme of the question.

(e) Some students write over-long introductions.

(f) Many candidates write a conclusion which does not add to the answer but only summarizes points already made. Marks are not awarded for repetition of information.

(g) Key instructions e.g. *describe, explain, compare,* are ignored.

(h) Some answers are presented untidily.

(i) There is an inadequate use of sketch-maps and diagrams.

(j) Many candidates make a poor choice of specific examples, locations are often imprecise e.g. *chalk cliffs on the south coast,*

These comments should underline the need for a carefully planned approach to the examination papers and to each question you answer.

Part II

---

1 Statistical methods

1.1 ASSUMED PREVIOUS KNOWLEDGE

In most GCSE syllabuses there is little on statistical methods so this is probably a new area of study. You may however have produced reports on fieldwork and done practical exercises in which you used some of the simpler techniques. The statistical work in A level Geography may also link with work you are doing in other subjects such as A level biology or economics.

1.2 THE PLACE OF STATISTICS

The various examination boards vary in the emphasis they place upon statistical methods at A level. There is a general recognition however of the contribution of statistical methods to present day work in geography. Statistical methods are not seen to replace traditional methods in geography but to complement them. By linking statistical work with fieldwork and with classroom-based interpretation of maps and diagrams, the processes of collecting results, data description, analysis of data and the presentation of data become logical stages in the same scientific operation.

Examiners are generally looking for evidence of:
(a) knowledge of calculation methods of some of the basic statistical techniques and indices.
(b) an understanding of the principles and processes involved.
(c) application of available techniques.
(d) ability to choose appropriate methods to solve problems.
(e) ability to interpret results.

1.3 SOME BASIC TERMS

Normal distribution This is a bell-shaped or symmetrical frequency curve. Observations equidistant from the central maximum have the same frequency. The three values of the plotted data – the average, the median and the mode (see below) all coincide at the same central point in this distribution.

Given a normal distribution curve it is possible to postulate the number of occurrences at any given value or between given values (Fig. 1.1). Approximately 68 per cent of the values lie less than one standard deviation from the mean and 95 per cent less than two standard deviations from the mean (a definition of standard deviation is given later in this unit).

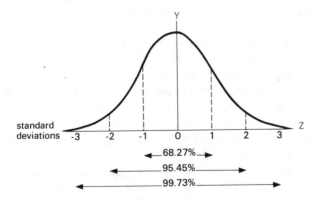

Fig. 1.1 The Normal Distribution Curve

The normal curve is also known as the *Gaussian distribution*. It is important to note that statistical normality is a property of a distribution *not* of individual values. So one value cannot be said to be normal but a set of values may be said to be normally distributed.

Skew (or skewness) (Fig. 1.2) This term is used to describe the extent to which a frequency curve is asymmetrical. When the modal class, i.e. the class containing the largest number of values, lies off-centre to the left when the data are plotted the distribution is said to have a *positive skew.* A *negative skew* exists when the modal class lies towards the upper end of the range, i.e. off-centre to the right when the data are plotted. Generally speaking the greater the skew the less representative is the average (the arithmetic mean value). Skewness may in fact be defined as the extent to which the mean differs from the median.

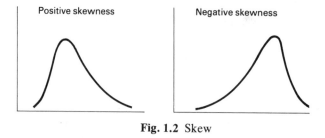

Fig. 1.2 Skew

Error The difference between observed and calculated values. See, for example, *sampling error* below.

Significance When statistical difference is *significant* it is extremely improbable that it occurred by chance. A geographer has to be concerned with probability – the probability that a particular conclusion based upon the interpretation of collected data is correct; the probability that a hypothesis is justified. So it is necessary to establish tests of significance.

A particular hypothesis may be assumed to be true and data are collected to test it. If the results observed in a random sample are markedly different from those expected in the context of the hypothesis and are not the result of pure chance, it is said that the observed differences are significant. The original hypothesis would then be rejected. If on the other hand, the results correlate with those expected, the observed correlations are significant and the hypothesis is accepted. Tests of significance are therefore procedures which enable us to judge whether hypotheses should be accepted or rejected.

Dependent and independent variables A variable is an item which can have several values. If to each value that can be assumed by a variable X there corresponds one or more values of a variable Y, we say that Y is a *function* of X i.e. $Y = F(X)$.

X is the *independent variable* because it may vary freely. Y is the *dependent variable* because its values depend on the values of X.

Correlation Correlation is a mathematical association between two sets of values. The measure of the degree of association between two paired variables may be established by the calculation of a *correlation coefficient* (e.g. Spearman's rank correlation coefficient, Section 1.8). It is important to remember that a correlation which is shown to be statistically significant does not imply that there is a *causal* relationship between both sets of data. The mathematical association between the two sets of data may have been caused by a third factor.

1.4   DATA COLLECTION

Sampling This is a means of obtaining a set of data of the smallest size that is representative of the total population or the whole area being studied. The set of data is within a desired degree of reliability.

When the data which has been collected to test a particular hypothesis has been identified, it may be found to be so large in volume that there are practical problems of time, cost or effort (workload). It therefore becomes important to select a sample representative of the total information available. A sample is a subset of the total population (population – a set of items or phenomena). It is the correct choice of a representative sample from the total population because the total population is in practical terms beyond reach.

Random sampling Random sampling techniques are used to obtain as true and representative a cross-section of population as is permitted by the size of the sample. Random number tables are usually used to select the sample in a way which ensures that each member of the population has as much chance of being selected as part of the sample as any other. This makes it possible to generalize from the characteristics (mean, standard deviation, probability etc) of the sample, i.e. the statistical inferences made from the data are valid for the total population.

Stratified sampling At times it may be best to collect and analyse data in a less general way than random sampling allows. For example, instead of assessing the characteristics for the whole population or whole body of data it may be preferable to examine individual groups separately. When data are grouped and a sample is randomly picked from within each group the process is known as stratified sampling, i.e. each group is known as a stratum.

Systematic sampling In systematic sampling an item is selected at some regular interval e.g.

every tenth item on a list. It is important however that the sample interval does not coincide with any periodic repetition of conditions. For example, a systematic sample of firms arranged in alphabetical order is more acceptable than a systematic sample for the analysis of climatic data which fluctuates periodically.

Sampling error Provided that a sample is truly random it is possible, given the sampling mean, to assess the limits within which the true mean falls with a known percentage probability. The value which controls these limits is known as the *standard error of the mean*. The formula for calculating it is:

$$SE\% = \sqrt{\frac{pq}{n}}$$

p = % of items in a given category
q = % not in this category
n = number of points in the sample.

This calculation not only provides an estimate of the limits of the true mean, it also emphasizes the limitations implicit in a sample mean. By calculating the sample error it is possible to calculate how an increase in sample size reduces the error. It is important in sampling because the art of sampling lies in choosing a sample size that will give an answer with the desired degree of accuracy and probability. It also shows that if a certain degree of accuracy is required a minimum sample size is essential.

Point sampling Random sampling can also be applied to data which has an areal distribution, e.g. farms. In order to achieve random sampling in a particular area the area under study is usually gridded and the grid numbered. The co-ordinates compiled from the numbers can apply either to a grid line or to the spaces between lines. A survey of farms is usually made using a sample of points. Land-use sampling is better based on areal sampling i.e. the spaces between the lines (Fig. 1.3).

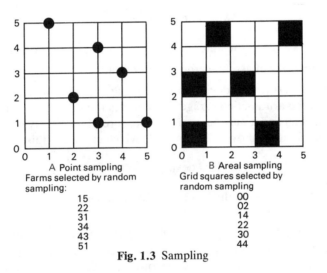

A Point sampling
Farms selected by random sampling:
15
22
31
34
43
51

B Areal sampling
Grid squares selected by random sampling
00
02
14
22
30
44

Fig. 1.3 Sampling

1.5   STATISTICAL MEASURES OF CENTRAL TENDENCY

The *central tendency* of data is the tendency of values of individual items within a set of data to cluster about a particular value such as the arithmetic mean.

Mean The arithmetic mean or average is obtained by the formula

$$\bar{x} = \frac{\Sigma x}{n}$$

Σx = sum of values making up set
n = number of values being considered
$\bar{x}$ = mean

Median the central value of an ordered series i.e. the value which, when the items have been ordered (ascending or descending) has an equal number of values above and below it.

$$\text{Median} = \frac{n+1}{2} \text{ where n is the number of values or occurrences.}$$

If there is an odd number of occurrences the median is one of the values. If there is an even number of occurrences the median lies between two of the values recorded.

Mode The value or class which occurs most often. Distributions may be unimodal (with 1 modal class), bimodal (2 modal classes) etc.

1.6   MEASURES OF DISPERSION

The degree to which numerical data tend to spread about an average value is called the dispersion (or variation) of the data.

Range The range of a set of numbers is the difference between the largest and smallest numbers in the set. The range is often given by quoting the largest and smallest numbers e.g. 2–12.

Quartile range A percentile is the value below which lies a particular percentage of an ordered distribution of values. The percentiles which divide the distribution into quarters (the 25th, 50th and 75th percentiles) are called quartiles. The 25th percentile is the lower quartile, the 50th is the median and the 75th the upper quartile. The *inter-quartile range* is the difference between the 25th and 75th percentiles. This is a crude but useful measure of the spread of data. The smaller the inter-quartile range the more closely the data clusters around the median.

This range lies astride the median and if the values are normally distributed each of the quartiles (upper and lower) would lie half of the inter-quartile range away from the median. This value is called the *quartile deviation*. It is expressed as

$$\frac{\text{upper quartile} - \text{lower quartile}}{2}$$

This value gives an indication of the range of the central 50 per cent of the occurrences above and below the median.

Mean deviation This is a way of summarizing the difference of each occurrence in a set of data from the average (or another constant such as the median). The difference between the size of any one value and the average value indicates the *deviation* of the unit from the average. The mean deviation is the mean value of all individual deviations from a given value i.e.

$$\text{Mean deviation} = \frac{\Sigma \mid x - \bar{x} \mid}{n}$$

$\mid x - \bar{x} \mid$ is the difference irrespective of sign (+ or −).

It is therefore a simple way of assessing the scatter of data.

Standard deviation The standard deviation is the square root of the average of the squares of the deviations from the arithmetic average. It indicates the degree to which individual values cluster around the mean and it may be used as a measure of variability of a frequency distribution (see normal distribution above). The formula which expresses this is:

$$\text{Standard deviation} = \sqrt{\frac{\Sigma (x - \bar{x})^2}{n}}$$

where $(x - \bar{x})^2$ is the square of the difference between individual values
n is the number of occurrences in the set of data.

Running means Some sets of data (e.g. agricultural production) consist of values which change over time. A central concern with such data is to reduce or eliminate the detailed differences between one particular value and another in order to identify and understand the overall characteristics. The running mean is a smoothing device designed to eliminate erratic or short term movements in a time series. This succeeds in throwing into emphasis the major fluctuations in the data.

If a 5-year running mean is being used, for example, the first value will be the mean for years 1–5, the second value will be the mean for the years 2–6 etc.

Histograms A histogram is a graph which displays the frequency of items within classes. So it is a graphical representation of frequency distributions.

1.7   DESCRIPTIONS OF SPATIAL DISTRIBUTIONS

Location quotient See Unit 21 *Location of industry*.

Lorenz curve (Fig. 1.4). This curve is used to compare an uneven distribution with an even one. The location quotient is based on this curve. The curve is drawn on a square graph with the X and Y axes having comparable scales. An even distribution results in the curve being a straight line at 45° to the horizontal. The more uneven the distribution the more concave will be the curve. The unevenness of a distribution represented by a Lorenz curve can be indicated by expressing the area under the curve as a percentage of that under the perfectly even theoretical distribution (the straight line). The curve is not an exact device but is an approximate visual method of representing a distribution.

Mean centre The mean centre is a measure of concentration. It is used for example to establish

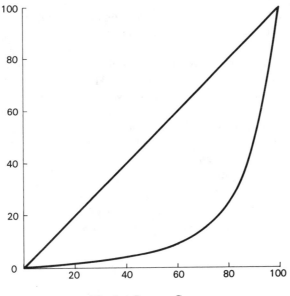

Fig. 1.4 Lorenz Curve

the position around which a set of factories is clustered. The location of a point can be defined accurately by means of two co-ordinates (x,y) which represent the distance of that point horizontally and vertically from a fixed reference point (e.g. the National Grid).

The mean centre of a point pattern is the point which has as its x,y co-ordinates the mean values of all the x and y co-ordinates in the distribution respectively. This is the mean centre of a spatial distribution (See Davis, P., *Data Description and Presentation,* OUP, 1974 p. 23).

Points are sometimes weighted according to their significance (e.g. the output of the factory) and the *weighted* mean centre is then calculated.

Nearest neighbour analysis This technique considers the location of individual points in a distribution in relation to others. The formula for the analysis is:

$$N.N.A. = 2\bar{d}o\sqrt{\frac{n}{A}}$$

do is the observed mean distance (i.e. the measured distance between each town and its nearest neighbour)

n is the number of points in the pattern
A is the area over which the points are distributed.

This method enables geographers to make simple objective comparisons between distributions.

Cumulative frequency graphs A cumulative frequency curve shows the number of occurrences or values above or below a particular level. The absolute numbers in each class are converted into percentages of the total. The number in each class is added to the classes above it until 100 per cent is achieved.

Measuring networks See the unit on networks.

Measuring shape Traditionally shapes have been described subjectively, often by comparison with other objects e.g. bell-shaped. Indices have been developed to provide a precise basis for comparison. The basic parameters used in measuring shape are—area, length of longest axis, radius of the largest inscribing circle, the radius of the smallest subscribing circle, the length of the perimeter. Ratios have been constructed so that a circle would have a value of 1.0 and the more linear the shape the more the ratio approaches zero. An example of a shape index is:

$$s = \frac{A}{0.282\,p}$$

A = area
p = length of perimeter

1.8   RELATIONSHIPS BETWEEN DATA

Scatter diagrams The drawing of scatter diagrams is part of the process of finding equations for approximating curves which fit given sets of data. The process is known as curve fitting. In order to express the relationship between two variables in mathematical form, the first step is to collect data showing corresponding values of the variables being examined. These points are then plotted on a rectangular co-ordinate system. The resulting set of points is a scatter diagram.

The drawing of a scatter diagram is used to investigate the relationship between two sets of data between which it is logical to assume there is a relationship. Usually there is a cause and effect association between the two sets of data which are plotted. If a causal factor (the independent variable) can be identified it is usually scaled along the horizontal axis. If values range widely the data may be plotted on a logarithmic scale. The diagram which results from plotting the data then allows the relationship between the two variables to be judged subjectively. If a definite trend in the distribution of points can be seen a relationship of some sort exists. The more closely the points on a scatter diagram conform to a straight line the stronger is the relationship (correlation) between the two variables. Objective tests of correlation then need to be applied (see Spearman's rank correlation coefficient below).

Spearman's rank correlation coefficient   This is a fairly quick method of assessing correlation. The coefficient is based not upon the actual values but on their rank order. It is useful because only rank order may be available for some data. Despite the fact that it is a crude index of correlation it makes a generalized estimate of correlation possible by using a simple formula.

The formula for its calculation is:

$$r_s = 1 - \frac{6 \Sigma d^2}{n(n^2 - 1)}$$

r_s = Spearman's rank correlation
d = difference in rank value of 2 sets of data
n = numbers of pairs being compared.

Chi-square   This is a non-parametric test. It tests whether the observed frequency of a given phenomenon differs significantly from the frequencies which might be expected according to the hypothesis which is being examined.

The data is processed in the form of frequencies and not in absolute values. In order that the test may be used it is necessary to set the hypothesis in precise terms. Usually this is done by formulating a *null hypothesis* (no) which postulates that two samples form part of the same population and that there is a high probability that the observed variations are the result of chance. The alternative hypothesis (H_1) being examined is that the observed differences are so great that they are unlikely to be the result of chance and the two samples must therefore be regarded as coming from different populations.

Observed values are values which actually occur (O).

Expected values are values which would occur if the null hypothesis applied completely (E).

The value of chi-square, χ^2 is obtained by the formula:

$$\chi^2 = \frac{(O - E)^2}{E}$$

This value can be referred to a table or graph and a probability value read off (see, for example, Gregory, S., *Statistical Methods and the Geographer*, Longman, 1979, pp 166–173).

1.9   MAPS AND GRAPHS

1.9.1   Cartographic methods

It is not possible to describe the methods of construction of the various cartographic methods and techniques listed in this section. It is important therefore that you read up construction methods etc. in one of the books listed at the end of the chapter.

The interpretation of diagrams and assessment of the advantages and disadvantages of using a particular method needs practice. You should look through the books you are using in geography identifying examples of a range of cartographic methods. For each of the diagrams you identify try to answer these questions:

(a) What is the particular technique used?
(b) Why was this particular method used?
(c) What are the advantages/disadvantages of this method in illustrating this set of data?
(d) What other method(s) might have been used?
(e) How may the diagram be interpreted? (i.e. what is the significance of what it shows?).

1.9.2 Methods of cartographic representation of distribution and spatial patterns

Methods	**Examples in this book** (by unit)
Isopleths	Weather maps; agricultural land use
Chloropleths	West Africa; The British Isles: regional problems; France

The *isopleth* map is composed of lines of equal values (isolines). The patterns of the lines illustrate the distribution of the phenomena mapped (e.g. contours, isobars). The *chloropleth* or colour patch map may be a density shading map. Given areas are coloured or shaded according to the values or densities relating to each area. *Dot maps* are maps on which values are represented by dots which are as precisely located as possible.

Advantages of the methods

Isopleth maps allow data to be plotted for a region without internal boundaries (e.g. of parishes) interrupting the pattern. They therefore illustrate general trends with changes in values shown smoothly rather than abruptly.

Chloropleth maps give an immediate impression of variations in values. They can also be quantitatively interpreted in terms of the areas on which the data is based.

Dot maps are simple to construct and enable values to be precisely located in space. They can be interpreted quantitatively.

Disadvantages of the methods

Isopleth maps may disguise abrupt changes which may occur in features of human geography from one locality to another. There is a degree of subjectivity both in deciding where to locate the values within areas and in the interpolation of the isolines.

Since the shading in *chloropleth maps* is related to specific areas, distributions may be shown in a disjointed way and gradual trends may appear as a series of steps. The shading relates to an average figure for each area so variations within the area are not shown.

With *dot maps* the selection of the dot value is critical – wrong visual impressions may be given by choosing too high or too low a dot value. If values vary widely the map can become rather confusing unless proportional circles are used to represent the highest values. Constructing a dot map accurately can be very time-consuming.

1.9.3 Magnitude symbols

Symbols/methods	*Examples in this book* (by unit)
Proportional circles	France
Pie graphs (divided proportional circles)	France
Proportional squares	not included
Bar graphs	Agricultural land use
	Problem regions of UK
Divided bar graphs	France
Population pyramids	Underdevelopment
Symbolic representation	not included

Proportional circles The radius of each circle is made proportional to the square root of the quantity it represents.

Pie graphs The quantities represented by the proportional circles are subdivided into component parts.

Proportional squares Squares are drawn instead of circles. The size of the square is proportional to the square root of the value.

Bar graphs Variable data are represented by bars or columns of different lengths.

Divided bar graphs These show the components which make up the total represented by the whole bar or column.

Population pyramids A special type of bar graph. Adjoined horizontal bars are placed on either side of a central vertical axis which represents age categories and is usually graduated in units of five years. The length of each column varies with either the number or the percentage of the total population in each group.

Symbolic representation Data are represented by symbols, e.g. car silhouettes to represent vehicle manufacturing. Variations in value are shown either by the number of symbols or by drawing the symbols proportionally.

Advantages of the methods

Proportional circles Because the square roots of the crude values are used, the size of the symbols can be kept within reasonable limits. Each symbol can be located precisely on a base map.

Divided proportional circles These share the advantages of the proportional circles with the added advantage of showing how total values are made up by different components. Pie graphs need not be located on a base map and can illustrate non-spatial data.

Proportional squares Again the use of square roots means that the size of symbols can be kept within reasonable limits. Comparison of data is easy because of the ease of comparing the areas of the squares.

Bar graphs An extremely versatile technique for representing data. The bars are simple to construct and give an immediate visual impression.

Divided bar graphs Share the advantages of the bar graph with the added bonus of showing the component parts. May be used for non-spatial data or located on base maps.

Population pyramids Can be interpreted quantitatively. Provide a good basis for comparing male with female population characteristics and for comparing age structures over time. The overall shape of the pyramid is indicative of a range of socio-economic factors (see Unit 23 *Less developed countries*).

Symbolic representation Gives a good visual impression.

Disadvantages of the methods

Proportional circles and *divided proportional circles* It is not easy to estimate the differences in value by comparing the areas of the circles.

Proportional squares These are not as easy to locate precisely as circles. If they are divided to show components the division is laborious and untidier in appearance than pie graph divisions.

Bar graphs, divided bar graphs and *population pyramids* When located on maps the symbols may be bulky and extended beyond the boundaries of the areas to which they relate.

Proportional symbols The major difficulty is that they are very difficult to draw accurately.

1.9.4 Graphs and flow-lines

Method	*Examples in the book* (by unit)
Line graph	Networks
Segmented bar graph (transect graph)	Rivers and river valleys
Flow-line	Soils, Mineral exploitation

For *line graphs* values are plotted on the vertical axis against (for example) time on the longitudinal axis. Points are connected by lines to indicate fluctuations in value through time. *Segmented line graphs,* e.g. the relief cross-section, represent variations in values or quantity along a line which may or may not be straight. A *flow-line* is a map in which lines vary in width according to the quantities of goods, vehicles etc. which move along the routes represented by the lines. Values are usually grouped into classes and the width of the lines then varies according to a regularly increasing scale.

Advantages of the methods

Line graphs give good visual impression of changes in value over time and identify peaks and troughs in trends. Composite data may be shown by the superimposition of a number of sets of data (i.e. a number of lines) on the same graph.

A great advantage of *transect graphs* is that they need not be straight lines so, for instance, the key features of a slope or of an urban area may be included in one diagram.

Flow-lines effectively relate quantity or volume to a direction of movement. They give an excellent visual impression and clearly show the relative significance of individual routes or paths. They may be used to show two-way flow and if multiple lines are used instead of solid bands it is possible to indicate the relative importance of components of the flow.

Disadvantages of the methods

Line graphs It is not possible to determine intermediate values from line graphs. It is not easy to interpret the fluctuations because the factors to which the fluctuations relate are not shown.

Transect graphs The decision as to what is shown on the diagram and where the transect is to be made is essentially subjective.

Flow-lines The choice of scale is critical, an evenly graduated scale may hide significant variations in volume or frequency. Too generous a scale may lead to the obliteration of key details of the map.

1.10   FURTHER READING

Guest, A., *Advanced Practical Geography* (Heinemann, 1980)
Gregory, S., *Statistical Methods and the Geographer* (Longman, 1979)
Matthews, M. H. and Foster, I. D. L., *Geographical Data; Sources, Presentation and Analysis* (Oxford, 1989)
McCullagh, P. S., *Data Use and Interpretation* (S.I.G. 4) (OUP, 1974)
Mowforth, M., *Statistics for Geographers* (Harrap, 1979)

2 Ordnance Survey maps

2.1   ASSUMED PREVIOUS KNOWLEDGE

Some GCSE geography examinations include a compulsory Ordnance Survey map question using an extract from either the 1:25 000 or 1:50 000 series. Students are expected to be familiar with these maps and if you have not studied them in detail it is essential that you do so before starting the A or AS level course. Map interpretation questions are sometimes provided as optional questions on the A level examination papers so there is every incentive to improve your skills in map reading and to develop your ability to interpret the evidence provided by maps.

One of the best ways to begin studies of OS maps is to buy the local sheets at a stationer's. Buy the 1:50 000 Second Series and the 1:25 000 Second Series maps and learn the symbols used on them. Measure distances, check compass directions and use the grid system to locate features on the maps. You can also borrow or buy mapwork books which include exercises on map extracts. Beware, however, of books which contain out-of-date extracts. The Boards normally use the OS maps Second Series which have different contour intervals, symbols and colours from earlier editions.

If you have never attempted GCSE questions based on actual map extracts used in the examinations you should obtain a copy of the book written by the authors, *Revise Geography*, Letts, 1987, or Wilson, J. G., *Language of Maps*, Schofield and Sims, 1975.

Interpretation of an oblique aerial photograph showing part of the map area may also form part of the map question. The photograph tests your skills in relating places and objects in the photograph with their counterparts on the map, and in identifying things shown on the photograph which are not evident on the map or vice-versa.

2.2   ESSENTIAL INFORMATION

2.2.1 The nature of A level questions

Only two A level Boards (AEB and Cambridge) set a compulsory map question, but this appears as part of an alternative paper which might not be taken. Nevertheless, most Boards expect candidates to be familiar with 1:25 000 and 1:50 000 Second Series maps. Questions may be based on evidence provided by an OS map extract. A level map questions are not concerned with testing your knowledge of the conventional symbols or your ability to measure distances. It is assumed you already have this information and these skills and can use them to describe, analyse, compare and comment on various aspects of the physical and man-made patterns discernible on the map extract.

Here are some examples of the types of questions which are asked.

(a) Describe in detail the physical features of the area west of easting 84 and suggest possible origins for these features.
(b) Analyse the pattern of settlement to be seen on the map extract.
(c) Compare the patterns of settlement and communications in the area north of northing 20 with that to the south.
(d) Comment on the nature and location of the land use types in the area of the map extract.

From these examples it should be evident that A level is concerned more with broad patterns and less with fine detail. Descriptions are required but they are linked with interpretation and as a result the map is used to test your knowledge of various aspects of physical and human geography as well as your ability to interpret OS maps.

2.2.2 Annotated sketch maps

Sometimes A level questions ask you to draw an annotated sketch map of all or part of the map extract to illustrate selected features such as physical regions and town sites. At the back of this book there is a 1 : 50 000 map extract of the Perth area of Scotland. It is for use with this unit and you will need to refer to it while reading the rest of this chapter.

Figure 2.1 has been drawn as an example of an annotated sketch map of the area east of easting 12 and south of northing 21.

It would be impossible in the time available in the examination to reproduce a sketch map as accurate as Fig. 2.1, but you should make your sketch as accurate and neat as possible. A few minutes spent drawing the grid lines before you sketch in the detail will help you to achieve a high standard of accuracy. All annotations should be clearly printed and, if necessary, arrows should be used to identify precisely to what the annotation refers.

Very often you will find it both convenient and time-saving to include small sketch maps or diagrams even though these have not been requested in the question. Settlement sites, situations and patterns as well as communications and land use can often be shown better by well-annotated sketch maps and diagrams than by lengthy descriptions.

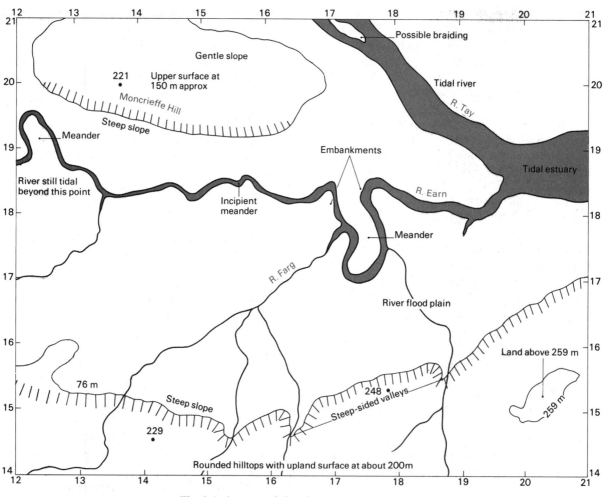

Fig. 2.1 Annotated sketch map of Perth

Sketch maps alone are not, however, an adequate substitute for a written answer. Examiners will only award additional marks for sketch maps if they contain extra material not already referred to in the written answer. But sketch maps will help to clarify the written answer.

Examiners often express concern at the quality of many of the sketch maps submitted by candidates. Untidily scribbled maps are the result of haste and lack of practice at map drawing. Good sketch maps can be drawn quickly and accurately only after continuous practice so take advantage of any opportunities available to develop this useful skill.

2.2.3 Aerial photographs

Sometimes the OS map question includes an oblique aerial photograph of part of the map. The photograph is used to test your ability in interpreting an aerial photograph and also to see how far you are able to relate photographic and map evidence.

Figure 2.2 shows an oblique view of part of the Perth map. When you are given questions about a photograph first locate the part of the map which appears on the photograph. Clues such as large buildings, bridges, rivers, roads and railways will help you. When Fig. 2.2 is compared with the map extract the river is an obvious starting point in locating the area. Other clearly defined landmarks such as the river bridges and the island in the foreground help to relate the photograph more precisely with the map.

Having identified the area, find other landmarks on the photograph which also appear on the map. On Fig. 2.2 the railway can be seen running through the city area with a branch crossing the river. Many other locations can be similarly compared. It should also be possible to identify the area of the map over which the aircraft was flying when the photograph was taken. On the Perth map it is the southern part of GR 1221. In which direction was the camera pointing? Do not expect the map to match the aerial photograph precisely. There could be several years' difference between the map survey and the taking of the aerial photo.

Fig. 2.2 An oblique view of part of the Perth OS map

2.2.4 Settlement

There are a number of features concerning settlement on maps which you must understand and be able to interpret.

Site The actual area upon which the settlement is built. The site of Perth, for example, is an area of level ground close to the River Tay at a point where the river is narrow and easily crossed.

Situation The position of the settlement in relation to its surroundings. Perth is a centre for the farmland of the Tay valley.

Pattern of settlement OS maps show the distribution of settlements in considerable detail ranging from individual houses in country areas to the built-up areas of towns and cities.

When examining settlement patterns on a map first distinguish the areas where settlement is sparse or non-existent. Find two such areas on the Perth sheet, then look for village sites. In the valley of the river Earn there are a number of villages on the south side below the high ground. What are their names? Other villages like Bridge of Earn may be near river crossings or route centres. Finally examine the urban distribution and note any distinctive features, for example, the development of New Scone (1326) close to Perth. You should summarize your analysis of the settlement pattern on a sketch map with the main features such as sparse settlement delineated and suitably annotated.

2.2.5 Land use

Ordnance Survey maps are not specialist land use maps but they do contain a great deal of information about land use. You should be able to distinguish a number of features which indicate how the land is used. At the same time you must appreciate that the information is limited and on no account should you attempt to fill in the gaps with guesswork. Use only the evidence supplied by the map.

The 1 : 50 000 map shows such types of land use as woodland, orchards, parks, quarries and heathland (although in the latter case the same symbol may also mean rough grassland). Route-ways and associated services such as railway sidings are also shown, as well as large industrial units, hospitals and schools. In urban areas it is possible to distinguish a limited number of land use types. Broad categories can be identified but no precision can be given to the boundaries of the morphological zones. For example, the Central Business District of Perth can be identified in GR 1123 between the railway station and the river where there is a focus of routes but the limits of the CBD cannot be accurately drawn (see Unit 18 *Urban Land Use*). Residential areas and small industrial plants are not identified separately but some large industrial sites such as the distillery (GR 098259) are named. Recreational areas such as golf courses are marked, as well as hospitals, prisons and cemeteries. Newer housing estates can be recognized by their layout and crescent-shaped road pattern. Find these and other land use types in Perth and then make a sketch map of the morphological zones of the city as far as you can identify them.

2.3 GENERAL CONCEPTS

The map is a record of distributions and patterns in space This concept is fundamental to your A level course and forms the basis for the type of map question which is set in the examination. To the geographer the map provides a two-dimensional record of spatial phenomena – it is a valuable tool which must be understood and interpreted. It is an essential part of the geographer's role to recognize, analyse and describe the distributions, trends and relationships which map evidence provides.

2.4 DIFFERENT PERSPECTIVES

The Ordnance Survey maps present a detailed survey of the main features in the areas which they cover. In this respect they are comprehensive rather than selective. As maps prepared by and for the government they provide basic information which is periodically updated so that it is possible to make comparisons between the different editions.

You should always examine maps critically to assess how far the techniques which have been used by the cartographer are successful and what limitations create difficulties for the person using the map. For example, the metric editions use contour intervals to the nearest metre but the vertical interval is retained as 50 feet. This produces irregular contour heights in metres.

The weaknesses and strengths of the more recent OS series are best analysed in conjunction with earlier maps on the same scale and with those issued by neighbouring countries in Europe.

2.5 RELATED TOPICS

Many questions relate your knowledge of physical and human geography to map evidence, for example *Describe and interpret the features of the physical landscape,* and *Comment on the nature and location of the land use types shown on the map extract.* It is therefore important to remember that although the interpretation of the OS map requires specialist skills, these skills need to be underpinned by a sound knowledge of physical and human geography with particular reference to Great Britain.

2.6 QUESTION ANALYSIS

1 With the aid of sketch maps and/or diagrams analyse the relief and drainage of the area.

(in the style of the Associated Examining Board)

Understanding the question The key word in the question is *analyse* and unless you fully understand what analysis of the relief and drainage of the area entails you may be content to confine your answer to a mere description. Analysis involves three distinct operations. In the first place the map area must be briefly described in terms of the distribution of high and low land and the drainage pattern. Secondly the area must be divided into its physical regions to identify the broad patterns of relief and drainage. Thirdly the distinctive features of the physical landscape require further elaboration. For example, the valley of the river Earn contains features of a mature river system such as meanders and an ox-bow lake (see the unit on rivers and river valleys).

The question asks for sketch maps and/or diagrams to illustrate the analysis. If these are well annotated they could reduce the amount of writing which is necessary to brief explanations of what they show.

Answer plan Write an opening paragraph describing in general terms the relief and drainage pattern on the map extract. The high land in the north-east of the map extends to the river Tay and is then continued to the south of Perth. A further area of high land extends across the southern part of the map with a steep drop to the valley of the river Earn. Perth spreads across low ground in the north-west of the area while the river Tay and its tributary the Earn form a widening flood plain to the east. The river Earn obtains most of its tributaries from the southern highland region while drainage from the highlands in the north-east is radial.

Draw and annotate a map to show the main physical divisions of the area (see Fig. 2.1). There are three main physical regions on the map.

(a) The flood plain of the River Earn and the adjoining estuary region of the River Tay.

(b) The area of upland north of the River Earn.

(c) The southern upland region.

The boundaries of each of these regions should be clearly distinguished on the sketch map. Describe the characteristics of the main regions e.g. heights and nature of the upland surfaces, dissection by streams and contrasts between the valleys of the river Tay and river Earn.

Take each of the regions in turn and with the aid of sketch maps and diagrams (cross-sections are useful), explain the physical features including drainage. Possible cross-sections might include one of the north-east highlands along part of easting 19, the valley of the river Earn along part of easting 14 and the Tay valley along part of easting 15. There is insufficient time in the examination to draw precise cross-sections on squared paper but reasonably accurate sketches can be made provided you use approximate vertical and horizontal scales.

Draw small sketch maps of some of the features such as the landforms of the mature valley of the river Earn and briefly explain their significance.

2 Describe the broad features of the settlement pattern and suggest from the evidence on the map some of the reasons for that pattern.   (*in the style of Oxford*)

Understanding the question By using the expression *broad features* the examiners wish to emphasize that they do not want an answer which analyses each settlement in turn with a detailed explanation for the site and situation of each one. The key word is *pattern* and unless you are able to identify patterns of settlement you should not attempt this question. Note also that the question requires *evidence on the map* which explains the pattern. The examiners would not give marks, for example, for the statement that settlement is absent from one area because the land is poorly drained, unless there was map evidence to confirm that this was the case.

Answer plan Draw an annotated sketch map of the map area and show the main patterns of settlement using such terms as *sparse, scattered, nucleated villages above the flood plain*. The built-up area of Perth should be shaded as an urban area.

Write descriptions of the settlement pattern you have identified and explain why the different elements within it occur. For example, the lack of settlement on the high ground in the north-east corner of the map can be accounted for by the height of the land, the steep slopes, extent of the rough pasture making farming difficult and the poor communications.

3 Comment on the nature and location of the land use types (excluding settlement) in the area of the map extract.   (*in the style of Scottish Higher*)

Understanding the question The question asks for comments on the *nature and location* of land use types. This is really an invitation for you to describe with as much detail as possible the land use which can be identified from map evidence and the reasons for the location of the different types. The best way to tackle the question is to identify the broad patterns of land use rather than attempting to write a minutely detailed description of the variety of ways in which the land is used in different sectors of the map. For example, deal with the broad pattern of the areas where heathland or rough grassland are to be found.

Although the question does not want you to include settlement in your answer, you should not exclude land use which has developed because of demand from the populated areas. The golf courses, parkland and the race course on the map extract are examples of a distinctive form of land use which can be classified as leisure facilities.

Answer plan Note down as rough work the main categories of land use which can be seen on the map, e.g. woods, heathland or rough grassland, marsh and recreational areas.

Write an introductory paragraph listing the main forms of land use and point out that large areas of lowland are probably given over to arable or pasture but there is no map evidence to support the assumption.

Describe the nature and location of each land use type in turn, in some cases drawing annotated sketch maps to illustrate the distribution pattern. Account for the land use types whenever possible, but do not consider each location in turn. For example, when describing the location of woodland keep to the broad pattern of distribution and avoid trying to account for each small patch of woodland.

2.7 FURTHER READING
Monkhouse, F. and Wilkinson, H. R., *Maps and Diagrams* (Methuen 1975)

3 Weather maps

3.1 ASSUMED PREVIOUS KNOWLEDGE

A number of the GCSE examining boards include a weather map in the examination. It shows the British Isles and features either a low pressure system crossing the country or a region of high pressure. You are expected to be able to recognize the symbols used and to describe the weather conditions recorded at certain locations. Simplified weather maps are published each day in *The Times* and *Guardian* newspapers with a survey of the weather to be expected in various parts of Britain. If you have never studied weather maps you should follow these simplified maps over a period of several weeks to appreciate how the pattern changes.

One interesting way of finding out about weather is to compare the forecast for your own area with what actually happens. Use the simplified weather maps and simplified descriptions already mentioned, or watch the weather forecast each evening on television. Make observations of cloud, wind direction and temperature in your own area. Recording instruments such as a barometer are not essential but you will need the guidance of a textbook such as Roth, G. D., *Collins Guide to the Weather* (Collins, 1981).

3.2 ESSENTIAL INFORMATION

Some of the information provided in the unit on *Weather and climate* will be of value while reading this unit.

3.2.1 Definitions – depressions

Depressions are areas of low pressure which can vary considerably in size. There are several different types.

Tornadoes Very low pressure systems only a few hundred metres across. They are common in the centre of continents in spring and summer.

Tropical storms These vary from 80 to 800 km across (50 to 500 miles). They usually originate

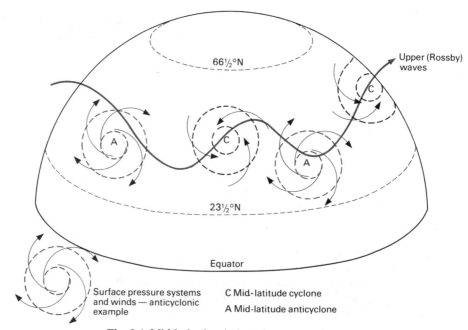

Fig. 3.1 Mid-latitude winds and pressure systems

over oceans and can do extensive damage. At their centres there are calm areas or 'eyes' where the sky is clear and the winds light. They occur most frequently in late summer or autumn and are most pronounced in the China Seas where they are called typhoons and in the Caribbean where they are known as hurricanes.

Frontal depressions These occur in middle latitudes where the westerly winds above the earth's surface move from west to east in a series of waves. Long waves (called *Rossby waves*) contain shorter waves, and it is the air streams of the shorter waves which produce the high and low pressure systems of the mid-latitudes. The location of the pressure systems in relation to the long wave pattern is shown on Fig. 3.1.

In a depression the pressure decreases towards the centre with air streams converging and revolving anti-clockwise around the centre in the northern hemisphere. The convergence of the air streams leads to a concentration of the isotherms between cold north-westerly air and warm south-westerly air with the boundaries between these air streams forming *fronts*. The convergence produces vertical movement of the air which results in cooling, condensation and precipitation.

Non-frontal depressions These can be formed by pressure changes brought about by local heating. They vary in size. Another form of non-frontal depression is the lee depression which is formed by air rising over mountains and descending on the lee side.

3.2.2 Ana and kata fronts

Meteorologists distinguish two main types of frontal depression; those in which the air in the centre is generally rising are called *ana-fronts*. Those in which the air in the centre is generally sinking are called *kata-fronts* (Fig. 3.2). In a depression with ana-fronts the rising air results in instability and precipitation. In kata-fronts sinking air is warmed by compression and produces

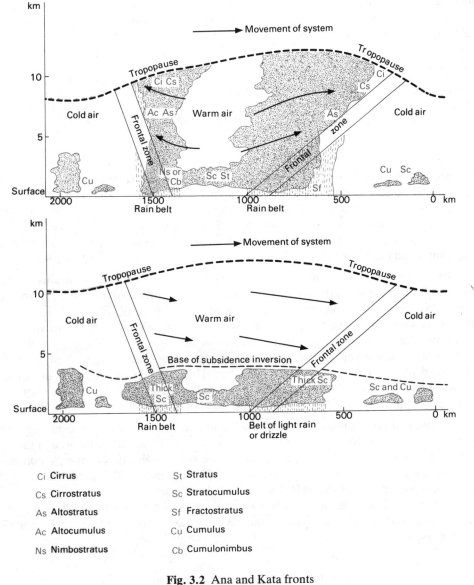

Ci **Cirrus**	St **Stratus**
Cs **Cirrostratus**	Sc **Stratocumulus**
As **Altostratus**	Sf **Fractostratus**
Ac **Altocumulus**	Cu **Cumulus**
Ns **Nimbostratus**	Cb **Cumulonimbus**

Fig. 3.2 Ana and Kata fronts

more stable conditions. An inversion layer forms, holding down the layer of cloud and, in the more stable conditions, producing only light rainfall.

3.2.3 Occlusion

In a depression the warm front and the cold front are separated by the warm sector. The area of this warm sector at the surface is gradually reduced as the cold front moves faster than the warm front and eventually catches up with it. The boundary between the two air streams is known as an occlusion. If the air behind the cold front is colder than the air ahead of the warm front it will cut under it and lift the warm air above the ground forming a cold occlusion. If the air behind the cold front is warmer than the air ahead of the warm front it will rise over the warm front to form a warm occlusion. Cloud and sometimes rain may result from the vertical displacement of some of the air in the occlusion.

3.2.4 Anticyclones

Anticyclones – regions of high pressure – usually cover large areas. Their winds are light and rotate clockwise and outwards from the centre. Anticyclones may be cold-centred or warm-centred. Cold-centred anticyclones are shallow and often move swiftly. Warm-centred anticyclones are deep and move slowly. The subsiding air near the centre of an anticyclone can give clear skies and much sunshine. However, if temperatures are low and humidities high, there may be condensation giving fog or low cloud and a temperature inversion may become established in the lower air layers.

3.2.5 Troposphere

The layers of the atmosphere nearest to the earth in which temperatures generally decrease with height are known as the *troposphere*. At about 8 km (5 miles) above Polar regions and 16 km (10 miles) above the Equator the *tropopause* marks the end of the troposphere and the beginnings of the stratosphere. In the stratosphere temperatures increase with height.

3.2.6 Jet stream

At the level of the tropopause wind speeds up to 500 km/h (310 mph) are reached. In the northern hemisphere there is one jet stream related to the Polar front which is discontinuous. The other is the sub-tropical jet lying between the Equator and approximately latitude 30°N.

3.3   GENERAL CONCEPTS

Air moves in a series of waves, known as Rossby waves, in mid-latitudes (see Section 3.2.1 and Fig. 3.1). Air flowing northwards at high altitudes from the Equator is deflected to the right in the northern hemisphere as the earth rotates in an anti-clockwise direction. This is known as the *Coriolis force*. Waves form in this air stream as the result of vorticity which is explained below. The ridges and troughs of these waves cause smaller waves to form on which originate the anticyclones and depressions which influence weather conditions in mid-latitudes.

Vorticity is the rotation or spinning of a column of air. It acts like water as it drains out of a bath. A column of air spinning of its own accord without regard for the rotation of the earth is said to have *relative vorticity*. A column of air can also spin as a result of the earth's rotation. This is called *global vorticity* and is strongest at the Poles and non-existent at the Equator.

The two forms of vorticity give the column of air an *absolute vorticity* – global vorticity increases and relative vorticity decreases towards the Poles. Relative vorticity increases and global vorticity decreases towards the Equator. This increase in one form of vorticity and decrease in another which is most intense in mid-latitudes sets up a wave pattern in the movement of the airstream, hence the development of the Rossby waves.

Convergence and divergence In mid-latitudes warm tropical air from the Equator meets cold Polar air and the two mix in vortices induced by the spinning earth. If the area in the centre of the vortice is relatively cold and pressure decreases with height, the airstreams converge near the ground, rise vertically and then diverge near the tropopause. This produces the *cyclone* or low pressure system which affects Britain and northern Europe.

If there is relatively high pressure and a warm core to the vortice the air flows will converge above the earth, descend and diverge at the earth's surface. This produces anticyclonic or high pressure systems.

3.4 DIFFERENT PERSPECTIVES

The science of meteorology is changing rapidly as a result of the enormous mass of data being collected by weather satellites, instrument-carrying balloons and surface stations. The introduction of computers has made possible complex calculations which enable predictions to be made more accurately and rapidly. The greatest advances have been made in our knowledge of the changes taking place above the earth's surface. It is the continuous monitoring of the atmosphere and checking on the data with the aid of satellite photographs which enables the changing pattern of winds and pressure to be analysed in depth and, when combined with recordings made on the earth's surface, used as the basis for three-dimensional models of the weather.

One result of this increase in knowledge has been a shift of emphasis on the significance of air masses (see Unit 11 *Weather and Climate,* Section 11.2.5. on page 75). The earlier idea that an air mass consists of uniform characteristics has been superseded by the knowledge that changes in the physical properties of air occur in air masses but are less intense than those which, for example, take place in fronts. The situation is more complex than was suggested by earlier theories which were formed on the basis of observations confined to the layer of air close to the earth's surface. This layer is more influenced by the properties of the earth's surface than are the layers above. However, the air mass concept with air streams displaying distinct characteristics does have a significant part to play in understanding weather patterns.

Our increased knowledge of conditions in the troposphere and the higher layers of the atmosphere above the earth's surface has also brought about new viewpoints on the processes which cause depressions. Fronts, which used to be considered as necessary components in the formation of a depression are now seen as secondary consequences of a trough in the upper westerlies.

3.5 RELATED TOPICS

Reference has already been made to the links between this unit and the one on *Weather and Climate.* Weather studies in connection with the synoptic charts used by the A level examination boards are mainly concerned with weather patterns over the British Isles and north-west Europe, although occasionally a North American synoptic chart may be used.

The day-by-day features recorded on weather charts include such phenomena as advection fog and thunderstorms which may form the basis for more general questions on weather and climate in the A level papers (see 3.6.2. below).

3.6 QUESTION ANALYSIS

1 The two synoptic charts (Fig. 3.3) are for 1800 hours on 14 May 1979 and for 0600 hours on 15 May 1979.

(a) Describe the synoptic situation at 1800 hours on 14 May and its development over the following 12 hours.
(b) Write an explanatory description of the weather experienced during this 12 hour period in (i) the South-West peninsula of England, (ii) North-West Scotland, (iii) East Anglia.

(Oxford and Cambridge, June 1980)

Understanding the question Examiners frequently provide two synoptic charts which show the changes which have taken place in the weather pattern over a period of time, usually twelve hours. The questions normally require answers which demonstrate your ability to describe accurately the changes which have taken place during the period covered by the two maps and your skill in interpreting and explaining the causes of the changes which have taken place and the weather phenomena recorded on the maps.

In this question part (a) requires two descriptions. One relates to the synoptic situation at 1800 hours on 14 May and the other to the developments in the weather pattern which have taken place during the following twelve hours. Part (b) asks for an *explanatory description* unlike part (a) which uses only the word *describe*. You are required to describe and identify the causes for the changes in the weather which have taken place in three regions of the British Isles.

When answering part (a) do not limit your description to the British Isles and the neighbouring coast of NW Europe. In the south the map extends to central Spain, southern Italy and the Adriatic coast of Yugoslavia, while in the north it includes much of Norway and Sweden, southern Finland and the western limits of the USSR. By contrast, when answering part (b) you must limit your answer to those weather stations which are within the three regions listed. In the case of the south-west peninsula there are two weather stations, one of which is on the Scilly Isles. In north-west Scotland there are two weather stations, one on Lewis, while in East Anglia one station is shown near to the Norfolk coast.

Answer plan Your first paragraph should be limited to a description of the general synoptic situation shown on the chart for 1800 hours on 14 May. Start by describing the main feature of the chart, i.e. the high centred over Eastern Europe with its extension westward to include central and southern England, Wales, southern Ireland and north-west France. Then describe the areas of low pressure, distinguishing between the shallow low pressure areas over southern Europe and the deeper depression south of Iceland. You must also describe the position and nature of the fronts between the high pressure system and the low.

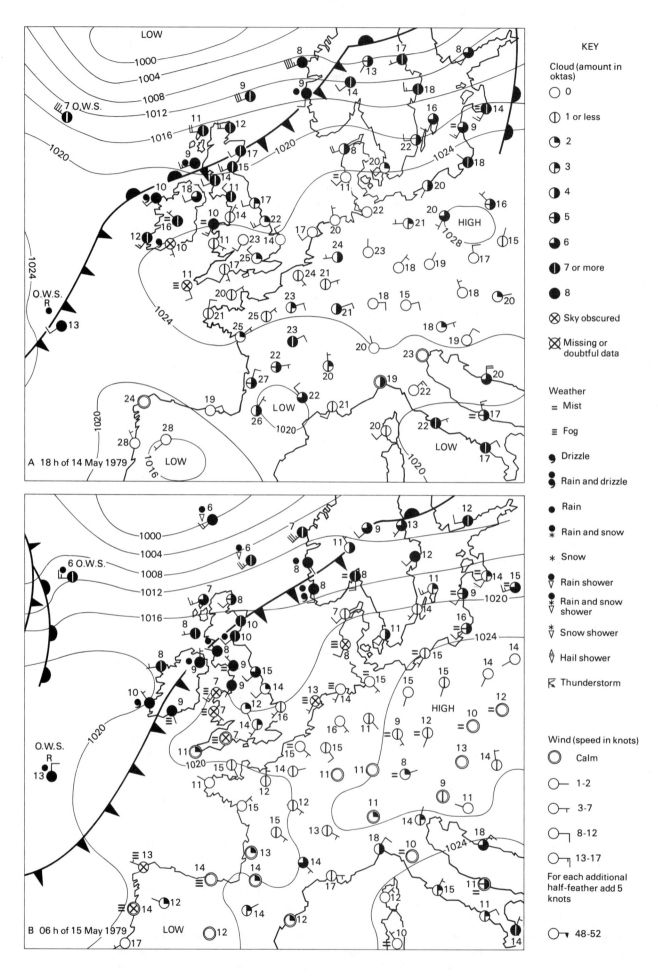

Fig. 3.3 Synoptic charts for northern Europe

The second section of your answer should describe the changes which have occurred during the twelve hour period. Use the same sequence as in the first paragraph, starting with the shift of the high pressure eastwards with evidence of its weakening, and concluding with the change in the frontal situation and the appearance of the fronts of another depression in the Atlantic to the west of Scotland.

When answering part (b) take each of the regions in turn and describe the weather experienced as shown by the symbols on both maps. Then give reasons for the type of weather and the changes which have taken place. Apart from explaining the local weather in relation to the general synoptic situation, remember to explain the causes for such things as the fog in the south-west peninsula and the fact that the temperature recorded at 1800 hours in East Anglia is low when compared with other neighbouring stations.

2 With reference to the synoptic chart for 0600 hours 15 May 1979 (Fig. 3.3), explain (a) the distribution of mist and fog; (b) the distribution and nature of the precipitation. Draw diagrams to illustrate your answer and include temperature/height graphs to explain the stability conditions likely to be found in the different regions covered by the synoptic chart. *(in the style of Oxford)*

Understanding the question Before deciding to answer this question you should make a note of the diagrams and temperature/height graphs needed for both part (a) and part (b), remembering that you must also include temperature/height graphs to illustrate the stability conditions in various parts of the map. If you are unable to draw graphs similar to Figures 11.1 to 11.5 in Unit 11 *Weather and Climate* (pages 74 and 75), you should not attempt this question. To understand part (a), look carefully at the map and find the distribution patterns for mist and fog. The mist areas are in the high pressure region and extend from southern Norway to Sardinia where there is very little wind, or none at all. Note that the synoptic chart is for 6 a.m. and in most places with fog and mist the sky was clear. These conditions will cause air near the ground to cool causing an inversion of temperature and the formation of radiation mist which will be trapped near the ground beneath the warm air of the inversion (see Fig. 3.1 on page 26). The fog symbols also make a distinctive pattern and when linked with the wind directions and temperatures of the areas concerned provide an explanation for the formation of these advection fogs. The precipitation distribution follows the fronts which gives the clue as to its nature. You will need to include a diagram to illustrate how the rainfall is caused (see Fig. 3.2).

Answer plan Start with a short description of the distribution patterns for mist and fog and then explain, with the aid of diagrams, how these will form given the conditions recorded on the synoptic chart.

Describe the distribution of precipitation and its nature, bearing in mind its close association with the cold front and, for one station, association with the warm front of the depression on the west side of the chart. Draw an annotated cross-section with temperatures included to show the conditions at the cold front.

To answer the last part of the question which requires temperature/height graphs of the probable stability conditions in various parts of the map area you must find at least one stable and one unstable area. The centre of the anticyclone is an obvious area of stability and a graph based on Fig. 3.2 will be required. Instability is associated with fronts and a graph based on Fig. 3.2 will illustrate this condition provided the temperatures are adjusted to those on the map. Explain what the graphs show.

3 How will movements in the upper atmosphere affect the surface weather conditions shown on the synoptic chart for 1800 hours on 14 May 1979. (Fig. 3.3)? *(in the style of the Joint Matriculation Board)*

Understanding the question This question is concerned with the air streams and conditions which will exist in the troposphere to bring about a high pressure system over central Europe with a low pressure system north of Scotland and a frontal zone across Scotland and southern Norway. In answering it is very important to avoid the detail on the synoptic chart. So long as you can identify the broad pattern as it has been described above you will have sufficient information on which to base your answer.

The conditions in the upper atmosphere which must be described in your answer are those which bring about the Rossby waves referred to in Section 3.2.1 and illustrated in Fig. 3.1. In addition remember that it is converging and diverging air streams which account for the high and low pressure areas (see 3.3).

Answer plan Describe, with the aid of a diagram, the general circulation of the air in the northern hemisphere with air streams rising at the Equator and moving northwards to meet Polar air moving southwards in the latitudes of north-west Europe. Then explain the Coriolis effect, how upper waves are formed (Rossby waves) which move from west to east and how high and low pressure systems occur in relation to these waves (Fig. 3.1). Draw a simple base map of the synoptic chart area and sketch in the approximate position of the Rossby waves in relation to the high and low pressure.

In your next paragraph use diagrams to describe convergence and divergence (see Chandler, T. J., *Modern Meteorology and Climatology,* Nelson, 1983, page 24), and relate the diagrams to the high and low pressure systems on the synoptic chart. Explain how the low and high pressure systems will develop as a result of the differences in pressure and temperature in the core areas of these systems.

3.7 FURTHER READING

Chandler, T. J., *Modern Meteorology* (Nelson, 1983)

Musk, L., *Weather Systems – Topics in Geography Series* (Cambridge, 1986)

Riley, D. and Spolton, L., *World Weather and Climate* (Methuen, 1981)

4 Weathering

4.1 ASSUMED PREVIOUS KNOWLEDGE

You may have studied the simplest forms of weathering for GCSE, for example, the process of 'onion-skin weathering'–the peeling off of outer layers of rocks as a result of heating and cooling in desert regions. In that case you will also have some knowledge of resulting land-forms such as screes. At A and AS levels you are expected to study the process in greater detail and to be able to relate observable features of the landscape to the different types of weathering, different types of rocks and to the climatic conditions under which the weathering takes place.

4.2 ESSENTIAL INFORMATION

4.2.1 Definitions

Weathering Weathering is the process of rock destruction. It is the breakdown (mechanical fracturing) or decay (chemical decomposition) of rocks *in situ* by natural agents. It is essentially a static process.

Weathering is therefore the first phase in the denudation of any landscape. Rocks must be weathered before there is debris to be transported and the effectiveness of the agents of transport (water, ice, wind) as agents of erosion depends upon the carrying of this debris.

There are three main types of weathering–physical, chemical and biological.

Physical weathering This is also called *mechanical weathering*. It is the process of the loosening of the surface of rocks and the gradual reduction of the rocks into fragments under the influence of atmospheric forces without chemical change taking place. The products of this process are usually coarse and angular.

Chemical weathering This is the process of the rotting of rocks. Minerals within the rocks are decomposed by agents such as water, carbon dioxide and various organic acids. Since minerals vary in their resistance to chemical agents this type of weathering attacks rocks selectively and may penetrate them deeply in places. The products of chemical weathering are generally 'finer' than those of mechanical weathering.

Biological weathering Flora and fauna increase the carbon dioxide content of soil and this increases the weathering potential of the biosphere. Various organisms may also cause reactions with minerals in particular rocks e.g. guano weathers limestones; chemotrophic bacteria oxidize minerals such as sulphur and iron.

4.2.2 Agents of weathering

In mechanical weathering there are two main processes at work: temperature change and crystallization. Mechanical weathering may also be assisted by the action of plant roots which penetrate and widen joints in rocks and expose a greater surface area to weathering.

Temperature changes produce the disintegration of rocks in a number of ways:

(a) Rocks are generally poor conductors of heat. The effect of daily heating and cooling is confined to surface layers of the rock. So the surface expands more than the interior and this sets up stresses which may lead to the fracturing of the rock in places roughly parallel to the surface. This is called *exfoliation*.

(b) Igneous and metamorphic rocks are made up of different minerals which expand at different rates when heated. Minute internal fracturing occurs within crystals and at their edges. Eventually the rock fractures.

(c) The *pressure release hypothesis* explains that many metamorphic rocks were crystallized under temperature and pressure conditions which were very different from those found at the surface of the earth. So minerals may be less stable at surface temperatures and pressures. As the rocks are exposed as the result of erosion, stresses are caused, fracturing the rock surface.

(d) When polycrystalline (many crystals) rocks are buried, grains of the rock may be deformed at the interfaces between them. In sedimentary rocks the cement between the grains may be affected. As the surface is eroded the load is taken off the rock and this release of energy can cause faulting which weakens the rock as it is exposed on the surface and weathering starts. So there is *granular disintegration*

(e) Temperature changes can also encourage *wetting and drying* weathering e.g., high temperatures cause evaporation of rock moisture. If rocks are alternatively soaked and dried they are more easily weathered.

Crystallization

(a) *Freeze-thaw* When water is turned into ice its volume increases by about 10 per cent. The freeze-thaw process is a very effective means of weathering in rocks which are fractured. The process cannot cause fractures but can widen them. The freeze-thaw process is especially effective in rocks such as cellular limestone in which the water collects in enclosed cavities from which it cannot escape as it expands.

(b) *Crystallization of salts* Salts are dissolved in the moisture which penetrates rocks e.g. sodium chloride (common salt), calcium sulphate (gypsum).

Chemical weathering

(a) *Hydration* Certain minerals take up water and expand. This causes additional stresses within the rock e.g. Anhydrite takes up water to become gypsum.

(b) *Oxidation* This is the process of taking up oxygen from the air. For example below the water table gault clay is blue or grey but above the water table where the water and clay are replaced by air it is oxidized into red or brown ferric compounds.

(c) *Hydrolysis* Felspars are important constituents of igneous rocks. Hydrolysis is a process which leads to the breakdown of felspars. It is caused by a chemical reaction with the water which involves H and OH ions.

(d) *Solution* This is not a very common process because few minerals are soluble. Solution may help weathering by removing the products resulting from other types of chemical weathering.

(e) *Carbonation* This is the result of the combining of carbonate ions with minerals. Carbon dioxide solution in the atmosphere converts calcium carbonate into the much more soluble calcium bicarbonate. This process is important in limestones and chalk.

4.2.3 Factors affecting the type and rate of weathering

The main factors are the hardness of the rocks (mineral composition), the texture of rocks (their crystalline state), rock jointing, relief and climate.

Rock resistance Rocks vary significantly in resistance to weathering. Resistance depends on the constituent minerals of the rock, the coherence of these minerals (how they are cemented together in the rocks), and the extent to which the minerals have been compressed.

The hardness of the minerals in the rock is measured by *Moh's scale of hardness*. This scale ranges from 10 (extremely hard) to 1 (very soft). Quartz for example is classified at 7, gypsum at 2. Most igneous rocks are hard. This is partly because of their mineral constituents e.g. quartz and felspar. It is also because as the minerals cooled and crystallized they were tightly bonded together. Sedimentary rocks tend to be softer because even those composed of hard quartz grains are often cemented together by a soft cement. If the cement is hard then the rock is very resistant to weathering e.g. quartzite which has a hard silica cement.

As far as the igneous rocks are concerned minerals which determine the rate of weathering may be divided into light-coloured (felspars and quartz) and dark-coloured (ferromagnesian minerals, e.g. mica). Light-coloured minerals have greater acidity.

Rates of weathering of minerals in igneous rocks

	Dark coloured minerals	Light coloured
Most susceptible	Olivine	Lime plagioclase
to weathering	Augite	Lime-soda plagioclase
↑	Hornblende	Soda plagioclase
↓	Biotite	Orthoclase
Least susceptible		Muscovite
to weathering		Quartz

The texture of rocks (i.e. the crystalline state) Under most conditions coarse-grained rocks are likely to weather more rapidly than fine-grained rocks which are composed of the same minerals. Although in fine-grained rocks the mineral grains have a greater surface area exposed, these surfaces are not open to weathering. Susceptibility to weathering of one of the minerals is a more important factor than the surface area exposed.

Usually one mineral in a rock is weathered more rapidly than others. The weathering of this mineral loosens the whole fabric of the portion of the rock exposed to weathering.

Rock jointing As far as both chemical and mechanical weathering are concerned the jointing of rocks is a vital factor influencing the nature and rate of weathering. Its importance is due to the fact that jointing increases the surface area exposed for attack by agents of weathering. It is very clear, particularly in limestone areas, that chemical weathering concentrates along joints and bedding planes. The joints allow acidic solutions, oxygen and carbon dioxide to enter the rocks and so encourage chemical rotting.

The pattern of jointing determines the character of the landforms produced. For example, plutonic rocks (the most common being granite) have a jointing system which divides the rock into rectangular blocks. As they are chemically weathered they are reduced to piles of partly rounded boulders such as the Tors of Dartmoor. Basalt, a volcanic rock, often has a well defined jointing pattern which forms vertical polygonal columns. Weathering of such rock has produced the Giant's Causeway in Northern Ireland.

Relief This is a factor which is often undervalued. If mechanical weathering is to continue, fresh exposure of the unweathered rock is vital. In areas of high land and steep slopes, such phenomena as landslides, slumps and solifluction result in the fresh exposure of bare rock. In lowland areas in contrast a thick layer of soil and weathered material protects the unweathered rock (although in others e.g limestone areas, soil accelerates weathering).

Climate The processes of weathering are dependent upon particular climatic conditions. For example, particular climatic conditions make the freeze-thaw a dominant weathering process. In the Tundra, for instance, the seasonal spring thaw and autumn freeze together with the likelihood of frost all through the year provide suitable conditions. In cool, temperate, humid regions there is also sufficient rainfall for water to penetrate joints and fissures and winters are cold enough to induce regular freeze-thaw.

Exfoliation and granular disintegration are most effective in regions with a large diurnal range of temperatures, for example, continental desert regions. On the other hand, wetting and drying needs sufficient precipitation to wet the rocks and temperatures which are warm enough to evaporate the moisture.

Chemical weathering is generally most effective in hot, humid climates. Equatorial climates provide ideal conditions for the rotting of rock masses.

4.3 General concepts

Weathering is the first phase in the denudation of any landscape.

The nature of the rocks themselves (hardness, mineral content, texture, jointing, etc.,) affects the nature and rate of weathering.

4.4 Different perspectives

At one time weathering was seen as a distinctive and separate process of landscape formation. In the same way mechanical and chemical weathering processes were regarded as separate types of weathering. Today the process of weathering is placed in the broader context of landform development and various types of weathering are seen as simultaneous and inter-related influences.

In the systems analysis of landscape formation known as the *rock material cascade* for example weathering is seen as contributing to the formation of rock waste which is the initial input into the system. The cascade is seen as being comprised of input (weathering and erosion), throughput (transportation) and output (deposition). Parts of the deposited output are cycled back into the crystalline rocks where the cascade begins again.

In recent years an important development in the study of geomorphology has been the study of slopes. Weathering and surface transport are now seen to be the two main groups of processes responsible for slope formation. The significance of the weathering process has therefore increased appreciably in the study of landforms.

4.5 Related topics

Unit 5 *Slopes* is very closely related to the work covered in this section. Since mechanical weathering is an important process operating in desert regions you should also read Unit 10 *Desert Landforms* in conjunction with this section. The units on glaciation and rivers and river valleys cover the transportation and deposition of weathered materials.

4.6 Question analysis

1 To what extent are the various processes of mass wasting influenced by climatic factors?

(in the style of Oxford and Cambridge)

Understanding the question There are two main parts to this question. In the first place you will be expected to outline what you understand by *various processes of mass wasting*. Then you are asked to judge how important an influence climate is upon these different processes. Since the first part of the question is the more descriptive section, it will probably be worth fewer marks (roughly one third).

There is a need to define clearly what you mean by 'mass wasting'. The wasting of rocks implies that the rock mass is broken down and rotted *in situ*. It is therefore a question about weathering and not about erosion.

Answer plan Begin by explaining what you understand mass wasting to mean. Establish very firmly that you see it as the process of weathering. It is then important that you demonstrate thorough knowledge of the various processes which are included under the umbrella heading of weathering:

MECHANICAL WEATHERING	CHEMICAL WEATHERING
temperature changes	hydration
pressure-release changes	oxidation
granular disintegration	hydrolysis
wetting and drying	solution
crystallization	carbonation
freeze-thaw	
dissolution of salts	BIOLOGICAL WEATHERING

You could present the above information as a table and write paragraphs on each of the main headings in which you define the various processes listed.

There is no need to relate each of the items listed to climatic factors – it would take far too long to answer the question. You can tackle the analysis by dealing with each of the main groups in turn. The purpose of this section is to establish that climate is a vital factor which influences both the nature and rates of weathering. See Section 4.2.3 on climate.

A final part to this section should emphasize that many of the weathering processes are totally dependent on conditions determined by the climate and that climate is therefore a fundamental influence.

In your conclusion you should qualify this by stating that although the climate creates the general conditions conducive to the operation of particular types of weathering, the extent to which that weathering is effective is determined by the chemical composition of the rocks. The resistance of rock to both mechanical and chemical weathering is a product of its mineral composition. So climate is not the only major factor in operation.

2 Discuss the relative importance in the evolution of landscapes of physical and chemical weathering.

(in the style of Southern Universities' Joint Board)

Understanding the question As long as you attempt to answer this question in an analytical and not descriptive way it should not cause any problems. You are not asked to describe how the two chief forms of weathering contribute to the creation of landscapes but to assess which of the two is the more important. This is a wide question. As it is worded it does not give you much guidance on how to structure your answer. The planning of your answer is therefore important, a good structure will earn good marks.

Answer plan A structure which would enable you to answer the question in the time available is:

(a) Introduction – definition of terms
(b) Relative importance of both types in relation to major climatic regions
(c) Relative importance of both types in relation to different rock types
(d) The place of weathering in the process of landscape evolution
(e) Conclusion – assessment of relative importance.

(a) *Definition of terms.* Mechanical and chemical weathering should be defined. You should also explain what you understand by the *evolution of landscapes*. This can be dealt with on two scales – the development of individual landforms, and the evolution of landscapes which are typified by particular landforms.

(b) Since climate is a prime factor in the weathering processes one way of attempting to assess the relative importance of both types of weathering is to examine their significance in different climatic regions. As far as weathering is concerned it is usual to divide the climatic belts into four main types:

Arctic regions Dominated by the freeze-thaw processes of mechanical weathering.

Humid temperate regions Chemical weathering is probably dominant since rocks are always moist at ground level. Chemical weathering is general, mechanical processes operate in a more limited way and are most effective in specific regions at specific times of the year.

Arid and semi-arid regions Mechanical weathering is dominant but chemical weathering is more important than once thought.

Tropical humid regions Hot, humid conditions are ideal for chemical weathering.

This simple pattern might suggest that both types are equally significant in the creation of landforms. Recent research however indicates that chemical weathering is much more important in Arctic and arid regions than previously believed. Carbon dioxide is more soluble at low temperatures so in Arctic regions meltwater absorbs the gas. Tamm now suggests that rock flour produced by glaciers may be the result of chemical weathering. Williams believes that snow patch erosion is also the result of that process. In arid areas the minute quantities of atmospheric moisture are important in processes such as exfoliation. So overall, chemical weathering is probably the more important.

(c) The effectiveness of mechanical weathering is related to the texture and jointing of the rocks. The mineral composition of the rock determines the effectiveness of chemical weathering. These factors influence the rate of weathering but the nature of the process is basically the result of climatic conditions. Given the distribution of the main climatic belts it would seem therefore that chemical weathering is the more widespread and active. In addition whereas mechanical weathering is confined mainly to rocks at and above ground level, chemical rotting may extend up to 100 metres below ground in suitable conditions. So again chemical weathering seems to be the main agent of rock destruction.

(d) (Read the section on the rock cascade in Section 4.4.) Both mechanical and chemical weathering provide rock fragments as the vital input into the sequence of denudation by processes of erosion. Since chemical weathering is the more widespread it is reasonable to assume that it provides the greater input. In view of the climatic conditions under which mechanical weathering is dominant, the products of mechanical weathering are likely to be transported by wind and ice and used to sculpture distinctive landscapes. The debris of chemical weathering on the other hand occurs mainly in humid regions where 'normal' erosive processes are dominant.

(e) Draw the main points you have made into a short conclusion. It would be a good idea to point out that the processes do not operate totally separately but are allied aspects of the same process of mass wasting.

3 Compare the nature and effects of the principal weathering processes of the humid tropics and the tundra.   *(in the style of the Associated Examining Board)*

Understanding the question This is a more complicated question than it first appears. It is easy to fall into the trap of just writing an account of weathering in two contrasting environments. In asking you to compare the examiner expects you to draw out factors common to weathering in both environments and to point out the differences and contrasts. It is also complicated because you are asked to deal with the nature *and* effects of the processes you outline. It would be easy to overlook the effects and so lose marks.

Answer plan The first part of the main body of your answer should be concerned with comparing the nature of the principal weathering processes operating in the two environments. Points to include:

(a) In both environments one of the two main types of weathering is clearly dominant. In the tundra mechanical weathering is dominant, and in the humid tropical environment chemical weathering is supreme.

(b) A feature common to both the tundra and the tropics is that weathering occurs at two levels – on the surface of the land and at subsurface levels:

	Humid tropics	Tundra
at the surface subsurface	on bare rock rotting of rocks to great depths	ice-wedging of bare rocks freeze-thaw above permafrost

(c) The subsurface weathering in both environments occurs above well-defined levels – in the humid tropics above the *basal weathering front* (or etchplain), in the tundra above the *permafrost*

(d) In both environments biological weathering occurs. This serves to complement the dominant weathering process. In the humid tropics rapid root growth forces the mechanical splitting of some rocks along fractures and joints. In the tundra flat waterlogged areas encourage the development of peaty masses. These masses produce organic acids which stimulate chemical weathering.

(e) In both environments the movement of weathered rock can take place. Chemical weathering in the tropics creates a deep mantle of rotted rock waste. Heavy tropical rainfall in hilly areas and on steep slopes leads to sudden landslides. In the tundra the soil and waste above the permafrost become waterlogged when snow and ice melt in summer. Vast quantities of liquid mud may then move down hill. This is called solifluction, which is not usually a rapid movement.

(f) In both environments the effectiveness of the weathering processes varies with the nature of the rock. Well-jointed soluble limestones are very susceptible to weathering in both environments. Other rocks, e.g. granite, may be fairly resistant to mechanical weathering but break down chemically in humid conditions.

(g) The rates of weathering fluctuate within the broad climatic belts according to the amount of moisture available. In equatorial forest areas the regular rainfall throughout the year produces more intensive weathering than in the savanna lands with their pronounced dry season. In the tundra the coastal areas with their longer springs and autumns and heavier precipitation are the most intensely weathered areas.

The second part of your answer should examine the contrasts between the nature of the processes. The main points are:

(a) How the dominant chemical weathering processes and subsidiary mechanical processes operate in the tropics should be contrasted *briefly* with the way in which weathering operates in the tundra.

(b) In the humid tropics the rate of chemical weathering is rapid (include some of the details listed earlier in this chapter). This is the area of the world where the rate of weathering is at its maximum. Contrast this with the factors in the tundra which restrict the rate of weathering.

The final section of the main body of your answer should then deal with some of the effects of the processes. Points which are relevant are:

(a) The transfer of material through landslides and solifluction causes natural hazards and disasters.

(b) The weathering processes expose fresh surfaces of bare rock which in turn are attacked by the weathering agents.

(c) The weathered debris in the tundra, since it is mainly mechanically derived, is angular and used by moving ice and snow to erode the land. The rotted material in the humid tropics is less 'useful' as an abrasive and cutting agent.

(d) There are also implications for man. Because the rotten rock may extend to considerable depths in the tropics, it is often easy to cut new roads. On the other hand it is difficult to quarry sound rock to build harbours, bridges etc. The freeze-thaw process may also affect building in the tundra. Concrete piles have to be sunk deep into the permafrost to support large buildings to avoid collapse when the ground thaws.

4.7   FURTHER READING

Clowes, C. and Comfort, P., *Process and Landform* (Oliver and Boyd, 1982)
Ollier, C., *Weathering and Landforms* (Macmillan, 1974)
Small, R. J., *The Study of Landforms* (CUP, 1974)
Sparks, B. W., *Geomorphology* (Longman, 1972)

5   Slopes

5.1   ASSUMED PREVIOUS KNOWLEDGE

For GCSE you will have studied rivers and river valley development. You will also have studied weathering. In both areas of work you are expected to be able to recognize particular landforms on the Ordnance Survey map and aerial photographs. You have also learned to draw diagrams to show the main features of individual or sets of landforms. Examination and essay questions usually asked you to name examples and to give some explanation of how they were formed. All this information and experience provides a good factual basis upon which to build up a more integrated and scientific view of the process of landscape evolution which you can gain from the study of slopes.

5.2   ESSENTIAL INFORMATION

5.2.1 Definitions

Slope form The shape of the land surface which makes up the slope.

Slope process The agents bringing about changes in the form.

Slope evolution The evolution of the slope is the change from past to present form.

Time dependent form Slope forms which depend on the stage of evolution they have reached e.g. a slope which becomes gradually gentler with time.

Time independent form Slope forms which remain the same through different stages of evolution e.g. a slope may be lowered but its steepness and shape remain constant. A retreating escarpment may retain the same form.

Slope retreat The wasting back of slopes by weathering and surface wash.

Regolith A covering of waste material composed of soil and weathered rock which lies beneath the soil. Its composition is constantly being changed by weathering as it moves down the slope.

Soil creep The gradual movement of the regolith down the slope. It is a slow process but it affects the whole slope. It is caused by the expansion and contraction of the soil combined with the effect of gravity (see *heave* below). Rates of creep vary with climate – maximum rates (5 mm per annum) being in humid tropical areas.

Solifluction A more rapid form of soil flow which occurs mainly in periglacial areas.

Through flow (sometimes soil throughflow) The downhill movement of water through the soil.

It is significant in humid tropical areas e.g. the clay particles are removed from slopes and concentrated in the valley floor.

Surface wash The transport of soil by water flowing across the ground surface. There are two processes involved.

(a) *Raindrop impact* – when raindrops hit bare regolith soil particles are detached. This produces miniature craters. In some instances this can lead to the formation of earth pillars when soil is washed away except for pillars protected by resistant caps.

(b) *Surface flow* occurs when rainfall intensity is greater than the rate at which the soil can absorb water. The ground is saturated and the rain no longer percolates into it. Surface flow also occurs at the base of slopes where the water table rises to the surface because the soil is saturated.

The rate of surface wash varies with rainfall, vegetation cover, slope angle, distance from the crest of the slope.

Mass movement The movement of soil and parts of the regolith down the slope. It can be classified as follows.

slow or gradual movements	*rapid mass movements (landslides)*
solution	rockfall
soil creep (heave)	soil slip
solifluction – frost creep, gelifluction	mudflow
surface wash – surface flow, raindrop impact	debris avalanche
	rotational slip

Heave (see Fig. 5.1) The heave mechanism is the process whereby rock waste moves downhill. It is caused by freeze-thaw processes, wetting and drying etc., which cause clay particles to swell and shrink and the expansion and contraction of loose rock fragments as a result of temperature changes. As material moves it shifts upwards towards the surface of the slopes and settles back vertically. The net result is downward movement.

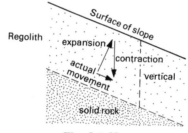

Fig. 5.1 Heave

Stable slope A slope or part of a slope on which rapid mass movements do not occur. The rock and regolith possess sufficient strength to resist the forces tending to cause mass movement. A stable slope has a smooth surface.

Unstable slope A slope on which mass movements occur because the rock or regolith is not strong enough to resist the forces which cause them. Active landslides and scars of past ones are evidence that the slope is unstable.

Landslide A particular kind of rapid mass movement on slopes. Landslides usually affect only a very small part of the slope and occur infrequently. A frequent cause is the saturation of the regolith by very heavy rain.

Free face A steep slope or part of a slope formed of bare rock.

Cliff A slope which consists mainly or entirely of a free face.

Scree An accumulation of rock fragments at the foot of a free face. The steepness of the slope of the scree depends on the *angle of repose* (usually $32° - 38°$). This is the maximum angle at which the fragments will accumulate without sliding further downhill.

Slope angle The angle made with the horizontal, it expresses the steepness of the slope. *Percentage grade* is the vertical rise in metres per 100 metres horizontal distance. Percentage grade $= 100 . \tan \theta$, where θ is the angle in degrees.

5.2.2 Slope form

The slope profile is the shape of the slope viewed as a cross-section at right angles to the hillside. For analysis it is divided into *slope units*. These are the straight and curved parts. Straight parts are called *rectilinear segments*, curved parts are either *concave* or *convex elements*. The straight parts are characterized by a constant slope angle. The *maximum segment* is the part which is steeper than the slope units above or below it. Below the maximum segment is the *concavity*, above the *convexity* (Fig. 5.2).

On simple slopes there is only one sequence of convexity — maximum segment — concavity. More complex slopes have more than one sequence.

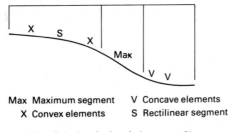

Max Maximum segment   V Concave elements
X Convex elements   S Rectilinear segment

Fig. 5.2 Analysis of slope profile

Slope angles Slopes may be roughly classified into the following categories.

Description	Angles	General features
Level/almost level	0°–2°	Drainage problems on impermeable rocks
Gentle	2°–5°	Most common slope in many landscapes
Moderate	5°–10°	In humid tropics soil erosion becomes a problem
Moderately steep	10°–18°	Problems for farming and building. In humid tropics soil erosion could be serious
Steep	18°–30°	Farm machinery cannot be used unless land is terraced. Building is expensive.
Very steep	30°–45°	The steepest land to carry a regolith
Precipitous/vertical	> 45°	Free face

5.2.3 Slope evolution

Three models of slope evolution have been developed (see Section 5.4). Not one of them is completely appropriate to all slopes. Structural and climatic factors determine to which of the three the retreat in slopes of a particular area most nearly corresponds.

The basic slope model (Fig. 5.3) This model is based upon the assumption that the caprock is strong enough to sustain a vertical face at its edge. This might occur, for example, when resistant sandstone or limestone overlies less resistant shale. The diagram shows that the slope has four components:

(a) *The waxing slope* is the curve over edge of the horizontal surface of the hilltop. It was called a waxing slope by Penck because, on a given vertical line, with time, it increases in slope. It is also called the *summital convexity*. The rounding off of the slope is the result of weathering.
(b) *The free face* See definition above.
(c) *The constant slope* is a slope with a uniform angle that did not alter as the slope developed through time. Many constant slopes only have a very thin veneer (cover) of rock waste.
(d) *The pediment* is solid rock. It has a concave shape with a decrease in slope angle in the downslope direction.

This model envisages slopes that are retreating.

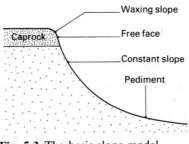
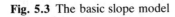

Fig. 5.3 The basic slope model

The nine-unit slope model (Fig. 5.4) This is a more complex model which relates slope morphology (form) to the processes of slope formation. The main processes in operation are weathering, throughflow and mass movements (landslides, slumping, rockfall). Units 1–3 in this model are comparable to the waxing slope and unit 4 is equivalent to the free face. Units 5–7 correspond to the constant slope and the pediment.

The main difference between the models is due to the fact that the basic slope model is based on semi-arid conditions while the nine-unit model relates to regions with humid temperate climates. Both models can be used as a static basis for comparison of areas and to identify areas in which similar units may be found. Neither model, however, incorporates the variety of forms that are to be found in different natural environments.

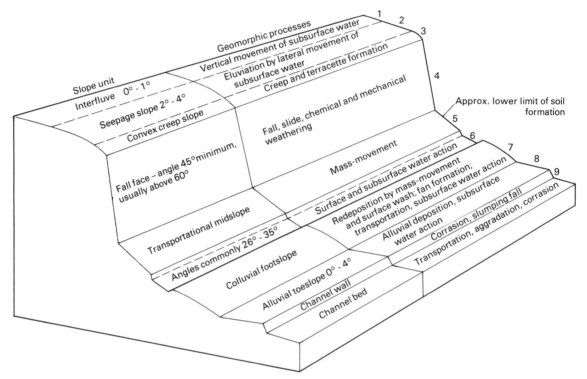

Fig. 5.4 The nine-unit model

5.2.4 The influence of structure on slopes

Simple slopes composed of one rock type Lithology influences profile form and angle. The strength, stability and permeability of a rock are important factors in determining slope form. As far as profile is concerned it has been found, for example, that convexities form the greater part of profiles developed on sandstones, about half on limestones and less than half on shales. Maximum slope angles also vary with lithology, for example, maximum slope angle on limestones is usually about 20°, on shales 9°, and on clays 5½°.

The nature of the regolith which is formed is also important. For example, the importance of surface wash as a slope forming process depends on the permeability of the regolith. Surface wash is more effective on less permeable regoliths such as clay.

Compound slopes Compound slopes are slopes composed of more than one rock type. Where beds of different degrees of resistance outcrop on a slope there is likely to be a number of convexity——maximum segment——concavity sequences. The maximum segment then corresponds to the most resistant bed.

5.2.5 The influence of climate on slopes
(see for example, A & D Youngs' book, pp 29–33)

Climate is a significant factor influencing slope form. It affects surface processes directly and indirectly. For example, it determines the significance of weathering processes such as frost shattering or surface run-off. Past climatic factors may also play a significant part in determining the nature of slopes e.g. the relict periglacial features of the present day temperate regions.

Vegetation The main effect of climate is achieved indirectly through its influence on vegetation. The main effect of vegetation is *the protection it gives the soil* from surface wash and rainsplash. Other effects are: *the action of roots* in holding the regolith on the slope, *the contribution to chemical weathering* by the products of organic acids and *the supply of organic matter* which improves soil structure.

5.3 GENERAL CONCEPTS

Form and process In the study of slopes as of any other aspect of the physical environment there is a distinction between form and process. Form is essentially the shape of the ground surface and is studied through techniques such as morphometry, slope profiling and contour surveys. Process refers to the agents which interact to produce the form. Processes acting on slopes include *weathering, soil creep* and *surface wash*. Both form and process existed in the past. The evolution of a slope is the succession of past forms created by interacting processes which have led to the form of the slope at present.

5.4 DIFFERENT PERSPECTIVES

Descriptive and genetic approaches It is possible to study slope form from a descriptive angle, e.g. by plotting isopleths of slope angles or plotting the frequency distribution of slope angles on a graph. The descriptive approach provides empirical data from detailed field observation. This makes it possible to test and refine existing theories and to develop new ones. The genetic approach is concerned with the origin of landforms. This approach is therefore concerned with past as well as present slope forms. So the main concern is the study of slope evolution.

Slope evolution theories

Slope decline (W. M. Davis) This theory is related to the Davisian cycle of erosion and implies that slope form is time-dependent (see Section 5.2.1). The movement of rock waste is seen as one stage between weathering on the one hand and transportation by rivers on the other. The forms of slopes change as the cycle of erosion advances. When slopes are first developed they are steep and covered with coarse material. Later in the cycle the graded slopes are gentler and are covered with a thicker layer of finer material. The slope decline is caused by the fact that the downwash of soil from convex upper slopes is faster than its removal from the slope base.

Slope replacement (W. Penck) This theory assumes that the surface of the slope is weathered evenly and crumbles – the fragments falling to the base. Thus maximum angle of the slope decreases and is replaced by a gentler slope. Evolution occurs by parallel retreat. This results in the greater part of the profile becoming a segmented concavity. In comparison with the Davisian theory the slope replacement theory is much more concerned with the ways in which the processes affect the forms of slopes and the ways in which the processes themselves are affected by the slope form. It is therefore a process – response model of slope evolution.

Slope retreat (L. C. King) In this theory the maximum angle of slope is held to remain constant as each of the upper portions of a slope retreats by the same amount. During the retreat the convexity, free face and constant slope (Fig. 5.3) all keep the same length while the pediment extends in length and becomes gentler in angle.

Dynamic equilibrium This is a modern theory of landform development. It is a general systems theory approach in which it is assumed that all aspects of landform geometry such as slope length, maximum slope angle and channel gradient are all closely related to each other.

The theory suggests that as long as the factors which control the processes of denudation remain the same the form of the land need not change i.e. the landforms are time independent (see Section 5.2.1). This is achieved when the system is in a steady state i.e. when there is an energy balance. The energy balance is such as to achieve the continuous and efficient removal of all detritus.

5.5 RELATED TOPICS

There are two sections of this book which are very closely related to the study of slopes – Unit 4 *Weathering* and Unit 10 *Desert landforms*. All three chapters should be studied as closely inter-related aspects of the denudation of the landscape.

5.6 QUESTION ANALYSIS

Questions on slopes and weathering are often combined. Before studying these questions it is worth looking at the unit on weathering again.

1 What are the main factors influencing the form of a slope? *(in the style of Cambridge)*

Understanding the question A vital factor in answering this question is that you have a good understanding of what is meant by the *form* of a slope. There are no hidden snags in the question. It is designed to test the quality of your knowledge and your ability to look at the issue from more than one angle.

Answer plan First, define what is meant by slope form. It is the morphology of the slope at any given moment of time. The form includes the thickness and composition of the regolith as well as the shape of the ground surface.

A very significant point which needs to be established early in the answer is that slope form is fundamentally affected by the environmental conditions in which the slope exists. There are three main environmental variables in this respect: structure, climate, and the stage of evolution reached by the slope.

Most of the rest of your answer then falls into three parts:

(a) *The influence of structure on slopes*–see Section 5.2.4 above. Elaborate on the points made in the section by quoting actual examples you have studied in detail especially any exercises you have carried out on slope form in the field. (See chapters 4 and 5 in Sparks' book which refers to examples.)

(b) *The influence of climate*–see Section 5.2.5 above and, for examples, A. & D. Youngs' book, pages 29–33. The answer to the following question is also relevant to this section.

(c) *The stage of evolution reached by the slope*–see Section 5.4 above. Emphasize the point that it is the structural and climatic factors which determine to which of the three models of slope evolution the form of a particular slope will conform most closely.

You must then establish the fact that slope form may also reflect past conditions. These may be grouped into two broad categories:

(a) *Inherited conditions:* features of slope form which are the products of previous erosional landscapes. For example, changes in land and sea levels may have caused rejuvenation of drainage systems. Consequently the landscape may now be characterized by 'valley in valley' features and in particular by *polycyclic slopes.*

(b) *Relict features:* features of slope form which date from a period when climatic conditions were different e.g. in cool temperate regions in Europe there are slope forms which are relict from an earlier glacial period. The slope profiles of those areas which experienced periglacial conditions were generally rounded and smoothed with net erosion occurring in the upper (convex) part and deposition on the concavity.

Finally, make the point that the slope form evident in a specific area is the product of the interaction of these factors.

2 Give an explanatory account of the processes by which weathered material is produced and transported in (a) humid cool temperate regions, (b) hot deserts.   (*in the style of the Joint Matriculation Board*)

Understanding the question This is a straightforward question as long as you realize that the key phrase is *an explanatory account*. It is not good enough to write a detailed description, you need to explain *why* particular processes operate and suggest reasons for their relative importance.

Answer plan

It is important that you should define *processes*. They are the agents which interact to produce the weathered material and move it downhill. Given the way in which the question is worded you should deal with the two climatic regions separately.

(a) *The humid cool temperate* It would be relevant to begin this section by pointing out that three main factors determine which processes operate and which of them is particularly significant:
 (i) the low intensity of rainfall which characterizes this climate
 (ii) the absence of a dry season so that rocks and regolith may be permanently damp
 (iii) the soil is well protected by a permanent vegetation cover.
As a result of the interplay of these factors, chemical weathering is the chief process of rock wasting and solution is an important weathering process. Even in the coldest spells the soil is not frozen to a sufficient depth to make solifluction important. Surface wash is not very important because of the low intensity of rainfall and the protection of the regolith by vegetation.

This is the only climatic belt in which soil creep is more important as an agent of transportation than is surface wash. Within this belt the rate of weathering is generally moderate. Transportation is slow for slopes of less than 30° angle have permanent vegetation cover. The general process of slope evolution is that of slope decline. The broad pattern is affected by structural factors e.g. a resistant caprock produces a free face on which mechanical weathering may be especially important and from which rapid movement of weathered material occurs.

(b) *Hot deserts* The main interacting factors in this climatic belt are:
 (i) the lack of vegetation cover
 (ii) the effects of high temperatures and of differential rates of expansion by and within rocks
 (iii) occasional storms which result in extensive sheet wash.

In this climatic region mechanical weathering is a very significant process. Because of the absence of vegetation cover lithological and structural differences in rocks may produce intricate micro relief features. Steep slopes lack a regolith and most slopes are either steep or very gentle. In mountainous areas slopes consist of cliffs and boulder controlled slopes of 30°–35°. The main transportation process is sheetwash following storms because weathered material is not protected by vegetation. Because of the absence of a regolith the retreat of steep slopes is wholly subject to control by weathering. Slope retreat is probably slower in deserts than in any other climatic region. Parallel retreat is the basic pattern of slope evolution.

3 Examine the factors which influence the rate and type of movement of weathered material on slopes.
                    (*in the style of Oxford and Cambridge*)

Understanding the question This is perhaps a more difficult question than it seems at first sight. It involves bringing together a variety of material on slopes and on weathering processes. The focus of the question is the 'rate and type of movement' so it is about *transportation* on slopes.

Answer plan The first point to make is that the types of movement which occur can be classified according to the rate at which the weathered material moves (see Section 5.2.1.).

You do not have time to describe each of these types of movement. What is important is to identify the factors which influence the rates at which these movements occur. The main factors are:

(a) the type and nature of the rock (see the unit on weathering).

(b) climatic factors – in addition to the information in the weathering chapter you could include: *soil creep rates* are related to climate (5 mm per annum in tropical rain forests, 1–2 mm in temperate deciduous woodlands). *Solifluction* is also dependent upon climate (the existence of a permafrost layer so that surface soil is saturated by melt water). *Surface wash rates* are directly related to the intensity rather than the amount of rain, and to the lack of surface vegetation in arid lands. *Saturation* is more rapid in tropical climates.

(c) depth of material in relation to the ground surface – the rate of movement in soil creep, for example, decreases with depth in the regolith.

(d) slope angle – the weathering of free faces results in rapid movements due to gravity. Surface wash is proportional to the angle of the slope and to the square root of the distance from the crest.

(e) the stage in evolution achieved by the slope. In a landscape characterized by slope decline the graded slopes become gentler and become covered with a thick layer of finer material. The slopes therefore become more stable and the rate of movement of material decelerates. In a landscape characterized by slope replacement the decrease of the maximum angle of the slope as it is replaced from below by gentler slopes also decelerates transportation of materials. This is a stimulus-response model in which transportation affects the slope but changing slope form in turn affects the rate of transportation. In a slope retreat landscape although the maximum angle of slope remains constant the extension of the pediment and the decrease in its angle may also reduce rates of movement.

(f) A final point to make is that if one accepts the concept of dynamic equilibrium (see Section 5.4) the landforms are time-independent. So the transportation (efficient removal) of debris is sustained as a result of the maintenance of the energy balance in the system.

5.7 FURTHER READING

Dury, G. H., *Environmental Systems* (Heinemann, 1981)
McCullagh, P., *Geomorphology* (OUP, 1978)
Sparks, B. W., *Geomorphology* (Longman, 1972)
Young, A., *Slopes* (Oliver & Boyd, 1972)
Young, A. & D., *Slope Development* (Macmillan, 1990)

6 Rivers and river valleys

6.1 ASSUMED PREVIOUS KNOWLEDGE

If you have not studied rivers, river valleys and the hydrological cycle in physical geography, you should know the following aspects of the subject as a basis for your A level reading: (a) The upper middle and lower courses of a river valley and the distinctive features of each of these stages. These include the types of erosion and their relative significance, the shape of the valley and its main physical features. You should be able to illustrate these points with diagrams (see Lines, C. and Bolwell, L., *Revise Geography*, Letts 1987). (b) How these features can be recognized on oblique aerial photographs and on Ordnance Survey map extracts. You should be able to give named examples of the physical features connected with the three stages, such as rivers with interlocking spurs, meanders, river cliffs and flood plains. British rivers display examples of these features (see Clowser, C. E., *Physical and Human Geography*, Blackie, 1963). (c) Other physical features which are found in some river valleys are gorges, waterfalls, cataracts and rapids, river terraces, deltas and estuaries. You should also understand the inputs and outputs of a river basin and the significance of rainfall interception.

6.2 ESSENTIAL INFORMATION

6.2.1 Definitions

Hydraulic action The mechanical work of flowing water in which loose fragments may be prised away from the bedrock.

Corrosion As the result of chemical action material is dissolved and removed in solution. Limestone is one of the rocks which will dissolve in this way.

Abrasion The erosion of the stream channel by material suspended in, or moved along by a stream. In the process the river's load is also abraded leading to the rounding of pebbles and fragmentation of material.

Cavitation The shock waves propagated by the collapse of bubbles in turbulent water which hammer any adjacent rock surfaces.

Bed load That part of a stream's load which is moved along the bed of the stream by sliding, rolling and saltation (hopping). It contrasts with the suspended load which can constitute about three-quarters of a stream's total load.

Stream velocity The speed of the flowing water. The velocity depends on the slope of the bed, the shape of the channel and the volume of water involved (discharge rate).

Interception The capture of raindrops by the leaves, branches and stems of plants, preventing some of it reaching the ground.

6.2.2 A stream's energy

The energy possessed by a stream will vary with the gradient and volume of water. There are a number of ways in which the stream loses energy.
(a) Energy is lost as a result of friction between the river and the sides and bottom of the channel. The most efficient shape for a stream channel is semi-circular. Bends increase friction and dissipate energy as heat into the atmosphere.
(b) The water in a stream with an uneven bed will be turbulent and lose energy as a result of shearing between turbulent currents.
(c) Energy is lost transporting material. Less energy is lost transporting material in suspension than in moving material along the river bed. When a stream has insufficient energy to transport its load it starts to *deposit* it. During a flood the enormous increase in the discharge results in greatly increased velocities and load. The additional energy can be used for extensive erosion.

6.2.3 The load carried by a stream

The amount of material carried by a stream will depend on its potential energy and also on the amount of material delivered to it down the valley slopes. This will depend on such things as the steepness and resistance of the rocks in the river basin to erosion, the nature of the vegetation on the valley slopes (since roots and vegetation can check movement downhill), and also on the amount of weathering to which the rocks are subject.

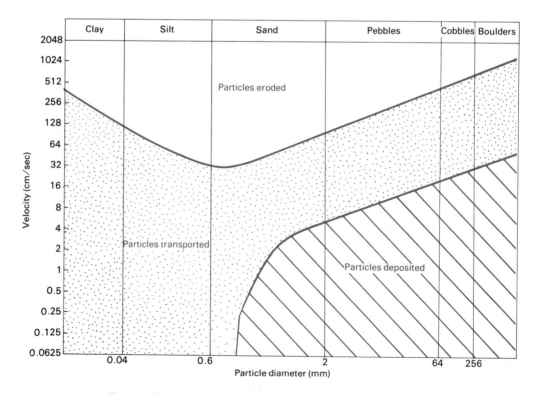

Fig. 6.1 The relationship between velocity and particle size

There is a distinct relationship between velocity of the water in the river channel and the particle sizes which can be eroded, transported and deposited. These relationships are shown in Fig. 6.1. The top line on the graph shows the lowest speeds at which particles of a given size which are loose on the channel bed will be moved. The section of the graph showing particles transported indicates the speeds at which particles of different sizes will be carried. Particles do not require such high velocities to be transported as they do to be set in motion. In general, the larger the particles the greater the velocity required to transport them. However, once the velocity falls below a certain point the particles are deposited. The velocity at which particles are deposited is higher for the larger and heavier particles than for fine clays and silts which almost float in the water (Fig. 6.2).

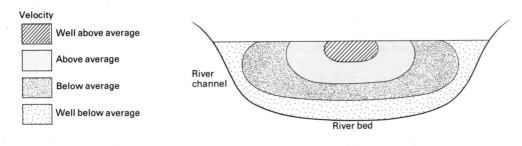

Fig. 6.2 Velocity distribution in a river channel

6.2.4 The long and cross profiles

The description of the long profile of a stream as a concave curve steeper in the headward section and flattening towards the stream mouth is an over simplification. It is rarely found in practice. This concept is based on the Davisian idea of a graded profile (see Section 8.3) which is set at variance with research findings involving the measurements of stream processes.

Stream channels in general do develop and produce a state of apparent equilibrium (quasi-equilibrium) between the channel characteristics and the movement of water and material through them.

It has been pointed out that there are eight inter-related variables involved in determining changes in river slope and channel form throughout the long profile of a river. These variables are *discharge, channel width, water depth, water velocity, amount of sediment load, load particle size, roughness of the channel bed* and *the slope (gradient) of the channel*. The significance of each of these in the long profile of a stream is shown on Fig. 6.3. Changes in one of these variables, for example an increase in the sediment load, may be compensated for by adjustment of one or more of the other variables such as an alteration in the depth of the stream and width of the channel.

The cross profile of a river valley includes the shape of the valley as a whole, the valley floor and the river channel. Slopes on the valley sides will be steep if a stream is actively downcutting

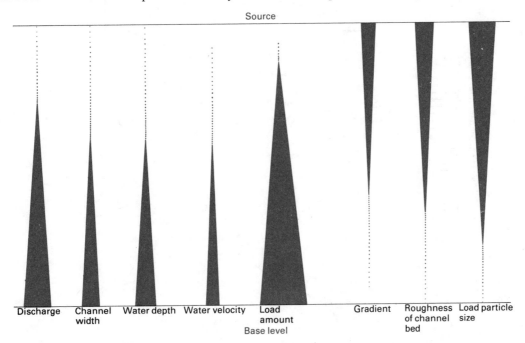

Fig. 6.3 Variations in stream channel characteristics from the upper reaches to the mouth

or the rocks are resistant. Floodplains and meanders occur where the stream is tending to erode laterally rather than downwards.

Over a period of time migrating meander belts will widen the flood plain leaving a low cliff or bluff-line bordering the meander belt. During floods deposition can occur on the sides of the channel to form levées. In the river channel dropped sediment can form shoals which cause the channel to braid. If the channel splits many times a network of distributaries develops.

Cross profiles of valleys can be asymmetrical. This may be caused by the structure of the rocks which may dip or contain faults. The stream tends to migrate down the dip (uniclinal shifting) taking the line of least resistance. Asymmetry also occurs where the structure and lithology do not exercise any control. Several theories have been put forward which relate asymmetry in these circumstances to periglacial processes. One group of theories is based on the possibility of periglacial processes steepening the valley slope while the other group assumes that frost action combined with solifluction would lead to a decline in the slope angle. The subject is very complex and a more detailed account can be found in Small, R. J., *The Study of Landforms,* CUP, 1970.

6.2.5 Meanders

The exact reasons why meanders develop in river channels which are straight are not fully understood (see Section 6.3). Once the current starts to swing it is most likely that meanders will occur. Meandering does not develop in sands where the shape of the channel changes with the amount of discharge.

The characteristic features of a meander belt are shown in Fig. 6.4. As the diagram shows, in a meander bend the channel cross-section is asymmetrical with erosion on the outer section of the bend and deposition on the inside of the bend.

Migration of meanders downstream 'planes off' higher land adjacent to the river leaving the low cliffs or bluffs mentioned in Section 6.2.4. During the migration, point bar accretion occurs on the convex bank where reduced velocity results in deposition. As the meander shifts it leaves cut-offs, swales and scars where the previous point bars occurred.

6.2.6 River régimes

The régime of a river is the pattern of the rate of discharge over a period of time. A number of factors determine the rate of discharge. They can be summed up as geology and soils, climatic conditions and miscellaneous.

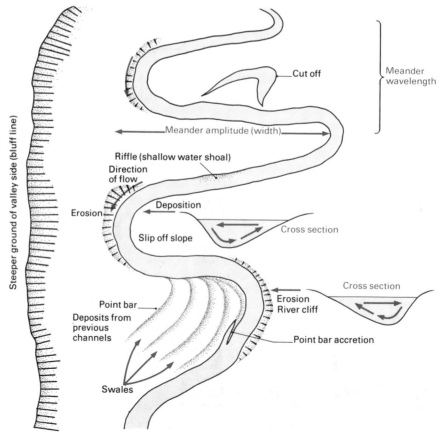

Fig. 6.4 Features of a meander belt

Geology and soils The amount of precipitation which finds its way into a river depends on such things as the gradients, permeability of the rocks, soils and plant cover in the catchment basin.

Climatic conditions The precipitation and its incidence throughout the year constitutes an important factor when linked with the annual temperature range and the evaporation rate. The nature of the precipitation, whether it falls as snow, torrential rain or as light showers, is also significant.

Miscellaneous Streams with many tributaries may have a different régime from streams with few. Tributaries coming from different climatic regions give rivers like the Nile distinctive régimes. Large areas of swamp land or mountain bogs in the catchment basin may even out rates of discharge by acting as reservoirs. A river rising at a high altitude may receive snow-melt when precipitation is low.

6.3   GENERAL CONCEPTS

Equilibrium (grade) In Section 6.2.4 reference was made to the long profile of a river being concave in shape. However, in most river valleys there are bands of resistant rock which erode more slowly than the other rocks and will appear as 'bumps' on the concave profile. The river will slowly wear away these obstructions and the long profile above the resistant rock outcrops will tend to become smooth as the temporary check resulting from the resistant rock is removed.

Eventually the resistant rock band itself will be smoothed and the profile will form a curve from source to mouth. When this profile has been achieved the stream's total energy is just sufficient to transport its load. The stream is said to be *graded* and to have a *profile of equilibrium* Any changes such as uplift, flooding, or diversion of some of the water, will upset the equilibrium and the stream will slowly adjust its profile until it is once more graded.

A definition of a graded stream has been put forward by J. H. Mackin.

> 'A graded stream is one which, over a period of years, slope is delicately adjusted to provide, with available discharge and with prevailing channel characteristics, just the velocity required for the transportation of the load supplied from the drainage basin.'

This definition probably places too much emphasis on channel slope and it is now recognized that rivers which are graded will not have identical long profiles. The shape of the curve will depend on such things as the lithological changes in the valley, the amount of material available for the river to carry and the number and size of the tributaries. The variables concerned were described in Section 6.2.4 and Fig. 6.3.

6.4   DIFFERENT PERSPECTIVES

W. M. Davis, writing in the 1890s, introduced the concept of the development of landforms as part of a cycle which he called the cycle of erosion. The cycle evolved from youth through maturity to old age and these terms became associated with river valleys as well as with landscapes.

The model assumed initial uplift with drainage running down the uplifted surface. Streams cut into this surface producing V-shaped valleys and interlocking spurs. After the youthful stage the valleys widened and in this mature stage streams developed flood plains and meanders. In old age, as the rivers approached base level, the flood plains became more extensive, the meanders more pronounced and much deposition occurred.

Davis stated that the landscape and the landforms within it were a function of structure (geology and lithology), process (types of erosion) and stage (length of time the agents had been active upon it). After a long time the Davisian landscape would be reduced to a peneplain unless there was a sea-level change to reactivate the process.

Davis presented geographers with an overall scheme to explain landform development and in this respect his theory is very important. It does, however, make a number of assumptions which cannot be justified e.g. that rapid uplift is followed by a long period of stability.

More recent geomorphologists, such as W. Penck, have emphasized that landforms depend on the relative rate of uplift of the land and the efficiency of the forces of removal acting upon it. Penck believed that landforms were wearing back rather than wearing down (see the unit on slopes).

However, the Davisian theory must not be dismissed as 'old-fashioned' and unrealistic. A cyclic framework may be appropriate over a long period of time when surveying the landscape as a whole. It is less significant in the short term when examining one aspect of it.

A number of theories explaining the origins of meanders have been put forward. The subject is complex and no single theory is satisfactory (see Smith, D. I. and Stopp, P., *The River Basin,*

CUP, 1978). Experiments with models suggest that meandering is promoted by gentle gradients, large discharges and small bed loads. It would seem that meanders, whether in rivers, the jet stream of the atmosphere, the Gulf Current of the North Atlantic, or elsewhere, are part of the nature of flow, but why this is so, no one is sure.

6.5   RELATED TOPICS

Erosion in a river valley is an aspect of physical geography. It is closely related to erosion by other agents such as ice and the sea. There are some similarities between deposition by a river and deposition by the sea or by a glacier. The coarser and heavier material is deposited first with a gradation of the deposition until the finer material is deposited. A number of landforms derive their origins from deposition and their shapes bear some similarities. For example, the moraines formed by glaciers, the spits and bars of marine deposition and the point bars and levées found in river valleys are all ridge-shaped landforms of transported material.

In the unit on desert landforms there was reference to the landforms to be found in deserts as a result of earlier pluvial periods. These features include dissected slopes, valley systems, rock fans and spreads of water-borne material. They are to be found for example in the Sahara and are the fossil remnants of the work of rivers.

River régimes are closely related to the study of water resources in different parts of the world (hydrology) and also to climatic patterns which affect river discharge rates.

Finally, the slopes to be found in river valleys and the concepts involved in slope profile development can only be fully understood by a study of slope development. This aspect of physical geography is considered in some detail in Unit 5.

6.6   QUESTION ANALYSIS

1 Study the diagram below which illustrates some components of part of the hydrological cycle.

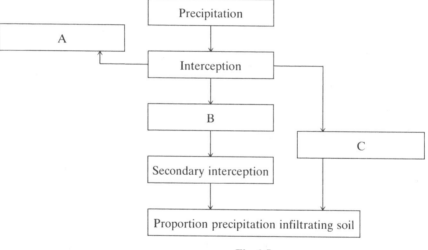

Fig.6.5

(a) Identify the processes labelled A, B and C.
(b) With the aid of a specific example, explain what is meant by secondary interception.
(c) Outline **three** physical factors which will cause the loss via interception to vary within an area of mixed woodland.
(d) With the aid of a sketch hydrograph, show how large-scale afforestation of a drainage basin in a granite upland might affect the river discharge characteristics following the passage of a storm.

(*Associated Examining Board, June 1986*)

Understanding the question This question is concerned with the inputs and outputs of a river basin and the discharge pattern following a storm. Provided you have understood the complex relationships between precipitation and the discharge of rivers, this is a very straightforward question.

Answer plan The answers to (a) are A – total evaporation loss from foliage; B – throughfall; C – stemflow.

Part (b) Secondary interception is precipitation which has been intercepted by trees, dripping off leaves and branches (throughfall) to be intercepted at a lower level by ground flora. For example, precipitation intercepted by coniferous trees may then be intercepted at a lower level by bracken and heather growing on the surface.

Part (c) Note that the question asks for three *physical* factors, so botanical factors cannot be included. The answer should be written in sentences containing a brief explanation. Among the physical factors are rainfall, wind speed, temperature, weather conditions (summer, winter).

Part (d) Note that the storm hydrograph is for an area of forest on a granite upland. Your sketch should be clearly drawn and carefully annotated. Obviously you cannot sketch a grid pattern but the values of the vertical and horizontal axes must be shown. As the question says *with the aid of . . .*, you should include a brief explanation under the sketch of such things as the relatively short time lag between the storm and the maximum discharge. This is due to the granite, which has a low permeability and is likely to have only a thin covering of soil in an upland region. This quick and substantial run-off will give the discharge curve its steep upward sweep.

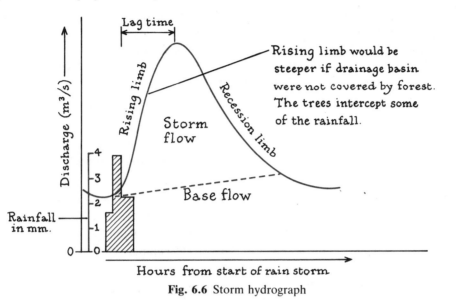

Fig. 6.6 Storm hydrograph

2 (a) Discuss the ways in which the amount and calibre of the load are related to the velocity and discharge of a river.

(b) Explain the variations in velocity which will occur in a cross-section of a river channel,

(i) within a meander belt;

(ii) along a straight stretch of the river.

Assume that the rate of discharge is constant.   (*in the style of Oxford and Cambridge*)

Understanding the question The two parts of this question are inter-related. Before selecting a question of this type make sure that you can answer all the parts adequately. Usually the first part is more straightforward than what follows, so read the whole question carefully.

Look also at the terminology and check that you fully understand such words as *calibre, discharge* and *velocity*

In part (a) the examiner expects you to be able to explain how the amount transported by a river, and the size of the particles carried, depend on the speed and amount of water in the stream channel.

In part (b) the movement and speed of the water must be related to specific cross-sections of the channel. Remember that in a meander bend the velocity changes at different points and your answer should make clear, preferably with a diagram, where the changes occur.

Do not assume that there are no variations in the velocity in a channel along a straight stretch of a river. This is the part of the question where an unsuspecting candidate loses marks because he or she ignores the significance of changes in the river bed.

Answer plan For part (a) draw a diagram like Fig. 6.1 and describe the relationship between the calibre and velocity of a river. Point out that an increased discharge rate will give the river more energy to carry additional material (see Section 6.2.3).

For part (b) draw a diagram to show the velocity of water in different parts of a river channel. Include a small diagram showing the flow of water in a cross-section of the river channel at a meander bend. Explain the reasons for the areas of high and low velocity you have shown on your diagrams (see Sections 6.2.5 and Fig. 6.3) and briefly explain how the changes in velocity will affect the shape of the meander over a period of time as some parts of the channel are eroded while in others deposition occurs.

Describe how the velocity along a straight stretch of river may be altered by the channel shape and size and roughness of the banks and bed. A simple diagram similar to Fig. 6.2 should be drawn.

6.7   FURTHER READING

Hilton, K., *Process and Pattern in Physical Geography* (UTP, 1979)

Small, R. J., *The Study of Landforms* (CUP, 1970)

Smith, D. I. and Stopp, P., *The River Basin* (CUP, 1980)

Sparks, B. W., *Geomorphology* (Longman, 1972)

Thornes, J., *River Channels* (Macmillan, 1979)

7 Drainage patterns

7.1 ASSUMED PREVIOUS KNOWLEDGE

You will probably have studied rivers and river valley development and you will have examined erosion by running water, valley form and the landforms which are produced by erosion. In addition you have normally been asked to recognize landforms and valley forms on oblique aerial photographs and the OS map; to draw simple diagrams to describe their main features; to name actual examples and to explain how they were formed. This is all useful background knowledge.

7.2 ESSENTIAL KNOWLEDGE

7.2.1 Definitions

Drainage geometry is concerned with the forms of the internal relationships of the drainage system itself.

Descriptive studies of drainage The classification of drainage patterns according to their appearance, e.g. trellised drainage, radial drainage.

Drainage morphometry The gathering of accurate data of the features of stream networks and drainage basins, e.g. stream order. The purpose is to compare the properties of individual basins in precise and meaningful ways.

Genetic stream classification Classification of streams and drainage patterns according to the way in which they were initiated and evolved, e.g consequent and subsequent streams (see Section 7.2.4).

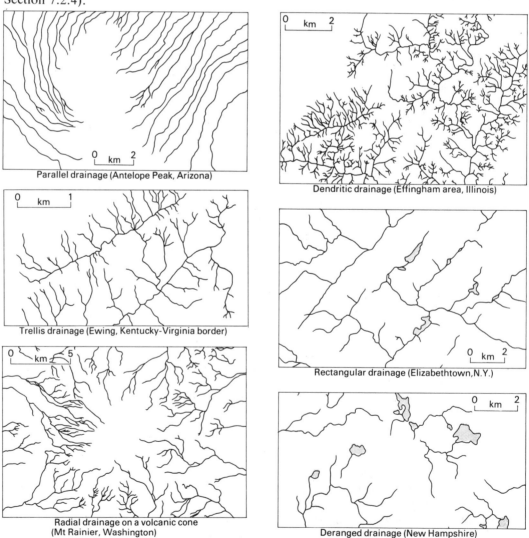

Parallel drainage (Antelope Peak, Arizona)

Dendritic drainage (Effingham area, Illinois)

Trellis drainage (Ewing, Kentucky-Virginia border)

Rectangular drainage (Elizabethtown, N.Y.)

Radial drainage on a volcanic cone
(Mt Rainier, Washington)

Deranged drainage (New Hampshire)

Fig. 7.1 A descriptive classification of drainage systems

7.2.2 Drainage systems

Descriptive studies of drainage Although drainage systems make many varied patterns it is possible to group many of them by means of a descriptive classification (Fig. 7.1) e.g.

Parallel or sub-parallel drainage develops on uniformly dipping rocks.

Dendritic (tree-like) drainage is associated with horizontal or very gently dipping strata and low relief. Structural control is very limited so streams are free to form many branches.

Trellised drainage often develops on eroded folded rocks. The main streams run along the fold axes and the tributaries flow down resistant ridges of rock. In areas of Jura folding, for example, the main streams either follow synclines or valleys formed by erosion of anticlines. In areas where cuestas are well-developed, dip-slope streams may flow across alternating outcrops of un-resistant and resistant rocks. The weak strata are eroded to form strike vales and tributary streams flow over the resistant beds.

Rectangular drainage is a very angular pattern based on geological controls, usually well-defined lines of weakness such as joints or faults.

Radial drainage Streams flow out from a high point or what was once a high point. This pattern is usually associated with domes such as volcanic cones or laccoliths.

Deranged drainage This is an 'immature' pattern where the drainage network has not had time to organize itself properly to create an integrated system. This is often found on a landscape which has been blanketed by glacial deposits.

7.2.3 Drainage morphometry (network geometry)

Many of the indices used in this approach are in the form of ratios or numbers. This makes it possible to make comparisons irrespective of scale.

Stream order This is the basic concept of network geometry and is the means whereby streams are located in a ranked hierarchy. A headwater stream with no tributaries belongs to the *first order* (the lowest order). When two first order streams unite they form a *second order* stream. Two second order streams join to form a *third order* and so on.

This ranking shows how a particular stream is related to the total network and how the total network fits together (Fig. 7.2).

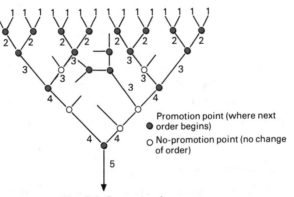

Promotion point (where next
● order begins)

○ No-promotion point (no change
of order)

Fig. 7.2 Stream order

Stream number The number of streams in each order for a given drainage basin (network). The table shows the stream numbers for the two fourth order basins in the diagram (Fig. 7.3).

Stream order	Stream number
1	133
2	33
3	8
4	2

Bifurcation ratio In the table above the ratio between the number of streams of one order and that of the next is a constant, e.g.

$$\text{Ratio order 1 : order 2} = \frac{133}{33} = 3.97$$

$$\text{Ratio order 2 : order 3} = \frac{33}{8} = 4.13 \qquad \text{average} = 4.03$$

$$\text{Ratio order 3 : order 4} = \frac{8}{2} = 4.0$$

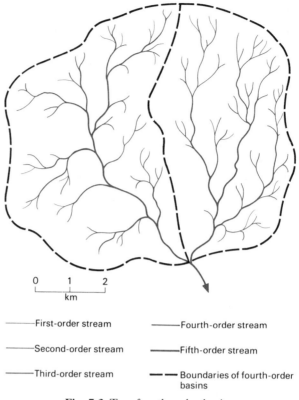

First-order stream   Fourth-order stream

Second-order stream   Fifth-order stream

Third-order stream   — — Boundaries of fourth-order
                            basins

Fig. 7.3 Two fourth-order basins

This ratio is the bifurcation ratio. The average of 4.03 is the ratio for the whole drainage basin. Bifurcation ratios usually range from 3.0 to 5.0. This constant value means that when the stream number is plotted against stream order on semi-log paper the points would approximate to a straight line. This straight line is known as an *exponential curve.*

Stream length Average length of stream in each order, e.g. the total length of all the first order streams divided by the number of first order streams. In the figure the stream length for first order streams is 0.36 km. Again there is a systematic relationship between stream length and stream order and once more it is represented by the exponential curve.

Drainage density Total channel length divided by the total area of the basin. This is a measure of the texture of the drainage net i.e. the dissection of the land surface. In the South Downs the drainage density is 2.8 miles of dry channel (valley) per square mile of basin area. In the badlands of Arizona the network is very fine and density is 200 to 900 miles per square mile.

Basin area or area of catchment In a drainage network the mean basin areas of the order occur in a roughly geometric sequence.

7.2.4 Drainage basins in relation to structure

The initiation and evolution of any drainage system are determined by (a) the nature of the surface on which the stream flows, and (b) the geological structure of the land over which it flows.

In the first instance the drainage pattern develops in response to the nature of the surface. For example, a gently tilting surface will encourage parallel drainage. In a heavily folded area consequent streams develop along synclines.

Genetic stream classification (see, for example, Sparks, pp. 129–130)

Consequent streams are streams whose courses are determined by the initial slope of the land. Longitudinal consequents develop in folded areas in the axis of the depression; lateral consequents develop down the sides of the depression.

Subsequent streams are developed by headward erosion along lines of weak structure. Most follow the outcrops of weak strata.

Resequent streams are also called secondary consequents. They flow in the same direction as consequents but belong to a later generation.

Obsequent streams flow in an opposite direction to the consequent streams.

The adaptation of streams to structure

Differential erosion results in the selective lowering of the ground surface on weak rocks and weak structures. Resistant rocks form areas of high relief. This differential erosion involves differential growth of stream systems with especially large streams or those following lines of weak structure developing as master streams.

These processes may result in

(a) the inversion of relief

(b) river capture

(c) antecedent drainage – the drainage system maintains its direction by cutting through folds rising across its path.

(d) superimposed drainage – a drainage pattern which originally evolved on an overlying unconformable rock cover since removed by erosion maintains its direction.

Read for example Sparks' book (p. 131) which provides examples of drainage patterns and their relationship to structure.

7.3   GENERAL CONCEPTS

The study of rivers may be seen as the study of channelled surface flow.

The drainage system is a major feature of the physical landscape The form of the drainage system, the spacing and orientation of the component streams is one of the key factors determining the character of the landscape. There are two main inter-related concepts.

The form of the drainage system can be described and classified in subjective and objective (numerical) ways.

The origin and development of the drainage system may be approached from a descriptive or genetic perspective.

7.4   DIFFERENT PERSPECTIVES

Earlier work in drainage patterns related closely to Davis' concept of the cycle of erosion. Discussions of drainage patterns were also considered in relation to the structure and lithology of the underlying rocks. Both descriptive and genetic approaches were used in the study of drainage.

More recently drainage patterns have been dealt with quantitatively. They are treated as geometric patterns. The approach is analytical and the purpose is to identify relationships between the different components of the drainage basin which may be applied universally.

Today all aspects of environmental science have a common basis in systems theory. Drainage patterns are then seen as surface morphological systems. These ideas have been developed in Dury, G. H., *Environmental Systems*, Heinemann, 1981 (chapter 7).

7.5   RELATED TOPICS

This unit is very closely related to the unit on rivers and river valleys. It is also linked with the units on weathering and slopes.

7.6   QUESTION ANALYSIS

1 Among the measurements used to analyse the physical characteristics of drainage basins are: stream order; drainage density; bifurcation ratio; area of catchment. Explain the meanings of each of these terms and state what relationships appear to exist between them.   (*in the style of Oxford and Cambridge*)

Understanding the question Anyone with a good knowledge of network geometry should have little difficulty in answering this question. It is important though that the answer is organized effectively so that it can be written in the time allocated. The two parts are of roughly equal significance and time should be allocated accordingly.

Answering the question
Start by drawing a diagram like the one on page 51 which will help with your explanations. Write a good paragraph on each of the four types of measurement (see relevant sections above).

Network geometry is normally systematic so there is an orderly relationship between the components of the drainage system. This relationship may be diagrammatically expressed as an exponential curve i.e. as a straight line relationship when values are plotted on semi-logarithm paper.

From the study of streams ordered in the ways outlined above certain tendencies or 'laws' may be identified which describe the relationship between them. Thus when the number of streams in a particular order is plotted against that order, the plotted values for the system lie on or near a straight line. When this straight line relationship exists the bifurcation ratio is constant. The number of streams in each order in a basin expressed by means of the bifurcation ratio indicates the form of the drainage system.

Within the same network the basin (catchment) areas of streams in the different orders also have a straight line (exponential) relationship which is known as the 'law of basin area'. The area of catchment needed to sustain a unit length of stream channel is called the 'constant of channel maintenance'. This is the inverse of drainage density.

2 Describe and account for the drainage patterns and landforms associated with limestone uplands.

(in the style of London)

Understanding the question You need to be certain of what this question requires before you begin to answer it. You could, for instance, spend a great deal of time cataloguing all the limestone uplands you know only to earn about one-third of the marks allocated to the question. There are three subdivisions; describe the drainage pattern, describe the landforms, account for what you have described. The question is specifically about limestone *uplands* – this means that you do not need to include chalkland landscapes as well.

Answer plan Since the question asks about limestone *uplands* say that you intend to deal with massive limestone areas where the rock occurs in considerable thickness and is hard enough and compact enough to form extensive moorland and highland areas as, for example, in the Pennines and the Dalmatian Mountains of Yugoslavia.

It is probably easier to begin the main part of your answer by describing the landforms of such areas. To save time and to keep your answer to a reasonable length this is best achieved by including an annotated diagram. You should add any additional landforms or technical names you have learned to the diagram.

In describing the drainage pattern the following general points should be included:

(a) drainage is partially vertical and partially underground

(b) the surface therefore lacks an integrated drainage system

(c) the drainage system may be classified into four zones.

The zone of aeration is characterized by the downward movement of inflowing water. Then there is *the zone of seasonal fluctuation* of karst water. When the water table falls this zone becomes part of the zone of aeration. When the water table rises it becomes part of the zone of full saturation. In *the zone of full saturation* drainage is not downward but lateral along the local drainage network. This movement occurs all the year around irrespective of fluctuations in surface drainage flow. In *the zone of deep circulation* water moves slowly and is not influenced by a drainage network.

In describing the pattern of surface drainage it is also possible to classify the valleys which have been formed by fluvial processes:

(a) *Through valleys* are cut by rivers which rise on impermeable rocks beyond the area of limestone.

(b) *Blind valleys* are river valleys closed at their lower end because of the disappearance of the stream from the surface.

(c) *Dry valleys* have no watercourse or are only temporarily occupied by one.

The second part of your answer is concerned with accounting for the features you have described. Include the following factors:

(a) The distinctive features of massive limestone areas are due to the solubility of the rock in water containing humic acids and carbon dioxide, strongly developed jointing systems and permeability of limestone.

(b) The basic process at work is chemical action. Most landforms are the result of such action. Relief forms in limestone areas seem unconnected because the land is essentially shaped by the vertical movement of water and the major sink holes tend to be isolated.

(c) The actual landforms and drainage patterns in a particular area depend upon *the location of the water table* (if it is well below the surface it allows the full development of the vertical movement of water), *the jointing pattern of the rock, the precipitation* (karst scenery develops most fully in areas of heavy precipitation or regions with heavy seasonal rainfall), *the thickness of the limestone mass* (the full range of underground forms and processes can only develop in areas of massive, thick limeston**e), *the strength of the limestone* (softer forms of the rock crumble and sink holes, caverns etc. collapse), *the purity of the limestone* (impurities can form deposits which inhibit the development of true karst scenery), and *the stage reached in the cycle of erosion* (see, for example, Sparks, pp 197–8).

(d) The denudation history of the region is relevant. For example, the pattern of surface dry valleys may be a remnant of a drainage pattern developed on an impermeable rock cover which has now been completely removed (superimposed drainage).

(e) Not all limestone uplands have karst characteristics. If the water table is permanently near the surface then karstic landscapes develop which are intermediate between the drainage patterns developed on impermeable rocks and true karst scenery. In arid lands karst processes and landforms are at a minimum unless they are fossilized features preserved from a previous climatic period.

3 Examine the statement that the form of drainage patterns reflects the sensitivity of rivers to variations in geological structures.

(in the style of Cambridge)

Understanding the question This question is not as simple as it looks. It does not ask you to prove that drainage patterns depend upon underlying structure. In asking you to *examine* the examiner is asking you to look at the statement critically. It is useful therefore to first provide evidence supporting the statement and the most conclusive evidence is reference to actual examples which illustrate the points you make (both Sparks and Small contain relevant examples). The discussion should then be broadened by looking at ways in which drainage patterns may not be responding to structure and other factors which drainage patterns may reflect.

Answering the question Make it clear at the start of your answer that you appreciate that the question is about structure and not relief. The influence of rock type and the effects of the geological processes of uplift, tilting, folding and faulting are therefore central to the answer.

A useful early general point to make is that the question is concerned with the principle of *differential erosion* – the principle that erosional forces will selectively attack weak rocks and weak structures and resistant rocks are consequently left as areas of high relief. Drainage patterns reflect the operation of this principle.

In examining the relationships between drainage patterns and structure it is relevant to distinguish between initial drainage patterns and accordant drainage patterns. *Initial drainage patterns* are described in Section 7.2.2 above and examples are shown in the diagram. These patterns develop in response to the nature of the initial surface.

Accordant drainage patterns demonstrate the operation of the principle of differential erosion (see Section 7.2.4 above). Examples reveal how minor variations in geological structures affect the evolution of the pattern and this justifies the statement in the question.

There are also examples of drainage patterns which tend not to conform sensitively to the structure. Drainage patterns discordant to structure may be the result of:

(a) river capture

(b) antecedence – an existing drainage system maintains its direction by cutting through folds rising across its path

(c) superimposition – the maintenance of a drainage pattern which evolved on an overlying unconformable rock cover which has since been removed by erosion.

In each of these three cases it may be argued that the drainage pattern which exists was originally developed in response to structural features. River capture is the result of differential erosion. Antecedent drainage reflects a previous structural pattern as does the superimposed drainage. It can be argued however that in none of these cases can it be held that the rivers have been sensitive to the variations in geological structure over which they now flow.

Geological structure also involves the nature of the rock itself. Drainage patterns developed on limestone (see Section 7.6.2) provide examples to support the statement contained in the question. Clay vales also provide relevant examples of the relationship of drainage patterns to lithology.

In broadening the discussion in a final section you should look at the influence of climate as a factor affecting the form of drainage patterns. Examples to which you could refer include

(a) the significance of the level of the water table in limestone areas in determining the nature of the drainage pattern (see Section 7.6.2). The water table level is directly affected by climatic factors.

(b) distinctive patterns of drainage in arid lands

(c) the effects of past climates upon present drainage patterns e.g. glacial diversions.

Your conclusion should be related to the weight of conflicting evidence you have presented in your answer. You could say that drainage patterns do not only reflect the sensitivity of rivers to variations in geological structure although this is clearly a prime factor in determining how a pattern evolves.

7.7   FURTHER READING

Dury, G. H., *Environmental Systems* (Heinemann, 1981)
McCullagh, P., *Geomorphology* (OUP, 1978)
Small, R. J., *The Study of Landforms* (CUP, 1970)
Sparks, B. W., *Geomorphology* (Longman, 1972)

8 Glaciation

8.1 ASSUMED PREVIOUS KNOWLEDGE

GCSE studies of glaciation concentrate on the distinctive landscape features which resulted from the Ice Ages. You should be able to give brief explanations of how these features were produced. You should also have studied glacial landforms on Ordnance Survey maps and oblique aerial photographs. There are two broad categories of glacial landforms – those found in highland regions such as pyramidal peaks, U-shaped valleys, cirques and arêtes, and those found in lowland regions such as kettle holes, drumlins and outwash plains. If you have not studied these elementary aspects of glaciation, read the appropriate units in Lines C. and Bolwell L., *Revise Geography*, Letts, 1987 or a similar book for GCSE candidates, such as Terry, A. G., *The Physical Landscape*, McGraw-Hill, 1969.

8.2 ESSENTIAL INFORMATION

8.2.1 Definitions

Nivation A complex process of weathering which deepens hollows by freeze-thaw action and removes material by solifluction (movement of soil and rock fragments down slopes).

Ablation The melting and evaporation of a glacier.

Abrasion The wearing away of soil and rocks by ice, wind or water.

Corrasion This takes place when material carried by ice, water or wind wears away underlying rocks.

8.2.2 Features of mountain and lowland glaciation

Knowledge is also required of the distinctive features of both mountain and lowland glaciation, some of which you may have studied for O level. The main features are:

Mountain glaciation Cirques (corries), arêtes, pyramidal peaks, trough (U-shaped) valleys, truncated spurs, hanging valleys, ribbon lakes, roches moutonnées, and moraine deposits.

Lowland glaciation Moraine deposits, gravel veneered terraces, outwash plains, underfit (often referred to as *misfit*) streams, erratics, drumlins, till plains, lacustrine deposits, varves, kames, eskers, urstromtäler and kettle holes. Whenever possible you should be able to name examples or precise locations where these features can be seen. The examiner will not consider that locating a moraine in East Anglia is sufficiently precise. You will be expected to give a more exact reference, for example, the Holt-Cromer moraine in north-east Norfolk.

8.2.3 Cirques

The 'armchair' shape of the cirque caused by the lengthening, widening and deepening of the hollow which contains a glacier has been the subject of debate. It was thought that frost-

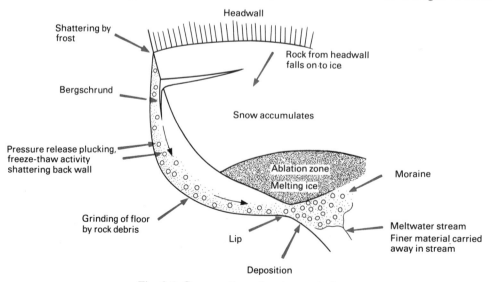

Fig. 8.1 Cross-section of a cirque glacier

shattering at the base of the bergschrund was the cause, but not all cirques have this crevasse. The shape of the cirque probably reflects a number of processes which have been summarized in Fig. 8.1. (See also Hilton, K., *Process and Pattern in Physical Geography* UTP, 1979).

The essential factors which help to produce a cirque are:

(a) *Sufficient snowfall to build up the cirque glacier* but not so much that the whole area is covered with an ice cap. Complete coverage by ice would check freeze-thaw weathering of the headwall.

(b) *Easily shattered bed-rock* which is strong enough to maintain the steep headwall and sides required to give the cirque its shape.

(c) *Daily seasonal flushings of meltwater* which can freeze on the backwall, prising away rock.

(d) *Pressure variations* which result in melting and refreezing. Water melted under a heavy weight of ice could refreeze when weight is reduced and then prise away rock.

(e) *Pressure release cracking in bedrock* which makes the cracks required for freeze-thaw activity.

(f) *Abrasion by loose rock* in the lower part of the cirque and the build up of material at the cirque lip.

8.2.4 Trough valleys

There is still conjecture about how the trough valley becomes over-deepened and the longitudinal profile of the valley becomes stepped. See Fig. 8.2.

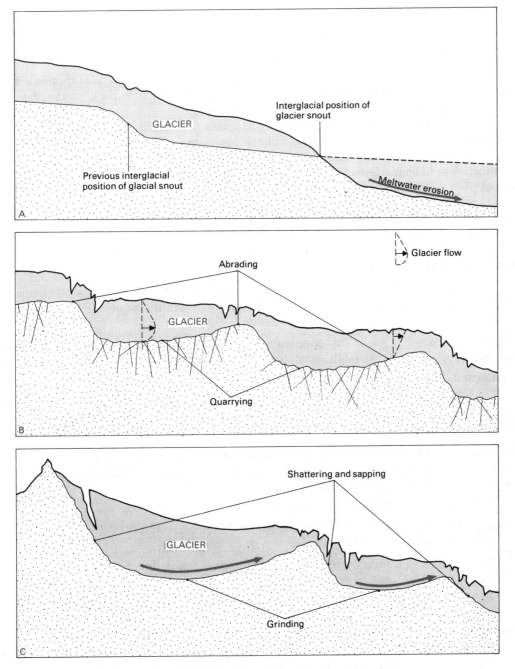

Fig. 8.2 Possible methods of valley step formation

One theory for stepped valleys advanced by Demorest and Streiff-Becker was that the maximum velocity of the glacier is at its base where the pressure of overlying ice is greatest. Less pressure and therefore less erosion takes place where the ice is thinner, so erosion over an uneven surface would produce hollows and steps.

It is now considered that valley steps may be due to different causes in different locations. In some places jointed bedrock may have been plucked and removed to make an irregular surface. Tributary glaciers may have increased erosion, or a narrowing of the valley may have had the same effect. Irregularities in the original valley floor may also have been accentuated by ice action. This action includes (c), (d) and (e) in 8.2.3 above, which are collectively known as sapping (Fig. 8.2).

The depth of trough valleys may have been caused by a number of the factors mentioned above combining to scour an existing valley.

8.2.5 Drumlins

These are typically composed of drift, but there are also pure rock drumlins and some which are veneers of drift over rocks. The processes by which drumlins were formed are not fully understood. Irregularly distributed patches of till in the ice may have been deposited when the ice retreated. Drumlins are found in highland regions, often in swarms, where the glaciers would have some velocity and this may account for their egg-like shape, which could have been caused by pressure.

8.2.6 Other information

You should know:—
(a) How drainage patterns may be modified by glaciation, with examples from specific areas.
(b) The Ice Ages effected changes in sea-level which caused changes on the coastline, e.g. rias, raised beaches, fiords; and river valleys, e.g. river terraces.
(c) *Periglacial conditions* existed beyond the ice sheets. You should know the processes which operated in these conditions and the associated landforms and physical features. The term 'periglacial' is applied to those areas which are marginal geographically and climatically to the ice sheets. The main features of these conditions are patterned ground (circles, nets, polygons, stripes), dry valleys, underfit rivers, mass wasting, loess deposits and valley asymmetry. You should be able to give locations for these features, many of which are to be found in southern England (see Hilton, K., *Process and Pattern in Physical Geography*, UTP, 1979 and Small, R. J., *The Study of Landforms*, CUP, 1978).

8.3   General concepts

To appreciate what happened during the Ice Ages and understand how this period helped to shape some of our present-day landforms it is necessary to identify the general concepts which underpin the physical details relating to glaciation.

During the Pleistocene glaciations, temperature conditions fluctuated The ice retreated in the interglacial periods and then readvanced, tending to destroy earlier landforms of glacial origin.

Continental ice sheets moved south across North America, northern Europe and parts of Asia Northern ocean areas were frozen and there was a world-wide lowering of sea level.

Glaciers also formed on higher ground Their remnants can be seen today in such ranges as the Rockies and Alps.

Areas beyond the edge of the ice sheets were subject to frost action There was an extensive zone of permafrost (ground permanently below freezing point) which has shrunk since the Ice Ages but still covers 26 per cent of the earth's surface.

Water expands when it freezes This makes it a formidable destructive force.

The erosive impact of glaciation is evident in changes to existing landforms, although the process of erosion by ice is not fully understood.

Meltwater is also an erosive force, as well as being responsible for deposition in sub-glacial channels and the formation of outwash plains.

The melting of the ice sheets at the end of the Ice Ages resulted in uplift of some of the land as the weight of the ice was removed (isostatic uplift) This uplift is a slow process which still continues. Meltwater has increased the height of sea-level, leading to the drowning of the original coastline in some areas.

8.4   DIFFERENT PERSPECTIVES

Reference has already been made to unresolved problems which relate to the processes involved in the formation of cirques, trough valleys and drumlins. Earlier ideas that glaciers merely carry material which has been loosened and then fallen from the slopes above did not fully explain the excavation of rock basins. Moreover, rock flour carried away from glaciers is evidence that corrasion and abrasion take place beneath the ice, probably aided by carbon dioxide weathering at low temperatures.

Recent research has emphasized that stress concentrations, particularly horizontal stresses, might be the reason why glaciers can cut deep rock basins in hard rock. Research has also been concerned with periglacial processes since these can be observed at first hand in northern Canada and the USSR. As a result we can now appreciate more fully the significance of periglacial conditions in shaping the landforms of such areas as lowland Britain. We now recognize that the ice sheets had a considerable influence on the processes at work beyond their margins and that the effects of the Ice Ages extend beyond the outwash plains.

8.5   RELATED TOPICS

Any study of soils must take into account the drift deposits which cover large areas such as East Anglia. These deposits, like soils in other areas which were once covered by ice sheets, are the direct result of glaciation and may bear no relationship to the rocks which they cover.

When studying coastal landforms remember that they may have been influenced by the raising of the land, or show evidence of 'drowning' as the result of the raising of the sea-level at the end of the Ice Ages. Depositional material such as sand and shingle may have been derived from glacial deposits on other parts of the coastline and the material forming sea cliffs may be derived from glacial till, as for example, on the coast of north Suffolk.

Glaciation and the subsequent meltwater made considerable changes to many drainage patterns as well as to the shape of river valleys. Your studies of the British Isles should give you the opportunity to appreciate the changes which have occurred in particular areas as a result of the Pleistocene glaciation.

In a similar fashion weathering and slope formation have links with glaciation in that some of the processes such as frost-shattering are common features. Studies of periglacial conditions suggest that many slopes in southern Britain may owe their origins to this period of glaciation.

8.6   QUESTION ANALYSIS

1 Write an explanatory account of landforms which are typical of lowland glaciation.

(*Southern Universities' Joint Board, June 1977*)

Understanding the question When studying examination questions look carefully to see if the question contains a key word which could guide the way you structure your answer. In this question the key word is *explanatory*. If this word were removed your answer could be limited to a description of each of the landforms, supplemented by sketch diagrams and location references. The inclusion of the word 'explanatory' makes it essential for you not only to describe the landforms in detail, but also to explain the processes by which these features were formed.

The question is limited to lowland glaciation and therefore no mention should be made of highland features. Of course, you must include those landforms such as moraines and erratics which are to be found in both highland and lowland zones.

Although you are asked to write an explanatory account, do not hesitate to include good, well-annotated diagrams whenever possible. You may prefer to show all the features on a composite block diagram near the beginning of your answer.

Answer plan Start with a brief introductory paragraph explaining that lowland areas, unlike mountain regions which have been glaciated, are essentially regions of deposition, although, as on the Canadian Shield, there is some evidence of erosion.

Then write a paragraph describing the features of a lowland area caused by erosion, together with explanations of how these features were formed (see Sparks, B. W., *Geomorphology*, Longman, 1976). The main landforms are ice-scoured hollows containing irregularly shaped lakes and smooth, rounded, steep-sided rock hummocks. Don't forget that, whenever you describe a landform in this question, you should give a precise location. If you have studied one region in detail, such as the Welsh borderlands, then use this information and add a suitable sketch map or diagrams to illustrate your written description.

Your second paragraph should deal with depositional landforms such as terminal moraines, erratics, gravel terraces, till sheets, drumlins, eskers, kames and varved clays. Give explanations for these landforms. Where there is conjecture, as for example in the formation of drumlins, outline the theories which have been put forward in as much detail as possible.

Then finish by describing landforms connected with changed drainage patterns, such as underfit streams, moraine-dammed lakes, diversion of drainage and lake shorelines.

2 How far do you agree with the statement that there are few landforms in the British Isles which do not show the effects of the Ice Ages? Illustrate your answer with specific examples.   (*in the style of Oxford*)

Understanding the question You are asked how far you agree with a statement which may or may not be true. Short statements of this kind tend to be generalizations. There may be a great deal of truth in what is said but do not accept the statement without questioning its validity.

Note that the question refers to the British Isles, not Great Britain or the United Kingdom. Examples can, therefore, be drawn from the Irish Republic as well as the rest of Britain. Do not fall into the trap of limiting your answer to those regions which, at some time, were under an ice sheet. The effects of the Ice Ages extended beyond the limits of the ice to those areas which experienced periglacial conditions.

Before deciding to answer this question check that you know a variety of landforms which are not the result of the Ice Ages. An answer which contains only examples of the many landforms linked to the Ice Ages is inadequate. The question states that *few* landforms do not show the effects of the Ice Ages. You must attempt an assessment, based on actual examples, as to whether the word 'few' is the most appropriate one to use.

Answer plan Write an introductory paragraph explaining that since the furthest southerly extent of the ice sheets was approximately from the coast of N.E. Essex to the Severn estuary and across the southern counties of Eire, the whole of the British Isles was either directly or indirectly affected by the ice because the southern areas were under periglacial conditions.

Give a summary of the landscapes of the highland zones of the British Isles, with brief descriptions of the most prominent features such as the arêtes, trough valleys, cirques, ribbon lakes and moraines. It should then be pointed out that some elements of the scenery of these regions were not connected with glaciation. These include the rounded upland surfaces of the Grampians and Southern Uplands, the remnants of volcanoes and lava flows such as the Eildon Hills near Melrose and the coast of Antrim; the fault scarps of the Highland boundary faults and the remnants of older rock formations such as Charnwood Forest and the Malvern Hills.

Write a paragraph about the landscape of the lowland areas of the British Isles which display evidence of glacial features such as till, drumlins, moraines and changes to the drainage pattern. Periglacial features such as combe rock and patterned ground are also present. By contrast, the scarplands of southern England, the granitic outcrops of Devon and Cornwall and the lowlands of southern Eire show little, or no evidence of the Ice Ages.

Conclude with a paragraph pointing out that evidence of the Ice Ages is never far away in the British Isles but other factors such as the nature and characteristics of the rocks, the work of rivers and the sea and the effects of weathering on such rocks as limestone and granite, have produced landforms which are not related to the Ice Ages. Taking these factors into account the statement can be regarded as accurate.

3 'The most characteristic landforms of upland areas which have been subjected to glaciation are the cirque or corrie and the trough valley.' Describe the processes involved in the formation of these landforms and suggest why there is still controversy over the way in which they were formed.

(*Oxford and Cambridge, July 1979*)

Understanding the question The first thing to notice about this question is that, although it contains a statement, you are not asked to discuss it. In this case the statement is used as the 'hook' on which to hang the question. The question is in two parts. It first asks for a description of the processes involved in the formation of a cirque and a trough valley. It then asks for reasons why there are still uncertainties about the way these two features were formed. The second part of the question does not ask for the various opinions which have been put forward as attempts to resolve the controversy but you would gain marks if you could describe some of the theories propounded, since it would reinforce your understanding of the problem.

Answer plan With the aid of an annotated diagram (see Fig. 8.1), explain the various processes at work in the formation of a cirque (see Section 8.2.3).

Then explain that controversy surrounds the 'arm-chair' shape of the cirque and that the earlier theory of frost shattering is not now considered adequate. Describe how the shape of the cirque probably reflects a number of processes.

Next describe the erosive action of a glacier in forming a trough valley. It should be explained that the processes at work are those which are also found in a cirque. Your explanation can be limited to avoid repetition.

Explain that controversy still surrounds the stepped long-profile of trough valleys as well as the over-deepening of these valleys which is evident in the cross-profiles. Put forward some of the possible explanations (see Section 8.2.4), and differentiate between earlier and more recent theories.

8.7   FURTHER READING

Clowes, C. and Comfort, P., *Process and Landform* (Oliver and Boyd, 1982)
Hilton, K., *Process and Pattern in Physical Geography* (UTP, 1979)
McCullagh, P., *Modern Concepts in Geomorphology* (OUP, 1978)
Small, R. J., *The Study of Landforms* (CUP, 1978)
Sparks, B. W., *Geomorphology* (Longman, 1972)
Sugden, D. E. and John, B. S., *Glaciers and Landscapes* (Edward Arnold, 1976)

9 Coastal landforms

9.1 ASSUMED PREVIOUS KNOWLEDGE

A level coastal studies take off where GCSE work ended and you will be expected to have covered the basic work on coastlines and the processes involved in their formation. At this basic level physical geography normally relates to the coastline around Britain. It is more familar than some other landscape types and very interesting to read about using one or more of the standard text books, or more general books such as Steers, J. S., *The Sea Coast* (Collins, 1969) or Eddison, J., *The World of the Changing Coastline* (Faber and Faber, 1979).

9.2 ESSENTIAL INFORMATION

9.2.1 Definitions

Longshore drift Waves usually approach a beach at an angle and material carried by the waves is moved obliquely up the beach. Depending on whether the waves are constructive or destructive (see Section 9.2.2) the material is deposited or moves down the slope of the beach to be pushed up at an angle by the next wave movement determined by the dominant wind.

Swash The rush of water up the beach from a breaking wave.

Backwash The flow of water down the beach after the swash has reached its highest point.

Fetch The oceanic distance over which the wind blows and generates waves. If the fetch is large and the time during which the wind has been blowing is long, for example from Cape Cod to Cornwall, the waves are likely to be large also. Waves in the North Sea are the product of a limited fetch, and however strongly the wind blows, the waves will have a limited height.

9.2.2 Waves

Waves are formed by the transfer of energy from air to water by wind blowing across the surface of the sea. Waves can travel long distances so the waves that break on a beach may not be the result of local winds. *Dominant waves* are those which affect the coast most in terms of erosion and deposition. Waves breaking on a beach carry material up the beach and may deposit it, building up the beach in a *constructive* process. Waves can also be *destructive* (Fig. 9.1). They are responsible for *marine erosion* in the following ways:

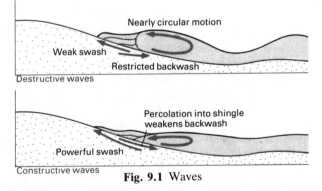

Fig. 9.1 Waves

(a) The swash and backwash of large waves cause abrasion of material carried by the waves, as well as erosion of cliffs and other coastal features.
(b) Rocks such as chalk are slowly dissolved in the water.
(c) Rocks are fragmented by hydraulic action i.e. the compression of air between the waves and the surface it hits. *Storm waves* have a formidable power, tearing away sections of beach and breaking through ridges of shingle and sand previously deposited.

When waves in a tidal sea erode an area of hard rock the result is a *wave-cut platform*. Between high and low water the rock is exposed to the action of waves which erode by abrasion and in some cases by solution, to form a gently-sloping platform of rock which may still be partly submerged at low tide.

9.2.3 The beach

A beach normally consists of unconsolidated materials such as sand, mud or shingle. The nature of the beach depends on the origin of the material which has been deposited on it. *Longshore drift* may deposit material a considerable distance from the original source.

A *shingle beach* is more mobile than a sand beach. During storms a ridge of shingle is formed above the normal spring tide level. Smaller ridges may exist marking the level reached by the high tides of the spring and autumn equinoxes. Further down the foreshore ridges may mark the last spring and neap tides (Fig. 9.2). *Sand and mud beaches* are much flatter than shingle because most of the swash sand returns down the beach with the backwash. These beaches do not have tidal ridges except occasionally when a small ridge appears along the spring high-water line. Only rarely is a *storm beach* to be seen.

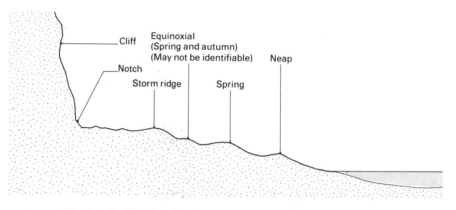

Fig. 9.2 Profile of a shingle beach at the time of neap tide

Beach types result from both the supply of rock debris and climatic conditions. Many rocks break down to sand or mud quickly, others, like flint pebbles, are long-lasting. The climatic factor is more significant than the nature of the rock in determining beach types. In the tropics the most important sediment is mud. Pebbles are most common in high latitudes due to the storm waves of these regions. Where coral flourishes the beach is likely to be composed of coral sand with a high calcium carbonate content.

Where there are dominant on-shore winds the dry sand may be blown inland to form a ridge of *dunes*. Apart from the formation of sand dunes, all the beach ridges described in this section are the result of short-term processes such as a storm or a spring tide.

9.2.4 Constructive action by the sea

A number of coastal features are evidence of the deposition of material which has been transported by longshore drift or by currents. The features caused by long-term processes are offshore bars, spits and tombolos. Features formed by short-term processes are beach cusps and sand bars.

Features formed by long-term processes The life-cycle of an *offshore bar* is shown in Fig. 9.3. The development of the bar leads to a lagoon on the inland side which gradually fills in. In time the bar is pushed inland and the marsh is gradually eroded leaving the remnants of the bar as coastal dunes. Much of the material from which the bar is formed is eroded from the sea floor and is not supplied by longshore drift. The most fully developed offshore bar in Britain is Scolt Head Island on the north Norfolk coast. *Spits* are built up with material mainly from longshore drift and extend from the coastline (Fig. 9.4). There are two kinds of spits, those that leave the coast at a marked angle and those that follow the coastline. Some spits growing from the mainland may

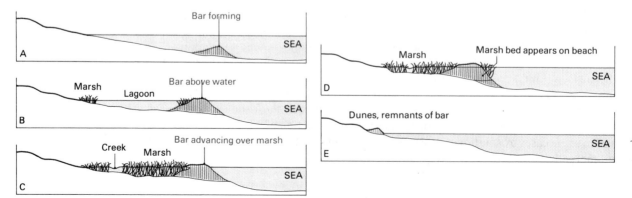

Fig. 9.3 Life cycle of an off-shore bar

Fig. 9.4 Spurn Head, Yorkshire. What is this feature called? What processes are responsible for its development?

connect with islands to form a shingle bar between the island and the mainland known as a *tombolo,* e.g Chesil Beach in Dorset. A complex area which has resulted from deposition is the cuspate foreland at Dungeness (see Sparks, B. W., *Geomorphology,* Longman, 1972).

Features formed by short-term processes *Beach cusps* are small, seaward facing peninsulas of shingle on the beach linked by curving bays. Their origins are not fully understood (see Fig. 9.5). *Sand bars* (not to be confused with offshore bars), are the result of the concentration of sand in ridges at the point where a wave breaks. They occur in tideless seas like the Mediterranean.

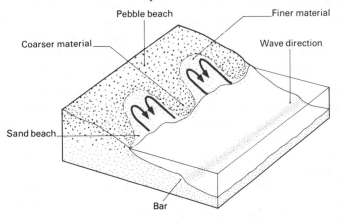

Fig. 9.5 Beach cusps

9.2.5 Changes in sea level

Many parts of the earth's surface are unstable and movement takes place which may raise or lower the land relative to sea level. For example, during the Ice Ages the increase and decrease in the area under ice occurred a number of times as the ice cap spread and retreated. This resulted in changes in the sea level. These are known as *eustatic changes* because they were widespread and therefore found in many parts of the world. Movements in relative sea level may also occur as a result of warping or faulting of part of the continental crust, or due to an increase or decrease in weight when an ice-sheet is formed or melts. This type of movement which affects the crustal balance is termed *isostatic*.

Any movement where the land rises relative to the sea is described as a *negative movement of sea level*. When the land sinks relative to the sea it is described as a *positive movement of sea level*. Positive movements of sea level result in some of the land being 'drowned'. The most obvious signs of positive movement along the coast are drowned river valleys called *rias*. There are many to be seen around the peninsulas of south-west Ireland, Cornwall, Brittany and north-west Spain.

Low-lying areas along the coastline such as the Fens were also covered by the sea when the level rose and layers of marine silt were deposited. Inland, swampy vegetation gradually built up a layer of peat. Small islands remained above this poorly drained area.

When hills and valleys lying parallel to the coastline were invaded by the sea, as for example along the longitudinal coast of Yugoslavia, long, narrow islands separated by inlets were formed.

Negative movements of sea level produce a number of distinctive landforms. Along coasts raised beaches and marine terraces are formed, in river valleys terraces, incised meanders and nick points can be found.

Raised beaches consist of platforms of rock with or without beach deposits. If the land form occurs above approximately 50 metres it is called a marine terrace and beach deposits are rarely found. There are many raised beaches in western Scotland resulting from isostatic uplift.

Marine terraces occur at different heights. They have bench-like shapes and are backed by steep slopes. Identification of marine terraces is not always easy since other landforms which look like marine terraces can result from a warping (bending) of the strata. For a detailed account of marine terraces and the problems associated with them see Small, R. J., *The Study of Landforms*, CUP, 1970 or B. W. Spark's book mentioned above.

9.2.6 Sea cliff erosion

Cliffs are subject to both marine and sub-aerial weathering.

Marine erosion is caused by waves and salt crystallization (the formation of salt crystals in cracks and pores causing the breaking off of mineral particles). Its effects depend on the nature of the rock forming the cliffs. Hard rocks with little jointing erode slowly. Most rapid erosion takes place in uncemented rock such as sands and clays of the Eocene and Pleistocene periods (see Section 9.2.2.).

Sub-aerial weathering depends on the nature of the rock – whether there are joints and bedding planes, whether the rock can be dissolved easily, its hardness and mineral composition, and whether the beds consist of different rock types. The main processes involved are:

(a) Mechanical weathering This consists of freeze-thaw action by water trapped in the rock.

(b) Chemical weathering There are a number of processes involved. These include *hydration* in which certain minerals absorb water and expand; *oxidation*, especially of clays when they dry out; *hydrolysis*, for example when felspar in igneous rocks break down to form clay minerals; and *carbonation* in which rocks such as chalk and limestone decompose as a result of the action of acid in rainwater on calcium.

(c) Biological weathering This is the action of plants and animals living on the cliffs. Visual evidence is the rabbit burrows which riddle many cliff tops but more destructive action results from the increase in the carbon dioxide content of the soil and reaction between organisms and the minerals in the rocks.

The nature of cliff profiles depends on the *composition of the cliffs* and on the *balance* between the marine and sub-aerial processes at work. *The dip of the strata is also important.* Blocks cannot break off easily from beds which are vertical, horizontal or dip inland, so cliffs with such strata arrangements tend to be nearly vertical (see Small, R. J., *The Study of Landforms*, CUP 1970, 444–449).

Sub-aerial processes are most effective on cliffs formed of drift material. Excessive rainfall or snow melt causes the cliffs to slump onto the beach to be washed away by waves and currents. Marine processes erode notches in the base of hard cliffs and exploit weaknesses in the rock, eventually causing sections to collapse.

9.3   General concepts

Energy transference An example is the action of waves along a coastline. Energy is transmitted from the wind to the waves and this same energy is then used along the coastline to erode and transport material. When the remaining energy is insufficient for transportation the load is deposited.

Coastal landforms are determined by a combination of short-term and long-term processes See Sections 9.2.3, 9.2.4, 9.2.5 and 9.2.6.

The nature of landforms is largely determined by their geological and lithological characteristics. The soft sands and clays of the East Anglian coast have associated landforms which contrast with those to be found along the Cornish coast where the rocks are hard and resistant.

9.4   Different perspectives

A number of processes in the formation of coastal landforms are not fully understood and different theories have been put forward to explain these processes.

The formation of wave-cut platforms and their effects on marine erosion of the cliffs behind the platforms There are two viewpoints. (a) The energy of the waves is used up on the wave-cut platform and by the time the waves reach the cliffs little erosion can take place. (b) Erosion continues at the cliff line and the wave-cut platform does not break the force of the waves. It would seem likely, however, that the rate of growth of the wave-cut bench slows down as it becomes wider. The problem is complicated by the fact that there have been changes in sea-level in the recent geological past.

The relative parts played by waves and currents in longshore drifting Experiments with such things as fluorescent materials suggest that currents may help in the movement of finer material, but shingle is only moved as a result of wave energy.

9.5   Related topics

In order to understand how coastal features are formed you require a knowledge of processes such as *weathering, mass movement* and the *work of ice and running water,* as well as an appreciation of *the physical and chemical properties of different rock types*

On a global scale, knowledge of the climate of the land area adjacent to the coast is also important since differences in climate will affect the nature of beach material and the sub-aerial processes at work on sea-cliffs.

The study of coastal landforms is therefore linked to a number of other aspects of physical geography which form part of A level syllabuses. These include weathering, the formation of slopes, the nature of rocks and world climatic variations. A unique feature of the study of coastal landforms is the possibility of carrying out field work which involves measuring, recording and analysing the rate of change.

9.6   Question analysis

1 What changes would you expect on a beach during a period of twelve months?
(Southern Universities' Joint Board, June 1978)

Understanding the question The question refers to the short-term processes which are at work on a beach. In this case the short term is a maximum period of a year. This allows for any changes brought about by spring and neap tides as well as by storms. One important word in the question is *during,* especially since changes to be seen at one time of the year may be removed later. These temporary changes should be included in your answer because the question is concerned with changes over time. It is not asking for a comparison of what the beach is like at the end of the year compared with twelve months earlier.

The question is limited to changes on the beach, and reference to the adjacent land area, such as cliffs, should only be made in so far as material from them may fall on the beach during the year to be added to the profile.

Although the question does not specifically mention the processes involved, a mere description of the changes would be inadequate. Some mention of why and how the changes take place must, therefore, be included.

Finally, remember that beaches made of different materials respond in different ways and it is important for your answer to distinguish what would happen on sandy beaches as well as on pebble beaches.

Answer plan Define briefly the limits of a beach, explaining that it is the strip between high and low water marks. Then explain that in the short-term period of a year there are not likely to be any major changes in the formation of the beach. However, at different times during the year, such as after storms, changes will be apparent. These changes may be only transient features to be destroyed during the next violent storm. Draw a diagram like Fig. 9.2 and explain what it shows.

Briefly describe the processes at work during the year on a shingle beach between the high and low water marks, and the effects these processes will have on the appearance of the beach. For example, longshore drift will shift material in one direction, and shingle on one part of the beach may disappear to be replaced by material from elsewhere (see Sections 9.2.2 and 9.2.3).

There may be considerable changes in the longitudinal profile especially after storms. Where there are groynes or rock outcrops, there can be much accumulation of material on the windward side and extensive scouring on the lee. The longitudinal redistribution of deposits may well cause an alteration in the course of a stream traversing the beach.

Describe the process which will add material to the beach from adjacent cliffs and give examples of beaches where material is added regularly from neighbouring cliffs.

Conclude by emphasizing that the changes on sandy beaches will be similar but there will be little evidence as the sand forms a fairly flat beach.

2 'Many factors, other than the work of waves, affect the form of sea cliffs.' Discuss this statement with reference to specific examples. *(Associated Examining Board, June 1978)*

Understanding the question Although the question mentions the work of waves it is so worded that there is no need to include in your answer any references to the effects of wave action on the form of sea cliffs.

The question is concerned with those aspects of geology, lithology and sub-aerial erosion which affect sea cliffs. Do not forget, however, that some of these factors are more active because of wave action.

You are asked to discuss the statement with reference to specific examples and unless you are able to give a number of precisely located examples you will fail to obtain good marks. The examples should not be referred to by such expressions as 'on the coast of East Anglia', or 'the chalk cliffs of the south of England'. Instead give a named location such as 'the cliffs at Cromer in Norfolk' or 'the chalk cliffs near Birchington in Kent'.

This particular statement, unlike some which appear in A level questions, cannot be considered contentious, but remember that the work which can be accomplished by waves will be partly determined by the lithology of the rocks forming the sea cliffs.

Answer plan In the introduction recognition should be given to the validity of the statement. The other factors involved, excluding the work of the waves, which affect the form of sea cliffs should be briefly mentioned.

This should be followed by a description of how physical weathering affects rocks of different types, with examples and, if possible, annotated sketches (see Section 9.2.4). Write a description of how chemical and biological weathering have helped to determine the form of sea cliffs.

Explain the inter-relationship between the lithological factors such as the dip of the strata and wave action to illustrate that lithology plays a part in the form of sea cliffs. Give examples of well-jointed cliffs, those with horizontal strata, cliffs made of soft clays and so on, and explain, with diagrams, how the profiles of these examples are likely to differ.

Conclude by pointing out that the nature of the cliff profile depends on the material forming the cliffs and on the balance between the sub-aerial and marine processes which shape it.

3 Make a detailed examination of the inter-relationships between erosion and deposition along coastlines.

(in the style of London)

Understanding the question This is a question about the processes involved in marine erosion and deposition. But the examiner does not want two straightforward descriptions of these processes. The key word in the question is *inter-relationships*. To explain the significance of this concept fully it is important to demonstrate the links which exist between the material eroded from one area and its ultimate deposition in another.

One of the best ways to bring out this inter-relationship is by describing the processes involved in erosion and deposition and then to refer to specific stretches of coastline where evidence of erosion and deposition can be seen. Examples are the coast of Yorkshire and Lincolnshire, the Ayrshire coast or the coast of Northumberland and Durham.

Any coastal field work which you have personally undertaken, or read about, which is relevant should be included in your answer.

Note that the word *erosion* is used in the question, not *marine erosion*. So it is worth pointing out in your answer that some erosion on the coast is sub-aerial and consequently part of the material deposited has been obtained in this way, usually from rock falls from cliffs on to the beach.

Answer plan Write an introduction explaining that coastlines receive the material of both marine and sub-aerial erosion and in this respect the inter-relationship is more complex than it would be if only marine erosion were involved. Also, emphasis should be placed on the relationship between the nature of the eroded material and the resultant depositional features e.g. beach shingle formed from flint eroded from cliffs some distance away.

Draw a map of the stretch of coastline which you will use as a case study, showing such things as the direction of longshore drift, dominant wind, cliff areas and their rock types, main areas where erosion is evident and the areas where deposition is taking place.

Use the map as the basis for a written description of the physical and geological background of the area, the nature of the coastal erosion and the areas where the material is deposited. Describe the formation of any spits, offshore bars or other depositional features, the location of which should be shown on the map. If possible include large-scale sketch maps of one or two of these features.

As a concluding paragraph point out that although the question has been answered by reference to a particular section of coastline, similar inter-relationships exist along other coastlines. In some cases research has shown that material is carried several hundred kilometres, but the principle remains the same. Give one or two examples of other stretches of coastline where erosion in some places is linked with deposition in others.

9.7 FURTHER READING

Bradshaw, M. J., Abbott, A. J. and Gelsthorpe, A. P., *The Earth's Changing Surface* (Hodder and Stoughton, 1978)
Clayton, K., *Coastal Geomorphology* (Macmillan, 1979)
King, Cuchlane A. M., *Physical Geography* (Blackwell, 1980)
McCullagh, P., *Modern Concepts in Geomorphology* (OUP, 1978)
Ollier, C. D., *Weathering* (Longman, 1975)
Pettick, J., *An Introduction to Coastal Geomorphology* (Macmillan, 1979)

10 Desert landforms

10.1 ASSUMED PREVIOUS KNOWLEDGE

Most GCSE syllabuses include desert landforms in the physical geography section. The main features included are landforms caused by wind such as the yardang, the deflation hollow, the barchan and the seif dune; and those formed by water, for example the wadi, mesa, butte and pediment. If you have not studied these features of desert scenery it is important for you to do so as an introduction to your A level work (see Cain, H. R., *Physical Geography*, Longman, 1975). You should be able to recognize these landforms from photographs or sketches and know in which parts of particular deserts they can be found. You must also know the processes by which these features were formed and their value, if any, to man.

10.2 ESSENTIAL KNOWLEDGE

10.2.1 Definitions

Exfoliation The peeling off of thin layers of rock from a surface. It is caused by the heating of the surface during the day and cooling at night. The resulting alternate expansion and contraction leads to the weakening of a thin rock layer which may be further weakened by the freezing of dew trapped in it.

Deflation Removal by the wind of fine products of weathering such as sand. The lightest material is blown away as dust and heavier material is blown along the surface. Some desert hollows are mainly formed by deflation.

Saltation The movement of particles by a series of jumps. This can happen in a stream when the current moves small stones or in desert landscapes where wind moves the sand.

10.2.2 The nature of present-day weathering

Mechanical or insolation weathering includes (a) *exfoliation*, (b) *block disintegration* – the breakdown of well-jointed rocks into boulders, (c) *granular disintegration* – resulting from the varying capacity of the rock minerals for absorbing heat, and (d) *salt crystallization* leading to expansion.

Chemical weathering leads to rock disintegration which occurs even in the presence of very small quantities of moisture.

10.2.3 Present-day wind erosion and deposition

Erosion The material carried by the wind is responsible for the etching of *yardangs* and for *deflation hollows* (Fig. 10.1), leaving a stone-strewn plain known as *hamada* from which smaller material has been removed.

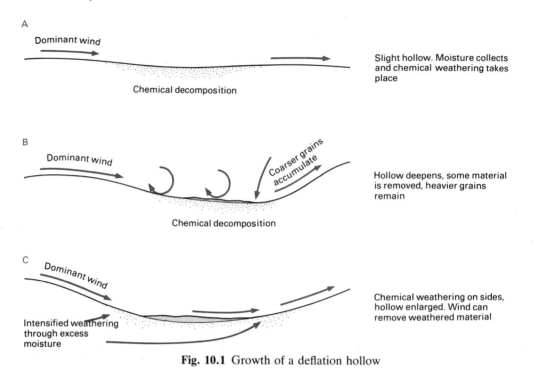

A

Dominant wind

Chemical decomposition

Slight hollow. Moisture collects and chemical weathering takes place

B

Dominant wind

Coarser grains accumulate

Chemical decomposition

Hollow deepens, some material is removed, heavier grains remain

C

Dominant wind

Intensified weathering through excess moisture

Chemical weathering on sides, hollow enlarged. Wind can remove weathered material

Fig. 10.1 Growth of a deflation hollow

Deposition A number of depositional forms are derived from the wind. They are *sand ripples, ridges, barchans* (Fig. 10.2) and *seif dunes* (Fig. 10.3).

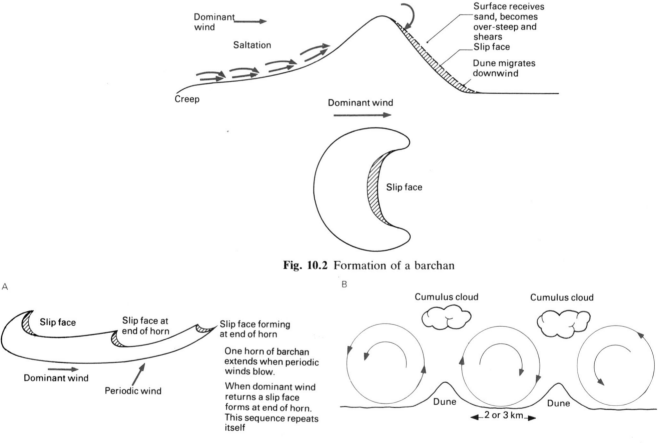

Dominant wind

Saltation

Creep

Surface receives sand, becomes over-steep and shears
Slip face
Dune migrates downwind

Dominant wind

Slip face

Fig. 10.2 Formation of a barchan

A

Slip face

Slip face at end of horn

Slip face forming at end of horn

Dominant wind

Periodic wind

One horn of barchan extends when periodic winds blow.

When dominant wind returns a slip face forms at end of horn. This sequence repeats itself

B

Cumulus cloud   Cumulus cloud

Dune   Dune

2 or 3 km

A Formation of a seif dune from a barchan   **B** Development of dunes by longitudinal roll vortices

Fig. 10.3 Two explanations of seif dune formation

10.2.4 The present-day work of water

Debris in wadis is moved during floods and acts as an erosive agent. Fans, cones and spreads occur where streams emerge onto open ground.

Sheet floods occur during the rare rain storms These are shallow films of water washing across the flat surface. Very little erosion takes place.

Dew forms at night providing the moisture necessary for chemical reactions. Hair-line cracks are filled with dew and chemical decomposition takes place, especially in granites and sandstones.

Moisture promotes crystallization in sodium chloride and sodium carbonate dust. In crevices the crystallization results in expansion, breaking down the rock.

10.2.5 Evidence of wetter periods in the past

During the Pleistocene period and earlier there were pluvial periods which account for many desert landforms today. The evidence for these wetter periods is:

(a) Extensive valley systems in the Tibesti and Hoggar Massifs of the Sahara with valleys radiating outwards. These could not have been carved by the present limited rainfall.
(b) In the central Sahara between the Tassili Plateau and the Hoggar Massif archaeological evidence indicates that the area was occupied by palaeolithic people who hunted antelope, gazelle, rhino and elephant, animals which live in a savanna environment.
(c) Pollen analysis in the Tibesti proving that oak and cedar forests once flourished there.
(d) Lake Chad was once larger than it is now; shorelines exist more than 50 metres above the present lake surface.
(e) Desert varnish (a layer of iron and manganese oxides drawn to the rock surface by evaporation) has been formed. This varnish can only form in alternate wet and dry conditions.
(f) Plateaux such as the Gilf Kebir in the Libyan desert have been dissected by stream action.
(g) The duricrusts of some desert surfaces could only have been formed during past wetter periods. Duricrust is a hardened layer formed on or near the surface. In hot climates with wet and dry seasons salt solutions are drawn up by capillary action and after evaporation are deposited as hard nodules. The climatic conditions required – periods of high temperatures interspersed with periods of heavier rainfall than the present-day climate provides – indicate a different climatic régime in the past.

10.2.6 The formation of erosional plains

L. C. King has proposed a pediplanation cycle to explain the development of almost level plains which are to be found in many desert regions. Steep-sided hills called *inselbergs* rise from these plains and King's theory (1967), is based on the parallel retreat of slopes following river incision of the original surface (Fig. 10.4). This theory does not fully explain the vast areas of near flatness in semi-arid regions and further research is required. B. W. Spark's book, *Geomorphology*, Longman, 1972, explains the situation very clearly.

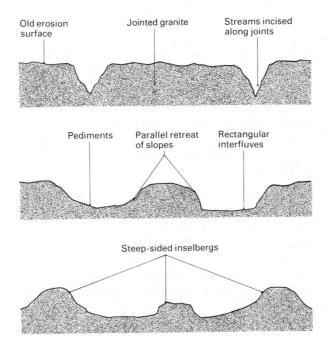

Fig. 10.4 Development of inselberg landscape

10.2.7 Inselbergs

As Fig. 10.5 shows, there are two forms of inselbergs. Their shape and the formation of the pediment are not fully understood (see Section 10.4). Beyond the pediment is the *bahada* made up of rock material from the inselberg. Further from the inselberg finer material may fill

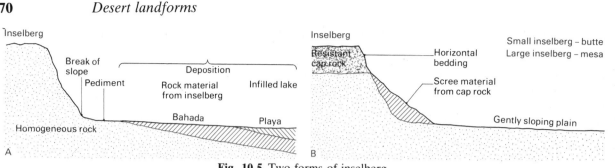

Fig. 10.5 Two forms of inselberg

structural basins. After rainstorms temporary lakes may exist and their sites are marked by flat plains called *playas* which are covered by salt after the lake water has evaporated.

10.3   GENERAL CONCEPTS

Some present-day landforms are the result of more humid conditions in the past Evidence for this concept is provided by the arid and semi-arid regions of the world. It is not possible to account for the formation of these landforms by reference to existing physical conditions. Some parts of desert landscapes are fossil forms, subject to little modification at the present time.

This concept has led to the abandonment of earlier theories that wind corrasion was responsible for the formation of flat desert surfaces. It is now recognized that there was over-emphasis in the past on wind corrasion as an important factor in the formation of desert landforms. Sand grains have little effect at heights exceeding two metres.

Mechanical disintegration and chemical action play a large part in desert weathering Most of the major erosional features of arid landscapes have been carved in the past by running water.

Erosional plains found in hot deserts are the result of parallel retreat of slopes by backwearing rather than by downwearing (see B. W. Sparks). The eroded material is spread out to give an almost level surface.

10.4   DIFFERENT PERSPECTIVES

The relative significance of the erosive process in arid and semi-arid lands is not yet fully understood.

The sharp break of slope (piedmont angle) at the foot of the inselberg (Fig. 10.5) and the pediment beyond before the zone of deposition begins is one area of contention. The theories are:

(a) D. W. Johnson in the 1930s suggested that, in the past, streams emerging from the lower slopes of the mountain front would undercut the foot of the slope. At flood times these streams would take different courses and rock fans would form. A number of these fans linked together would produce the pediment. The weaknesses in this theory are that very few rock fans exist, the hills show no canyon indentation and erosion should slow down as the hill mass becomes smaller, yet piles of boulders suggest it is still going on.

(b) More recently A. Wood and others have suggested that the mountain front has retreated parallel to itself as a result of chemical decomposition and the removal of waste material by the wind and by gravity. The pediment is formed by sheetflood which would remove debris but not cause any real erosion. The angle between the mountain front and the pediment represents a change in the process of erosion.

(c) In 1966 R. F. Peel suggested that the foot of residuals is kept sharp by the conservation of water from the hills at this point, accelerating the weathering. His views are partly supported by the presence of depressions at the foot where more weathering appears to have taken place.

The formation of seif dunes is not fully understood and two possible explanations are shown in Fig. 10.3. R. A. Bagnold (1941) suggests that a barchan swings round at one end when it experiences a strong wind from a direction different from that which normally prevails. The dominant and periodic winds produce the prolonged wing which continues to grow to form the seif dune (Fig. 10.3A). S. R. Hanna (1969) suggests that the seif dunes are the result of longitudinal helical roll vortices in the air. Once a prevailing wind has swept the sand to one side to form ridges, heating on the flanks of the ridges will cause the air to rise and circulate in a corkscrew-like movement. This rotation will move more sand up the ridges leaving the lower ground bare (Fig. 10.3B).

10.5 RELATED TOPICS

The study of desert landforms is largely concerned with assessing the effects of various weathering processes in an arid or semi-arid environment. The chemical and physical processes at work are identical with those which you have studied or will study in connection with weathering in other climatic régimes and in work on the formation of slopes. Obviously the emphasis is distinctive in deserts where rain is rare and there is a high diurnal temperature range.

The part played by running water as an erosive and depositional agent is negligible, but desert landscapes have, in part, been fashioned during previous pluvial periods. Pluvials occurred during Pleistocene times when the landscapes in temperate latitudes were being altered by glaciation. Consequently, areas once glaciated and some desert regions are fossil landscapes derived from the Ice Ages.

If you understand the processes which have helped to shape desert landforms you will be better equipped to study other landscapes and similar topics in geomorphology.

10.6 QUESTION ANALYSIS

1 Describe the conditions which lead to the development of the landforms shown in the diagram below.

(in the style of Cambridge)

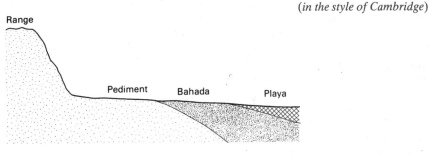

Fig. 10.6 Desert landforms

Understanding the question The words *Describe the conditions* are the key to what the examiner will be looking for in your answer. The landforms shown in the diagram are, of course, formed in arid and semi-arid environments and much of your answer must be concerned with the climate and weathering processes in these environments which have determined the landscape cross-section shown in the diagram. As the processes by which some of the features have been formed are not fully understood you must be prepared to explain in your answer the different theories which exist and their weaknesses, if any. The diagram shows three distinct zones, the hill range, the area of deposition (bahada and playa) and the pediment with its sharp break of slope. You must ensure that each one is dealt with thoroughly in your answer, not forgetting the range.

Answer plan First point out that the diagram shows a cross-section in an arid or semi-arid region where the landscape has been formed partly during more humid climates in the past, and partly from the processes at work at the present time. Give a brief summary of the present climatic régime in desert regions and how it affects mechanical and chemical weathering.

In a separate paragraph describe King's pediplanation cycle, (Section 10.2.6 and Fig. 10.4), and explain how the cycle leads to the type of isolated range shown on the diagram.

Explain the different hypotheses proposed by Johnson and Wood to explain the existence of the pediment, bahada, playa and break of slope (Fig. 10.5). Add the suggestion from Peel that the sharp angle is formed by the conservation of water from the hills at this point.

Distinguish between the areas of erosion and those where the material is deposited, and do not forget to mention that the playa was originally a wind-formed depression which has been filled by deposition.

2 Assess the parts played by wind and water in the development of desert landforms.

(in the style of Oxford)

Understanding the question At first sight the question seems very straightforward, concerned as it is with the work of wind and water in an arid or semi-arid environment. However, the word *development* indicates that the question is not confined to the processes operating at present and your answer would not be adequate if no reference were made to the development of some of the landforms as the result of pluvial periods in the past.

Water is another word which must be considered carefully. Do not assume that it refers only to rainfall and the work of streams. Water in its widest sense includes any form of moisture, such as dew which is a regular occurrence in desert regions. Your answer must, therefore, explain the rôle played by moisture in mechanical and chemical weathering. A further point to consider is the significance of the phrase *assess the relative importance of*. You are being asked to judge whether wind or water is the more significant in forming desert landscapes.

Answer plan Write an introductory paragraph pointing out that desert landforms have developed over many thousands of years and there is ample evidence to show that past climates have been more humid than the present one. Mention briefly some of the evidence which supports this statement (Section 10.2.5).

Describe the limited ability of the wind to erode, giving examples, including a sketch of a yardang (Section 10.2.3). Explain that the wind is more effective in transporting and depositing material. Make sketches of sand ripples, barchans and seif dunes without becoming involved in the problem of the formation of seif dunes. Explain briefly how these features were formed by wind action.

Describe the landforms which could only have been developed in wetter conditions than at present. Outline the present-day work of water (Section 10.2.4), including the part played by moisture in mechanical and chemical weathering.

Sum up by assessing the relative significance of wind and water in the formation of desert landscapes, pointing out that water in the past, and moisture at the present time, have both contributed a great deal to the erosion of the landscape. By contrast, the work of the wind is more evident in the depositional features than as an erosive force.

3 'Landforms in the hot arid areas of the world do not always show a simple relationship with the present-day climate of these areas.' Discuss.   (*in the style of London*)

Understanding the question The statement must be read very carefully. Your first impression might be that you are being asked to discuss only those desert landforms which were formed during the pluvial periods of the Pleistocene and earlier times. A second reading will show, however, that the stress in the statement is on the phrase *simple relationships with the present-day climate of these areas.* Your answer must, therefore, review desert landforms and identify those which are the result of present-day climate and those which are fossil landforms, remnants of earlier pluvial periods. Remember that the question is in the form of a statement, so, at some point in your answer, preferably in the concluding paragraph, you will need to state how far you agree with the statement, or whether you would wish to modify it in some way.

Answer plan Start with a brief introductory paragraph explaining what physical processes are helping to mould the landforms at the present time (Section 10.2.2).

Then describe the main landforms produced by present-day processes, e.g. the playa, yardang, barchan and seif dune. Sketches of some of these features should be linked with your explanations.

Explain that many of the landforms of hot deserts, such as wadis, inselbergs and the large, gravel strewn plains cannot be attributed to the present-day climatic régime. Give examples of areas of desert landscape which must have been formed when the rainfall was heavier than it is now (Section 10.2.5).

With the help of diagrams describe the pediplanation cycle introduced by L. C. King (Section 10.2.6 and Fig. 10.4). Show how this cycle is only possible if rainfall at earlier times was sufficient to erode extensively. You must also point out that the Peel theories attribute much of the erosion to chemical decomposition and the removal of waste material by the wind and gravity.

In your conclusion you should assess the validity of the statement. In this case there can be no doubt that it is accurate.

10.7   FURTHER READING

Goudie, A. and Watson, A., *Desert Geomorphology* (Macmillan, 1990)
Goudie, A. and Wilkinson, J., *The Warm Desert Environment* (CUP, 1977)
Hilton, K., *Process and Pattern in Physical Geography* (UTP, 1979)
McCullagh, P., *Modern Concepts in Geomorphology* (OUP, 1978)
Small, R. J., *The Study of Landforms* (CUP, 1978)

11   Weather and climate

11.1   ASSUMED PREVIOUS KNOWLEDGE

You should know about:

(a) Instruments used at weather stations and the recording of daily weather information.
(b) The interpretation of weather maps and meteorological information. Some examining Boards expect candidates to answer a question based on a weather map.
(c) Low and high pressure systems and their associated weather conditions. You should be able to understand and draw diagrams which explain the formation of a frontal system and the weather experienced as it passes.
(d) The nature and causes of rain, frost, fog and similar weather phenomena.
(e) The circulation of the atmosphere and the wind and pressure systems.
(f) The major climatic regions of the world, their temperature and rainfall regimes.

If you have not studied these aspects of weather and climate you should read one of the GCSE books, such as Hume Brown, J., *Weather and Climate* (Blackie, 1972) or Manley, G., *Climate and the British Scene* (Fontana, 1952).

11.2 ESSENTIAL INFORMATION

This section does not include work on depressions and anticyclones. These are dealt with in detail in the unit on Weather maps. Since the two units complement one another they are best studied together.

11.2.1 Definitions

Radiation The sun emits energy mainly as electromagnetic waves which travel through space and are converted into heat when absorbed by the earth's atmosphere. This emission of electromagnetic waves is known as radiation. About half the sun's radiation is in wave-lengths visible to us as light. The rest consist of slightly shorter (ultra-violet) and slightly longer (infra-red) wave-lengths.

The earth is warmed by the rays which are absorbed and in turn the earth re-radiates energy in smaller amounts than those emitted by the sun. Terrestrial radiation occurs in longer wave-lengths than solar radiation and much is absorbed by the atmosphere which acts as a blanket.

Heat passes from the earth to the atmosphere by turbulence and by latent heat transfer, that is, evaporation at the earth's surface resulting in the absorption of latent heat and condensation in clouds with the release of this heat into the atmosphere.

Conduction Heating by contact, for example, heating of the lower layers of air by direct contact with the earth.

Convection The upward movement of a liquid or gas, such as air, which has been heated. It expands, the density is reduced causing it to rise, carrying its heat with it. It is replaced in the lower layers by cooler fluid or gas.

Advection The transfer of heat by horizontal movement of air, e.g. movement of tropical air from low to higher latitudes.

Anabatic wind A local wind caused by the heating of slopes during the day resulting in warm air rising up the slope.

Katabatic wind This is the reverse effect, occurring when the hill slope is cooling. It cools the air close to it which becomes more dense and sinks down the valley slope.

Air mass A widespread section of the atmosphere whose temperature and humidity characteristics are similar horizontally at all levels above the earth's surface. When air rests over an extensive uniform surface for long periods it acquires the temperature and humidity characteristics of that surface. These characteristics will be gradually distributed vertically through the air mass. Those parts of the earth where air masses occur and acquire such characteristics are called *source regions*.

Relative humidity The actual moisture content in a given volume of air, expressed as a percentage of that contained in the same volume of saturated air at the same temperature. It can be calculated using the formula:

$$\frac{\text{relative humidity}}{100} = \frac{\text{absolute humidity}}{\text{saturation content at the same temperature}}$$

Dew point The temperature to which air must be cooled to become saturated by the water vapour it holds i.e. the relative humidity is 100 per cent.

Environmental lapse rate (ELR) The actual temperature decrease with height such as an observer might record ascending in a balloon. The actual lapse rate will depend on local air temperature conditions.

Dry adiabatic lapse rate (DALR) The rate at which rising unsaturated air cools, or subsiding unsaturated air warms. It is at the rate of 3°C per 300 metres.

Saturated adiabatic lapse rate (SALR) The rate of decrease in temperature in ascending saturated air, or of increase in descending saturated air. Rising, moist air will cool as it rises, but the cooling will be less than 3°C for each 300 metres.

Temperature inversion Normally air temperatures decrease as height increases, but sometimes the lower layers of air are cooler than those at higher altitudes. This reversal of the normal pattern is often produced by rapid cooling of the earth's surface (Fig. 11.1).

11.2.2 Water vapour in the atmosphere

Water may be present in the atmosphere as an invisible vapour, as water droplets or as ice. The source of water vapour includes the oceans and large areas of forest. There is a maximum

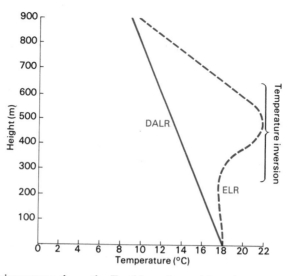

Fig. 11.1 Inversion: air is warmer above the Earth's surface with an increase in the difference between the DALR and the ELR

amount of water vapour that a given quantity of air can hold at a certain temperature (see relative humidity and dew point in Section 11.2.1). The process by which invisible water vapour is condensed and returns to the earth's surface as precipitation is as follows. The atmosphere contains a multitude of condensation nuclei, such as dust and sea salt. Condensation occurs around these nuclei as air is cooled. Cooling may result from (a) radiation, (b) movement up a slope, (c) convection, (d) advection (warm air crossing a cold surface), (e) mixing of air e.g. at a frontal boundary.

11.2.3 Stability and instability

Air is stable when, if forced to rise, it tends to return to its original position. This will happen when the air is cooler than the surrounding air. If, as the air rises, the temperature of the surrounding air falls more slowly than the temperature of the rising air, the rising air will be cooler and denser than its surroundings and tend to sink back.

If on the other hand, the temperature of the environmental air falls rapidly with height faster than the dry adiabatic lapse rate, the rising parcel of air will become warmer as it rises and its speed of uplift will increase. Such air is unstable (see Figs. 11.2–11.4).

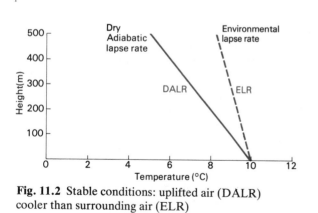

Fig. 11.2 Stable conditions: uplifted air (DALR) cooler than surrounding air (ELR)

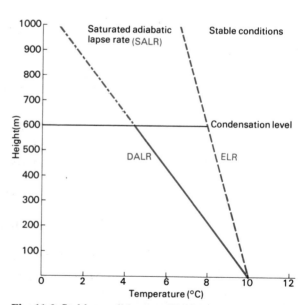
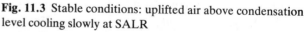

Fig. 11.3 Stable conditions: uplifted air above condensation level cooling slowly at SALR

Conditional instability occurs when moist air is forced upwards and is at first cooler than its surroundings. At some point condensation will occur and heat will be released into the rising air. It will then cool less rapidly, eventually becoming warmer than its surroundings and therefore unstable as it continues to rise (Fig. 11.5).

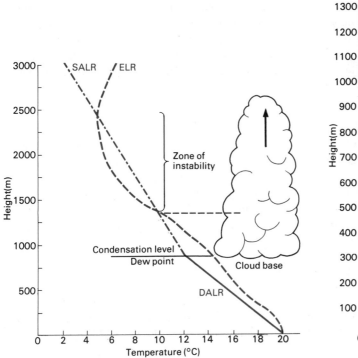

Fig. 11.4 Instability: ELR varies with height giving a zone of instability

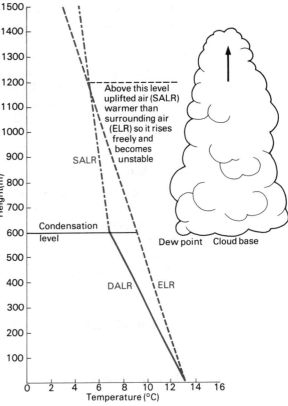

Fig. 11.5 Conditional instability: above 1200 in SALR air is warmer and less dense

11.2.4 Clouds

Cloud formation is associated with lapse rates. Clouds occur at the height at which dew point is reached and condensation takes place. In turbulent conditions layers of air are mixed so that water vapour is fairly evenly distributed throughout the air mass. Adiabatic changes in the rising and falling air produce conditions in which clouds form. *Turbulent cloud* is usually low and of the layer type. *Orographic cloud* forms when air, which has been heated from below, is forced to ascend to the level of its dew point. Under stable conditions *fair weather cumulus* will form. Under unstable conditions thunderstorms associated with *cumulo-nimbus clouds* of great vertical extent are possible.

Cloud will form when warm, moist air is forced to move above colder air. Condensation will occur under these *frontal* conditions.

Clouds can be classified according to the height at which they occur (see Gates, E. S., *Meteorology and Climatology for Sixth Forms and Beyond,* Harrap, 1978).

11.2.5 Air masses

Air masses (see Section 11.2.1), cover many hundreds of kilometres. They originate from source regions such as the sub-tropical high pressure zones, polar and continental regions, from which they migrate and affect the climate of areas over which they pass. Two air masses with different characteristics may meet. They do not mix easily and tend to have sloping boundaries called *fronts* between them.

Air masses are classified according to their source regions and according to the paths they take after leaving the source region. They are called Arctic, Polar and Tropical and depending whether they pass over maritime or continental regions are described as:

mA Artic maritime   *cP* Polar continental
cA Artic continental   *mT* Tropical maritime
mP Polar maritime   *cT* Tropical continental

Britain is affected by *mP, cP, mT* and *cT* air masses.

11.2.6 Local winds

There are three main types of local winds:

(a) Anabatic and katabatic winds (see Section 11.2.1).

(b) Land and sea breezes During daytime air flows from over the cooler sea to warmer land with a reverse flow at night when the sea is relatively warmer than the land. The cause of this flow is the reduced pressure over the land in the daytime and over the sea at night.

(c) Föhn wind This is experienced in the Alps. Air forced to rise over the mountains cools adiabatically and condensation may occur. On descending the other side of the mountains the air will warm up at the SALR until dew point is reached. However, this may be at a higher altitude if there is less moisture present. Then the air will heat up at the DALR, resulting in a relatively warm and dry wind. The chinook which is experienced in Alberta is a Föhn-type wind which has crossed the Rockies from British Columbia.

11.2.7 Micro-climates

Micro-climatology is the study of climatic differences which occur in a small area. Differences of aspect, slope, soil colour, vegetation and plant cover can produce distinctive climatic conditions. Man-made landscapes such as streets, buildings and reservoirs all produce local contrasts in climate when compared with the conditions which prevail elsewhere in the region.

Although people cannot control climate they can, either deliberately or accidentally, affect micro-climates by such actions as removing or changing the vegetation pattern, urban development and water control.

11.2.8 Climatic regions of the world

Although you will have studied climatic regions for O level, you need to appreciate how climates are classified and to know the details of one major classification such as *Köppen's*. An account of the major climatic regions is contained in E. S. Gates' book quoted in Section 11.2.4.

You must also know the distinctive climates which are caused by altitude. The climatic zones with their own vegetation patterns are caused by the decrease in air temperature as altitude increases, often associated with an increase in precipitation. Remember that a mountain situated in the tropics will display a different zone pattern compared with a mountain located in a temperate climatic region.

Moreover there are endless varieties of local climates within a mountain mass, determined by such things as slope, aspect and exposure. Weather changes are more rapid on mountains with precipitation usually heavy and winds which are partly determined by local relief features.

11.3 GENERAL CONCEPTS

Concepts essential to an understanding of weather and climate include the effect of water vapour in air, the transfer of energy in the atmosphere and the heat budget of the earth. These concepts are founded in physics and will not be described in detail here.

Other important concepts are:

Weather and climate over large areas of the earth's surface are determined by air masses The study of air masses is an important part of the A level syllabus and frequently appears in the form of a question on the examination paper (see Section 11.2.5).

The urban heat island This is caused by the construction of large areas of houses, roads and factories. The concept was developed in the research of T. J. Chandler and others in the 1950s and 1960s. You can read about it in Barry, R. G. and Chorley, R. J., *Atmosphere, Weather and Climate* (Methuen, 1968).

Human intervention as a means of determining micro-climates This intervention may be accidental or deliberate. As the results can harm the environment, there is an awareness that changes to the micro-climate must be closely monitored, and, if necessary, checked.

11.4 DIFFERENT PERSPECTIVES

Climatic and weather studies are largely concerned with explaining the physical basis which is responsible for local, national and world-wide climatic patterns. At A level the approach is factual and does not include areas of conjecture.

11.5 RELATED TOPICS

In your study of soils (see pages 79-85), you will have noted that climate is an important factor in soil formation. The physical and chemical processes are the result of climatic conditions and the biological actions are indirectly caused by climate which helps to determine the natural vegetation and animal life.

In physical geography, weathering, slopes and the subsequent development of landforms in different parts of the world depend to a considerable extent on the climatic regime of the region. In some cases the landforms (see the units on glaciation and desert landforms), are, in part, the result of different climates which existed many thousands of years ago.

Climate is also fundamental to an understanding of agricultural production in different parts of the world. The main climatic factors which affect agriculture such as temperature and rainfall distribution also affect natural vegetation. People have assisted cultivation in some regions by such things as irrigation, drainage and the introduction of new varieties of crops which can tolerate climatic variations.

As with agriculture, A level studies which include ecosystems (see pages 86–93), plant communities and changes to the natural vegetation by such methods as *slash and burn,* become more significant if you have a good knowledge of the climate of the region and the inter-relationships between the climate, soils and vegetation. Knowledge of weather and climate is also useful to the study of ecosystems.

11.6 QUESTION ANALYSIS

1 (a)

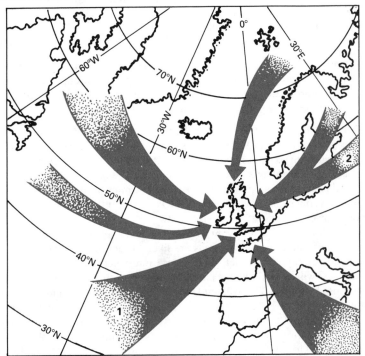

Fig. 11.6 Paths of principal air masses experienced over the British Isles.

(i) Give the names of the air masses that follow the paths numbered 1 and 2.

(ii) What temperature and humidity characteristics will be experienced over the British Isles when under the influence of air path 1 in winter?

(iii) What temperature and humidity characteristics will be experienced over the British Isles in winter when under the influence of air path 2?

(iv) What temperature and humidity characteristics will be experienced over the British Isles under the influence of air path 2 in summer?

(b) Figure 11.7 shows the lapse rate in the lowest 3000 metres of an air mass that follows path 1, when at its source.

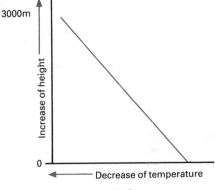

Fig. 11.7

(i) Sketch on Fig. 11.7 the lapse rate to be expected after poleward movement in winter.

(ii) How has the stability changed?

(c) Figure 11.8 shows the lapse rate in the lowest 3000 metres of an air mass that follows path 2, when at its source.

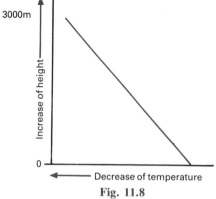

Fig. 11.8

(i) Sketch on Fig. 11.8 the lapse rate to be expected after equatorward movement during the winter.
(ii) How has the stability changed?
(iii) What are the weather implications of the change in stability?

(Oxford and Cambridge, June 1986)

Understanding the question The first part of the question tests your knowledge of air masses and their characteristics. In parts (b) and (c) you are required to apply your knowledge to specific examples. Do not be put off by the two graphs, they have been simplified as illustrations and do not require elaborate additions when you complete them as part of your answer.

Answer plan Air mass 1 is from the sea area of the tropics, so it is an mT air mass. Air mass 2 is from continental Europe's northern latitudes, so it is a cP air mass. These air masses have distinctive characteristics. The mT air mass will bring warm air to Britain in winter, raising temperatures. Humidity will also be increased, producing mist and low cloud.

The cP air mass will bring low temperatures to Britain in winter, as the air is very cold. Initially the air mass has a low humidity, but some moisture may be picked up over the North Sea, producing stratus cloud and possibly snow showers. Humidity otherwise is low.

In summer the cP air mass will bring warm air from the continent. It will be dry air with a relatively low humidity, and only fair-weather cumulus clouds are likely to develop.

Part (b) You are asked what will happen to the lapse rate when mT air moves polewards in winter. Remember the air mass will be cooled at its base as it moves northwards, and the layers of air above the earth's surface will also be cooled. This will tend to make the lapse rate line more horizontal. The stability resulting from the cooling of the lower layers will increase, i.e. the air mass will become more stable. This will result in warm and humid conditions with low layer cloud, and drizzle or sea fog near the coast.

Part (c) Polar continental air moving towards the equator in winter will be cold and dry at first. Lower layers will be warmed, making the lapse rate line slope more steeply. Originally stable, the warmed air will rise and become unstable. The weather implications are that the unstable cold air will give showers, probably of snow.

2 Explain the meaning of the term *micro-climate*. Illustrate the importance of micro-climate in both physical and human geography.   *(Oxford and Cambridge, July 1979)*

Understanding the question The first part of the question requires a definition (see Section 11.2.7). Your definition should be followed by some examples which clarify the difference in scale between micro and macro-climates. This question of scale is important because some studies of micro-climates, particularly when plants are involved, are confined to a few centimetres around and above the plant. By contrast, the study of urban heat islands which are formed by the built-up part of a city involves many hectares.

The second part of the question does not ask you to write down all you know about micro-climates. It is confined to describing the importance of micro-climates in physical and human geography, and the key word is *importance*. These two elements are best tackled as separate sections of your answer, using as many varied examples to illustrate your statements as possible.

Answer plan First write a clear definition of the word micro-climate, giving examples of the different scales which distinguish a micro-climate from a macro-climate.

This should be followed by a long section on the importance of micro-climates in physical geography. In this section you should explain how small areas and surfaces respond to temperature change and how different types of rock are affected by crystallization. You must also explain how aspect and local relief can produce micro-climates with anabatic and katabatic winds and frost pockets in valley bottoms.

Give examples from the Alps or other mountain regions which you have studied, or from areas in your locality where there are marked climatic changes over short distances attributable to physical causes. Do not forget the effects of the sea and expanses of water on adjacent land areas.

A similar section should follow on the importance of micro-climates in human geography. This should include references to settlement patterns in mountain valleys where aspect is important. You should also describe urban heat islands, with examples, to illustrate how man has influenced local climates.

Describe the changes made to the micro-climates of specific environments as a result of man's action. For example, the building of large power stations with cooling towers in the Trent valley; the cutting down of hedgerows in Eastern England; the formation of smog in cities such as Los Angeles and the planting of trees to check the mistral in the Rhône valley.

Write a short concluding paragraph pointing out that micro-climatology is important if we are to appreciate more fully the changes which can be brought about to local environments by human action.

11.7 FURTHER READING

Chandler, T. J., *Modern Meteorology and Climatology* (Nelson, 1983)
Gates, E. S., *Meteorology and Climatology for Sixth Forms and Beyond* (Harrap, 1978)
Hanwell, J., *Atmospheric Processes* (Allen and Unwin, 1980)
Money, D. C., *Climate and Environmental Systems* (Unwin Hyman, 1989)
Musk, L., *Weather Systems – Topics in Geography Series* (Cambridge, 1986)
O'Hare, G. and Sweeney, J., *The Atmospheric System* (Longman, 1984)

12 Soils

12.1 ASSUMED PREVIOUS KNOWLEDGE

Soils are not studied in any great detail for GCSE. However, students who have taken geography at this level should appreciate some of the properties of clay, sand and loam. They should also know that some soils tend to be acid while others are alkaline and that this factor can affect the vegetation associated with them. In regional or farm studies mention will have been made of lateritic soils in the tropics, the fertile loess (limon) of parts of north-west Europe and the rich silts and peaty soils of the Fens.

12.2 ESSENTIAL KNOWLEDGE

12.2.1 Definitions

Regolith The layer of weathered rock fragments which covers most of the earth's land area. It varies in thickness from place to place and the surface layers are called soil.

Texture This is determined by the percentages of sand, silt and clay which are present. Soils with a large proportion of clay are plastic, sticky and cohesive. Sandy soils are the opposite and feel gritty when rubbed between the fingers.

Structure This is very important to the soil's fertility since the structure affects aeration and workability. There are five types: structureless, platy structures, prismatic structures, blocky structures and crumb structures. All these result from the nature of the organic matter and the properties of the soil (Fig. 12.1).

Chelation Organic compounds washed down through the soil detach and remove plant nutrients and mineral ions such as iron and aluminium from the upper layers of the soil.

Leaching The dissolving and washing down of calcium and other bases through the soil as a result of percolating rain water. Bases are substances which react with an acid to form a salt and water solution. Some bases dissolve in water and are then called alkalis e.g. the hydroxides of sodium, calcium and potassium.

Eluviation The process of washing down material such as organic matter or minerals through the soil.

Illuviation The deposition in a soil horizon of minerals, humus and other materials.

Soil profile A section through the soil showing the different layers or horizons. The horizons are usually lettered from A to D with A as the upper horizon and D as the bed rock.

Soil catena The relationship of soil types to the local topography. The changes depend mainly on changes in gradient, hydrological conditions and vegetation.

Zonal soils Soils occurring over wide areas on well-drained land which have been there long enough for the climate and organisms to have expressed their full influence, e.g. chestnut soils.

Azonal soils Immature soils without well developed soil characteristics e.g. alluvial soils or lithosols (those at high altitudes on resistant parent material).

Intrazonal soils Soils affected by some local conditions not involving climate or vegetation, such as poor drainage, e.g. gleyed soils or those where parent material exerts a strong influence such as calcimorphic soils (soils which exhibit distinct features as a result of the parent rock being limestone).

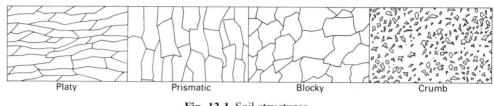

Platy   Prismatic   Blocky   Crumb

Fig. 12.1 Soil structures

12.2.2 Constituents of soils

There are four soil constituents. They are:

Mineral matter This is derived from weathering of the parent material and consists of particles of different sizes such as clay, silt and sand.

Organic matter This is formed by the decomposition and assimilation of plant tissues and animal matter. Decomposed organic matter is called humus.

Air This is usually saturated with water vapour and rich in carbon dioxide.

Water This can be acid, neutral or alkaline and is held as a thin film around particles. It is the medium by which plants are supplied with nutrients.

12.2.3 Soil-forming factors

There is a complex inter-relationship between the following factors:

(a) Parent material The nature of the parent material will have a marked effect on young soils. Its influence will become less as soil becomes older.

(b) Climate This is of major importance in soil formation. Rainfall, temperature and their seasonal and diurnal variations affect soil.

(c) Type and amount of organic life Although vegetation is usually dependent on climate, it can act as an independent variable, as the supply of organic material can be altered and interrupted if the vegetation is changed. Also organisms such as bacteria and earthworms have a marked effect on soil formation by helping the breakdown and incorporation of organic material.

(d) Relief Altitude can affect climate, aspect can influence solar warming and slope angle can affect run-off and soil erosion.

(e) Time Soils form over a period of time at different rates and gradually develop features of maturity.

12.2.4 Soil-forming processes

Podzolization This occurs in the cool, humid regions where leaching is dominant. Sesquioxides (oxides of aluminium and iron) and clay minerals are removed from the upper soil horizons. This produces the true podzol (see Section 12.2.8), particularly in association with heath or coniferous forest. Podzolic or leached soils also occur under a range of vegetation types including deciduous forest and pasture land (see Section 12.2.8).

Calcification This is characteristic of dry regions in continental interiors where leaching is slight and there is considerable evaporation (see Section 12.2.8).

Ferrallitization The accumulation in the humid tropics of sesquioxides in the B horizon (see Section 12.2.8).

Salinization This takes place in arid areas where drainage is impeded and salt accumulates, usually by upward leaching from a saline ground-water supply.

Gleying The reduction of iron compounds by micro-organisms in waterlogged soils.

Many of these soil forming processes are associated with *eluviation, illuviation, leaching* and *chelation* (see Section 12.2.1).

12.2.5 Some aspects of soil chemistry

The main products of chemical weathering in the soil are insoluble clay minerals. These are very small particles carrying a negative charge of electricity on their surface. These particles are dispersed evenly forming a colloidal state. Associated with the clay particles is humus. The two form a clay-humus particle which is negatively charged.

Also present in the soil in solution are electrically charged ions – atoms which have lost an electron (cation) or a proton (anion). The positively charged ions (cations) include calcium, sodium and potassium. The negatively charged ions (anions) include soluble silica and bicarbonate.

The negatively charged clay-humus particles attract the positively charged ions (cations) which attach themselves loosely to the clay-humus particles. They are then said to be adsorbed i.e. loosely captured and capable of being exchanged for others.

The amount of negative charge varies for different types of clay-humus particles and this affects the total amount of exchangeable ions. This amount is known as the cation exchange capacity of the soil. The interchange of ions takes place, for example, after a heavy rainfall. The rain washes away (leaches) cations of minerals such as potassium, calcium and magnesium and replaces them with hydrogen ions, increasing the concentration of hydrogen ions on the clay-humus particles and making the soil more acid.

In time the soil water acquires more calcium and other cations as a result of mineral weathering and plant decay and more ion exchanges take place. However, where rainfall amounts are consistently high the hydrogen ions predominate and the soil remains acid. Where the climate is drier and less leaching takes place there is an accumulation of calcium and magnesium ions, for example in chernozem soils.

12.2.6 Man's influence on soils

The main ways in which soils are modified by man are: altering the plant succession through grazing etc., removing the natural plant cover and replacing with crops, timber etc., ploughing and draining which changes the soil structure and the arrangements of the horizons.

12.2.7 Soil fertility

Soil fertility is dependent on the following factors.

The physical properties of the soil These are its depth, texture, structure, stoniness and drainage. A fertile soil should have a deep and well-aerated rooting zone.

The availability of organic matter (humus) This increases the chance of creating a fertile soil by improving the structure and increasing the moisture-holding capacity of sandy soils.

Suitable conditions for organic decomposition and the incorporation of organic matter in the soil These vary according to the amount and type of litter available, the nature of the soil and the climate. The richest soil forms where there is plenty of plant litter, aeration and drainage are good and the soil is neutral or alkaline and soil fauna such as earthworms mix the plant material with the soil minerals. Under these conditions the organic matter breaks down completely and the humus is evenly distributed in the upper part of the soil. This type of organic distribution is called mull.

Less fertile soils are called moder and mor (see Bridges, E. M., *World Soils*, CUP, 1978, pages 18–19).

The appropriate chemicals must be present in the soil Some 16 chemical elements are known to be essential to cultivated plants, though some are only required in trace amounts. Calcium is one such element – improving the structure of the soil.

The degree of soil acidity or alkalinity is also important as several nutrients become less available to plants at the extremes of pH values. (The concentration of hydrogen ions in solution is indicated by the pH scale. Neutral soils have a pH value of 7, higher values are alkaline, lower ones are acid.)

12.2.8 Soil profiles

Coniferous forest zone Pine needles and other litter from coniferous forests form an acid humus (mor). Litter accumulates during the cold winters and the spring thaw removes plant nutrients, iron and aluminium from the upper soil layers which therefore have a bleached colourless layer of silica. Lower down the iron and aluminium accumulate to form a darker illuvial horizon which may give rise to an impervious layer known as hardpan. This soil is known as a podzol (Fig. 12.2).

Prairie grassland In this area the main organic matter is grass and its roots. Precipitation is

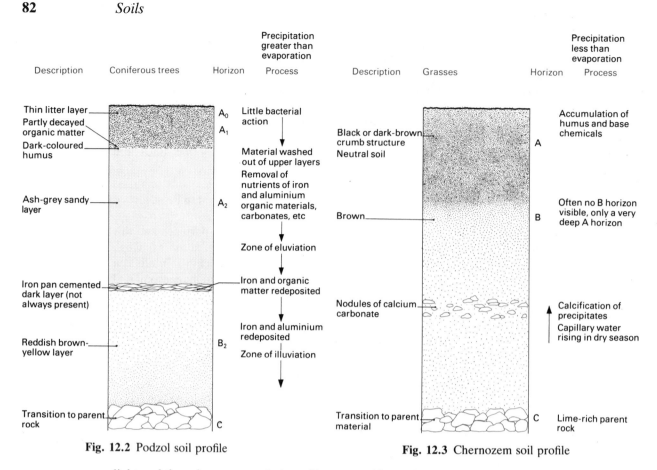

Fig. 12.2 Podzol soil profile

Fig. 12.3 Chernozem soil profile

light and there is an accumulation of humus and base chemicals near the surface. This produces a black earth or chernozem soil with a crumb structure (Fig. 12.3). In dry seasons, when precipitation is less than evaporation, capillary water rises and a calcic horizon of calcium carbonate forms.

Deciduous woodland Leaf fall accumulates in the autumn and decays into a less acidic humus known as mull. Precipitation is greater than evaporation so there is still marked leaching but the minerals are not broken down chemically. The soil is called a brown earth. It is fairly uniform

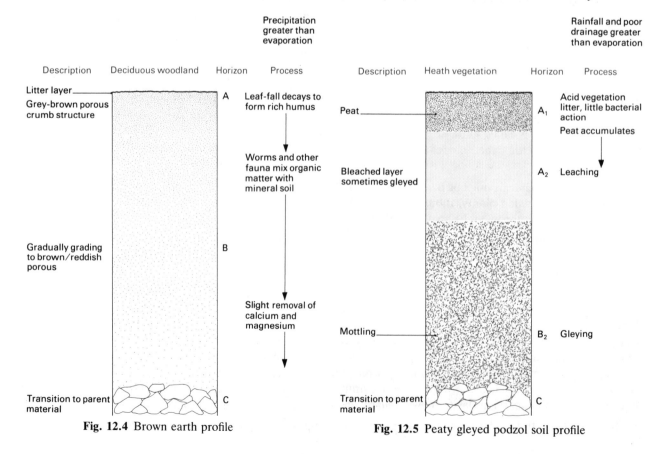

Fig. 12.4 Brown earth profile

Fig. 12.5 Peaty gleyed podzol soil profile

in colour, lacking the distinct horizons of the podzol. This is due to the greater number of organisms which turn over the soil (Fig. 12.4).

Heathland soils In upland areas, rainfall totals are high, drainage is often poor and the natural vegetation is heathland. In these areas a peaty gleyed podzol develops. The vegetation provides an acid litter, there is still bacterial action and peat accumulates. Heavy rain leaches the soil to form a bleached layer and a mottled B horizon, sometimes with an iron pan development (Fig. 12.5).

Tropical soils High temperatures and rainfall speed up rock and mineral weathering in the tropics. The leaf fall and its rapid decay keeps bases in rapid circulation. Leaching is heavy but, rather than silica, iron and aluminium oxides remain to give the soil its characteristic red or yellowish colour. Such soils tend to be infertile due to the lack of humus content and lasting bases. Where there is a marked dry season many of the soils develop lateritic crusts which may be as much as 10 metres thick and which are rich in iron.

12.2.9 World soils

In Section 12.2.8 descriptions were given of some major soil profiles which owe their characteristics partly to the climatic régimes in which they occur. Figure 12.6 gives a diagrammatic representation of the major soil profiles to be found between the Pole and the Equator in the northern hemisphere. It also shows soil-forming processes associated with the soil types. The diagram indicates that there are no natural boundaries between the different groups and that sub-groups also exist.

There are two major groups of soils, those with calcium carbonate present, called *pedocals*, and those with aluminium and iron present, but no calcium carbonate, called *pedalfers*.

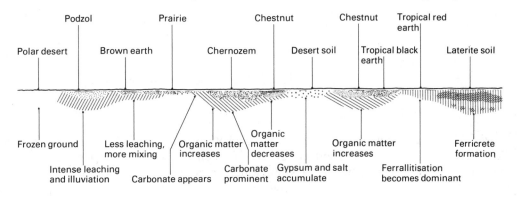

Fig. 12.6 Profile relationships of zonal soils in a transverse from Pole to Equator

12.3 General concepts

Soil formation can be regarded as a system Soil is formed as the result of the interaction of the five factors listed in 12.2.3. It is an example of a process-response system because it gains and loses energy as a result of a number of processes. Various inputs and outputs of mass and energy occur. For example, there are *inputs* of material when weathering occurs and of plant nutrients when organic matter in the soil decays. There are losses (*outputs*) as a result of leaching and when minerals and nutrients are removed in the drainage water.

Recycling occurs when nutrients absorbed by plants are returned to the soil as plant litter Therefore:

Soil is a dynamic medium which is changed by natural processes and develops over time

Soils can be grouped according to their energy balance sheets to provide broad classifications The soil groups display similarities of their horizons which develop over time in certain climatic and vegetation conditions.

Soil is a resource which can be used for a variety of purposes such as agriculture, forestry and recreation The extent of the resources can be influenced by man who can destroy soils or take measures to conserve them. In some cases human interference has affected soils by modifying or removing the vegetation cover. In these conditions wind and water erosion may be active. Wind erosion blows away the top soils producing such disasters as the 'Dust Bowl' in the dry western interior of the USA. Water erosion can also remove top soil and create gullies. On the positive side, land has been upgraded by drainage and the application of fertilizers. Large schemes such as the reclamation of the IJssel Meer increase the soil available for agriculture.

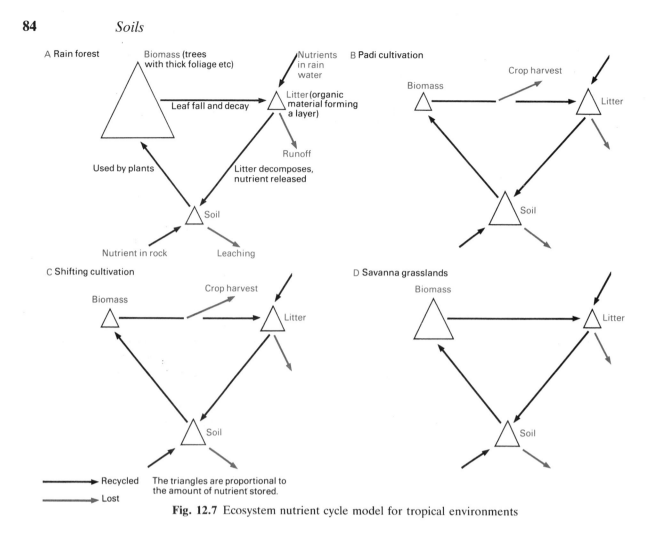

Fig. 12.7 Ecosystem nutrient cycle model for tropical environments

12.4   Different perspectives

One way of looking at soil is to regard it as some kind of working organization, a system through which there is a flow of materials and energy. A soil system can thus be considered as a series of inputs and outputs of energy and matter within which a recycling of nutrients occurs. A model of this nutrient recycling process can be designed consisting of (a) *a biomass store* – plants and organisms living near or above the surface; (b) *litter* – organic material which decays to make humus; (c) *soil* – derived from parent rock, deposition and weathering. Figure 12.7 is an example of an ecosystem nutrient cycle model designed for different tropical environments.

12.5   Related topics

Soil studies are closely inter-related with aspects of biogeography, climate, geology and physical geography. It is convenient to study different elements of the ecosystem separately, but you will find many instances when work on one aspect leads to a greater understanding of others.

   Studies of agricultural patterns and farming practice will inevitably be strengthened if you fully understand the soil systems, whether the area involved is one of highly commercialized farming or a backward area of the Third World.

   You will find that a knowledge of soils is of value when studying environmental problems such as soil erosion and pollution. Soil erosion tends to occur when farming methods developed in one environment are used in another for which they are unsuitable because of the nature of the soil, vegetation, climate and relief. Conservation schemes aim to restore the soil and prevent further deterioration, usually by the planting of suitable vegetation.

12.6   Question analysis

1 'The natural chemical properties of the soil are more important in determining its fertility than the physical properties'. Discuss.   (*in the style of Cambridge*)

Understanding the question   Before answering the question you must first make certain that you know what is meant by *natural chemical properties* and *physical properties*. List these properties to guide you in writing your answer. Make additional notes which support or refute the statement that the natural chemical properties are more important in determining the soil fertility than the physical properties. Remember that soil fertility is a complex subject which is dependent on a number of factors of which the natural chemical properties is one. It is significant that the question used the word 'natural' to describe the chemical properties of the soil. At the present time large quantities of artificial fertilizer are used to make soils fertile and the chemistry of the soil can be altered relatively easily. Mention this in your answer.

Answer plan Start by explaining what is meant by soil fertility and how it depends on the internal properties of the soil (see 12.2.7). Describe how the physical properties of the soil can affect its fertility, especially the structure of the soil since a fertile soil is one which provides a deep and well-aerated rooting zone. Explain how the introduction of plant nutrients can help to improve the structure as well as the fertility of the soil. Describe in detail the part played by natural chemical elements in soil fertility. Write a concluding paragraph which stresses the interdependence of the chemical and physical properties, but emphasizes the paramount importance to all soils of maintaining the correct levels of plant nutrients.

2 With reference to the areas covered by any **one** of the world's forest types, discuss the role played by physical, biological and human factors in soil profile development.

(Welsh Joint Education Committee, June 1979)

Understanding the question The forest types you are most likely to have studied in detail are the tropical rain forests, deciduous forests of temperate latitudes and the coniferous forests of northern latitudes. Look carefully at what else the question requires before deciding which one to select. The question asks you to discuss three factors which help to determine soil profile development in *one* forest type. You must therefore be able to apply *each* of the three factors to the forest type you have selected. Select the forest type which you understand most fully when the three factors mentioned in the question are taken into account.

You should note that the question asks you to discuss. Do not, therefore, limit your answer to three paragraphs, one for each of the factors. You must also comment on their relative importance in soil profile development for the forest type you have selected.

Answer plan An introductory paragraph is required describing the nature and distribution of the forest type you have selected. You must then describe the climate and other physical features of the region and how they will affect the soil.

Draw a soil profile diagram and give a full explanation of what it shows. Give examples of how man's activities have affected the forest zone and, in turn, the soil profile. If possible draw a second soil profile to show changes brought about by human factors.

Write a short concluding paragraph emphasizing the inter-relationships between the physical, biological and human factors in soil profile development.

3 A transect from the Tropic of Cancer to the Arctic Circle through a hypothetical continent shows the following succession of soil zones: semi-desert soils; chestnut brown soils; chernozems; brown forest soils; podzols.
(a) Explain why the soils are zoned in this way.
(b) Draw an annotated soil profile to describe one of the soil zones.
(c) Explain the changes which are likely to occur to the soil if the deciduous woodland in the brown forest soil zone is replaced by coniferous woodland. *(in the style of Oxford)*

Understanding the question The question describes the soil zones to be found in a transect from south to north across a hypothetical continent. In this sense it is a kind of model since in reality a transect would show variations within the zones due to topography, drainage, plant cover and other physical or biological factors.

The examiner needs to be satisfied that in part (a) you fully understand how soil zones are formed and the factors and processes which result in the particular zonal types listed in the question. Part (b) requires a diagram of a soil profile (see Figs. 12.2–12.5). The examiner will award low marks if the diagram is badly drawn and the annotations inadequate or illegible. Part (c) poses the problem of how a change in vegetation may affect the nature of the soil. In this case the change is from the litter provided by deciduous trees to that provided by coniferous woodland. Remember that changes will not occur suddenly. They will require time and could only become apparent after twenty or thirty years.

Answer plan Explain in an introductory paragraph that the zonal soils listed in the question are the result of the interactions of a number of soil-forming factors and processes (see Sections 12.2.3 and 12.2.4).

Describe briefly the processes which have been responsible for each of the zonal types listed in the question. For part (b) select a soil profile you know well and draw an annotated diagram similar to Figs. 12.2–12.5.

Part (c) requires an explanation of the differences of the litter layer under deciduous trees and that under coniferous trees and how this would affect the chemical composition of the soil (see Figs. 12.1 and 12.3).

12.7 FURTHER READING

Bradshaw, M., *Earth. The Living Planet* (Hodder and Stoughton, 1977)
Bridges, E. M., *World Soils* (CUP, 1978)
Courtney, F. M. and Trudgill, J. T., *The Soil: An Introduction to Soil Study in Britain* (Edward Arnold, 1987)
Hilton, K., *Process and Pattern in Physical Geography* (UTP, 1979)

13 Plant communities: ecosystems

13.1 Assumed previous knowledge

You probably did nothing about ecosystems for GCSE geography but if you took biology at this level, you probably have background knowledge to the topic. Some of your systematic and regional geography work will also provide useful background information, particularly work on the climatic regions and vegetation belts of the world. Work on soils is relevant, as are studies of distinctive areas such as the Amazon Basin, tundra regions etc.

13.2 Essential information

13.2.1 Definitions

Systems A system is a structured set of objects (i.e. components), or a structured set of attributes, or a structured set of objects and attributes combined together.

Set of objects means that a system has boundaries which separate it from other systems. *Structured* means that the system has internal order, that is, the components are arranged and inter-connected in some kind of pattern.

Attributes are the characteristics of the system and include appearance and behaviour. These attributes can be measured.

Ecosystem An ecosystem is a system in which both the living organisms and their environment form components (elements) of the system. These elements are linked together by flows and are separated from outside elements by a boundary e.g. a pond or forest is an ecosystem.

Food chains Within the biological part of the ecosystem there are food chains in which one living organism is dependent on another. The levels in the food chain may be seen as forming a pyramid with each step of the pyramid called a trophic level (Fig. 13.1).

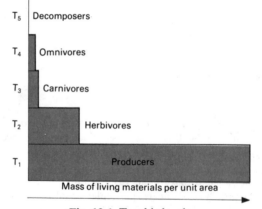

Fig. 13.1 Trophic levels

Ecological Community An assemblage (grouping) of particular species of plants and animals which are linked by the flow of energy, the cycling of nutrients and the regulation of population within a particular physical and chemical environment.

Biomes Major terrestial ecosystems of the world. They may also be called provinces, biochores or regions (see Fig. 13.2).

Biomass The total content of the organic matter. The higher the trophic level in a food chain, the less the biomass. The proportion between the biomass at a given trophic level and that at the next higher trophic level is called the *biomass ratio*. Biomass ratios vary within an ecosystem and from one ecosystem to another. The biomass is usually measured as dry weight per unit area of organism.

Ecological succession Bare ground is soon colonized by plants where growth is possible. Subsequently there is a series of sequential replacements as one set of dominant plants is replaced by another. This process of sequential replacement is known as *succession* (Fig. 13.3).

Ecological niche Within an ecosystem the place that a particular species occupies in the total system is called its *ecological niche*. The number of niches in a given ecosystem is a measure of

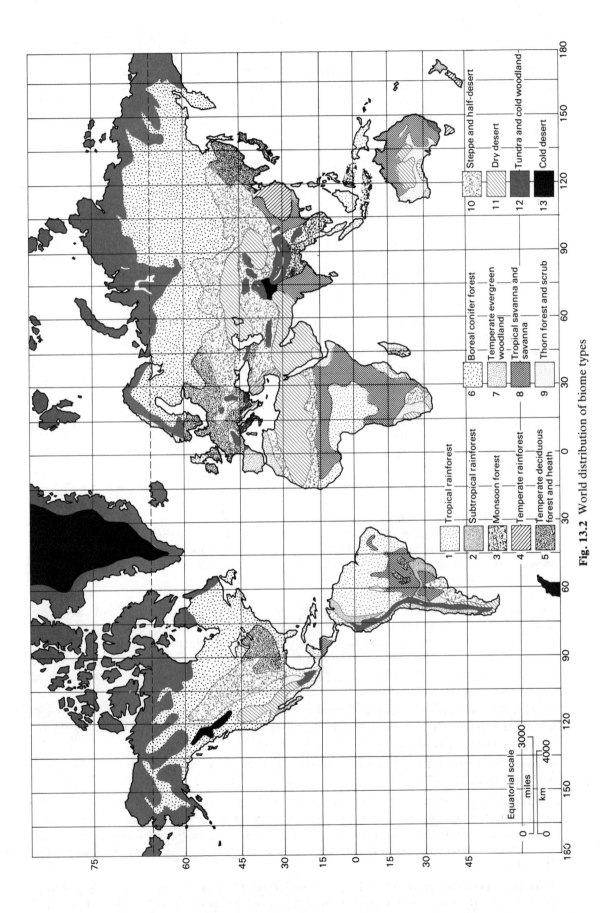

Fig. 13.2 World distribution of biome types

1	Tropical rainforest
2	Subtropical rainforest
3	Monsoon forest
4	Temperate rainforest
5	Temperate deciduous forest and heath
6	Boreal conifer forest
7	Temperate evergreen woodland
8	Tropical savanna and savanna
9	Thorn forest and scrub
10	Steppe and half-desert
11	Dry desert
12	Tundra and cold woodland
13	Cold desert

Equatorial scale

miles 0   3000
km   0   4000

how complex the system is. In polar regions there are few niches in the ecosystem, in the humid tropical areas the number of niches is enormous.

Climax community Like other systems, ecosystems move towards a state of stability. Each species alters its own environment over time and that of its associates. Factors such as competition and parasitism lead to vegetative change over time. Dominant species emerge and plant succession occurs whereby the dominants become larger and more complex. In Fig. 13.3 you can see the succession of plants in a coastal area of Australia with a humid sub-tropical climate. The end product of the process of succession (in the diagram, forest) is known as the *climate climax vegetation*. Unless climate or geological conditions change significantly this vegetation will persist.

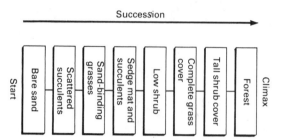

Fig. 13.3 Dominant plant succession in a humid sub-tropical coastal area

The nature of the climax is determined chiefly by the physical environment – the climate, rocks and soils. Forest is the usual climax in those places on land where there is sufficient light and where temperatures are not too extreme. In drier regions the climax is grassland. The normal climax may however be *arrested* by flooding, fires and human interference.

13.2.2 Ecosystems

An ecosystem includes both organisms and their environments. So a forest ecosystem includes all the living organisms of the forest (plant and animal), the soil in which most of the plants live, the moisture taken in by the plants and animals as well as the special micro-climate which a forest establishes itself.

Ecosystems are therefore very complex. There may be many components (e.g. the different species in a forest). The linkages between the components may be very intricate. So we isolate aspects of the ecosystem in order to study them, e.g. the food web.

Although there is a great variety of ecosystems in existence, all of them are characterized by general structural and functional attributes. Ecological relationships exist between *abiotic* (non-living) environmental substances e.g. water, carbon dioxide, and *biotic* components i.e. plants, microbes, animals.

Ecological relationships are fundamentally energy-orientated The basic source of energy for any ecosystem is radiant energy (sunlight). This energy is converted by *producers* (Fig. 13.4) by the process of photosynthesis into a chemical form by the production of carbohydrates. The producers are such chlorophyll-bearing plants as grass and trees and phytoplankton in the oceans as well as bacteria which oxidize inorganic compounds which are important in creating the movement of nutrients through the system.

One of the principal features of the ecosystem is a one-way flow of energy (see diagram). The energy moves to *primary consumers* such as herbivores that derive energy directly from the

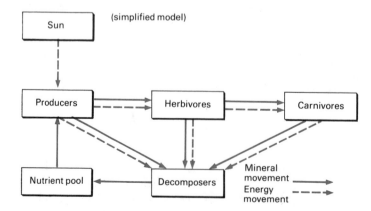

Fig. 13.4 Energy and mineral movement in ecosystems

plants (producers). Carnivores are *secondary consumers* which obtain their energy indirectly from the producers by way of the herbivores. Both groups of consumers are called *heterotrophic* (other feeding) because they get nutrition by feeding on other organisms.

A third group of heterotrophs are the *decomposers* – mainly bacteria and fungi. Through the action of enzymes the decomposers mineralize organic matter and make it available for re-use by producers.

From the diagram you can see that the movement of energy within an ecosystem is non-cyclic and undirectional. The movement of nutrients however is cyclical.

The processes of energy flow and mineral cycling are fundamental to the ecosystem. These processes occur through the vehicle of living organisms – animals, plants etc. Since each species has unique attributes, no two ecosystems are exactly alike. In analysing different ecosystems we are essentially concerned with the *interdependence* of life forms, with the relationship of particular life forms to the whole system, and with ecosystem stability.

The dynamics of population This term refers to the flows of energy and matter in the form of organisms, and the capacity of a species to alter its own environment and that of its associates together with the ability to adjust to its habitat. The number of individuals of a species is partially determined by the difference between birth rate and death rate. Low percentage death rates will produce high populations (see Unit 14 *Growth and distribution of population*). Migration also plays a part in the control of animal population. Populations do not expand indefinitely; the rate of growth levels off at a level called the *carrying capacity of the habitat*. This capacity is the product of a number of factors which are usually grouped as factors of *environmental resistance*. The process is known as *population self-regulation*.

A steady state Like other systems therefore the ecosystem moves towards a condition of stability – the steady state. It is the result of the dynamic interaction of all the forces operating within the system. This does not mean that the number of species within the system becomes constant. Instead there is a dynamic state of fluctuation around a mean. The steady state is characterized by the evolution of the climax community,

13.2.3 Ecological communities

The significant properties of ecosystems are energy flow, nutrient cycling and population self-regulation. These processes do not occur in isolation. They operate in particular environments in relation to particular assemblages of different species populations. These assemblages are known as ecological communities.

At both a micro (small scale) and macro (large scale) level it is possible to recognize zones or belts which are characterized by particular assemblages. For example, there is on one scale the world latitudinal zoning into biomes and on the other the summer zoning of plants from the shoreline into a lake, characterized by terrestial plants (e.g. elm trees) then sub-aquatic plants (e.g. willow trees) to reeds, water-lilies and finally totally submerged plants.

13.2.4 Biomes

Biomes are arranged latitudinally. The same biome is found within the same general latitudes in different continents (Fig. 13.2) e.g. the tundra stretches across northern North America, northern Europe and northern Asia.

In the mountainous areas of the world the distribution of biomes relates to altitude rather than latitude. The particular biome found at a particular altitude depends however upon latitude. In the northern hemisphere for example a given zone is found at progressively lower altitudes in mountainous regions as one moves northwards.

The latitudinal distribution of biomes reflects the prime influence of climate in determining the pattern. Temperature is largely dependent on the incidence of solar radiation. This is directly related to latitude. Wind patterns are similarly associated with latitude (remember the wind and pressure belts of the world diagram you learned for GCSE) and this strongly influences patterns of precipitation. Climatic factors are therefore vital. Soil is also an important regulatory factor in determining the distribution of biomes though the soil is itself partly the production of climatic conditions.

13.3 GENERAL CONCEPTS

Ecological studies Ecology is concerned with the inter-relationships of living organisms and their environments. Since the purpose of the studies is to discover the principles which govern these relationships, the ecologist investigates the totality of the living conditions of plants and animals; their reactions to the environment and to each other; the nature of their inanimate surroundings. So, ecological studies are concerned with seeking an understanding of the structure and dynamics of ecosystems.

Systems (see Section 13.2.1 for definition) Systems are analysed at four main levels of abstraction.

Morphological systems are defined in terms of their internal geometry – the number, size, shape and linkages of their components. In physical geography, for example, a map of a stream network represents a morphological system, for all the units are spatially linked and each unit may be classified in relation to all the others.

Cascading systems In cascading systems the relations between the individual components or units involve transfer of mass or energy from one component to another. Output from one becomes input to another. The transfer can be controlled by *regulators*. They may also provide storage. In assessing the behaviour of cascading systems scientists are therefore concerned with the *rate of throughput*

An example of a cascading system is the fall (cascade) of precipitation on the land or surface water moving along drainage channels to the sea.

Process-response systems combine the attributes of the two previous types. A process-response system changes its internal geometry and/or behaviour in response to cascading inputs. They involve at least one morphological system and at least one cascading system. The systems are linked together and often have some components in common.

An example is a lake in a desert basin. The lake is a morphological system with the components of its internal geometry being surface area, volume, depth, etc. When rain falls it is a cascading system which shares with the lake the amount of water which falls. The rain may cause the lake level to rise so its surface level increases. The increased surface area also increases the rate of evaporation (the regulator). The water in the lake represents storage. Since the volume of the lake, the surface area, etc. changed in response to precipitation then we can say that the morphological system has changed in *response* to a change in cascading input (rainfall).

Control systems Process-response systems can be affected by human intervention – by controlling the output and input flows for example. Environmental management is concerned with the operation of control systems according to the ideals and priorities of those responsible for the management.

All types of systems have a tendency to maintain themselves by moving towards stability.

13.4 DIFFERENT PERSPECTIVES

At one time the traditional approach to 'natural vegetation' was to deal with it as a distinct aspect of the physical environment. The close relationship of vegetation to climatic zones and soil types was emphasized. The basic vegetation pattern was used to provide a framework within which typical agricultural economies were studied. This is an outdated approach which will not earn good marks at A level.

The development of biogeography as an aspect of geography led to a change of perspective. The framework within which geographical studies of vegetation and animal life were made became the ecosystem. The ecosystem includes not only the living organisms but also the environment. Environmental scientists now have a common theoretical framework within which to work. This is the *General systems theory*. The systems idea provides a complete frame of reference for the physical, biological and social worlds. So the older divisions of physical and human geography can be brought together within the same framework.

13.5 RELATED TOPICS

Study of the ecosystem is closely linked with Unit 12 *Soils* and Unit 11 *Weather and climate*.

13.6 QUESTION ANALYSIS

1 (a) In the boxes provided indicate the six main components of the ecosystem.

(b) Show the inter-relationships between these six components by means of arrows.
(c) Outline four ways in which man can upset the balance of the ecosystem.   (*in the style of London*)

Understanding the question The first two parts of the question are straightforward. They are designed to check that you have accurate knowledge of the basic features of the ecosystem. Both parts together will probably account for half the marks. Part (c) is more complex. It asks you to relate your geographical knowledge to the concept of the balance of the ecosystem. Each of the four ways you state should be backed up by reference to actual examples.

Answer plan Use the diagram on page 88 to complete the six boxes. Place the sun in the top box and the decomposers in the lowest. Use the same diagram to help you show the inter-relationships. Indicate the energy movement flows and the mineral movement flows separately.

In beginning the final section you need to outline what you see as the balance of an ecosystem before stating ways in which man can upset this balance (see above). You need to spell out that the presence and persistence of a given life form in an ecosystem imply that its numbers and distribution are more or less stable for the time being. This does not mean that the situation is static but that there is a state of balance with adaptation occurring to maintain that balance. This internal balance is characterized by inter-dependence, competition and evolution.

There are four ways in which balance may be upset:

1 *By the establishment of sedentary agriculture* e.g. the nineteenth century establishment of plantations in tropical forest areas and the colonization of the temperate grasslands by grain farmers.

In this type of economy food chains within the existing system are subordinated to the production of crops which are either eaten by man (e.g. wheat) or which are fed to other herbivores to provide animal protein (meat, milk, cheese). As far as possible all competitors are removed. High inputs of energy and matter are needed to weed, to act as pesticides, to provide nutrients.

Much of the energy gained in this system can be cropped by man. Compared with earlier forms of agriculture however there is a great reduction of biomass. Forms of energy have also been transformed with work done by man replaced by the use of fossil fuels (through mechanical planting and harvesting). The cropping and manuring cycles have been replaced by industrial fertilizers. The system has also been affected by the materials man returns to it – wastes may include toxic products which can interrupt food chains.

2 *By affecting global energy supply* This occurs as a result of the emission of gaseous waste and waste composed of particles into the atmosphere. Since the nineteenth century there has been, for example, an accelerating rate of carbon dioxide release into the atmosphere. This allows an increasing amount of solar radiation to be absorbed by the atmosphere and therefore produces global warming. It has been calculated that if the carbon dioxide content of the atmosphere were doubled there would be an increase in global temperatures of 1.5 to 2°C. This would affect the distribution of the main climatic belts, cause changes in local climates, allow greater penetration of harmful radiation, encourage the melting of the ice caps and so cause a rise in sea level. Existing balances would therefore be significantly affected.

3 *By the deliberate introduction of new species into an existing system.* The introduction of predators and parasites can interrupt existing food chains and may also lead to the replacement of species which are unable to compete with the introductions. You should give an example of this.

4 *Through mineral exploitation.* Eighty per cent of the metallic ores and 95 per cent of the non-metallic minerals and rocks which are economically valuable are mined by open cast methods or by dredging. This disrupts balances within existing ecosystems. For example, the extraction of sand and gravel from bays has caused the erosion of beaches; the erosive power of streams has been increased by dredging; the increased amount of sediment suspended in the water as a result of dredging may kill flora and fauna. The refining of minerals releases products such as liquid acids and toxic gases which may devastate the flora and fauna of surrounding areas.

2 Show how climate vegetation and soils are interrelated in tundra regions. (*in the style of Scottish Higher*)

Understanding the question This is not a complicated question but it is not a purely descriptive one either. The higher marks will not be gained by showing how much you know about the tundra but by using this knowledge to draw out the inter-relationships between the three factors.

Answer plan Probably the best way to start is by defining and locating the tundra regions.

Since climate is the fundamental factor in the determination of the main features of both vegetation and soils it is rational to begin the main part of your answer with a description of the main features of the tundra climate. Main points are:

(a) It is characterized by very long cold winters and short cool summers. In only two to four months do temperatures rise above freezing point. Killing frosts can occur at any time of the year.

(b) The diurnal range in temperature is less significant than the seasonal variations in the length of day and night. Insolation is absent in midwinter, nearly continuous in midsummer.

(c) Although air temperatures may remain below freezing until June and winter begins in September, solid objects may become quite warm in the short summer. Even when air temperatures are at zero black bulb temperatures of 38°C have been recorded.

(d) Precipitation is low, much of it in the form of snow. Little evaporation occurs because of the low temperatures.

The zonal soil in the tundra is essentially a product of the climate. Because of the low temperatures permafrost exists beneath the soil, sometimes only a few inches below ground level. The summer thaw of snow and ice in the soil layer produces waterlogged conditions since the water is unable to penetrate the permafrost. So the typical soil is a gley, characterized by its pale colour. This is because shortage of oxygen in the waterlogged ground produces grey-coloured iron compounds. The low growing vegetation cover provides a thin organic surface layer which is largely undisturbed by processes of evaporation (temperatures are too low) or leaching (water cannot penetrate permafrost). You might include a profile of a gley at this point. Typical tundra vegetation is found in regions bordered by the isotherm for the warmest month of 0°C to the north and 10°C to the south. Important species found in the region are mosses, lichens and, in warmer areas, dwarf birches and willows. There are also dwarf shrubs of ling growing to 50 cm high. The vegetation is closely adapted to the climate:

(a) Only a few hardy species are able to survive in the climate. The ability to survive is not so much determined by the long, severe winter as by the short, cool summer growing season.

(b) Many plants have distinctive adaptive features e.g. many do not seed but reproduce vegetatively. Seeds which are produced are very resistant to frost and remain dormant for very long periods until sufficient warmth is received. Plants are low and often tufted or rosetted so that leaf shoots are protected from wind and frost when the snow cover disappears. Mosses and lichens in particular are resistant to temperature changes in summer.

(c) As in deserts the plants have to live in conditions of moisture deprivation. Rainfall is low, much water is unavailable because it is frozen. Some of the plants therefore have leathery, waxy or hairy leaves to reduce transpiration.

(d) As a further adaptation to the cold a large proportion of the tundra biomass exists as roots (76 per cent).

(e) Micro-climatic factors are also important e.g. aspect is significant with south-facing slopes capable of supporting a richer vegetation cover than northern slopes.

There are also vegetation – soil inter-relationships. For example, dead vegetation does not decay rapidly because of the low temperatures. Consequently the surface layer of the soil is peaty and acid.

In conclusion point out that in extreme climatic conditions such as those of the tundra and arid lands the inter-relationships between climate, vegetation and soils are very clear.

3 Figure 13.5 depicts a temperate grassland ecosystem in terms of its main compartments (column B), human activities (column A) and the changes caused in the ecosystem by those human activities (column C).

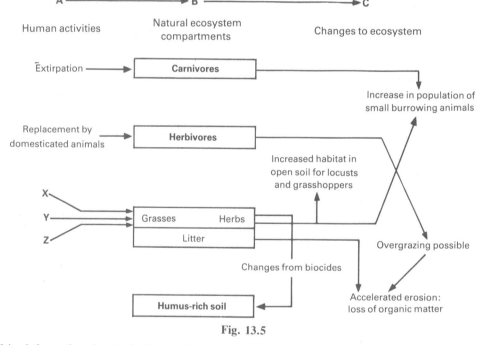

Fig. 13.5

(a) Using information given in the figure, give a reasoned definition of the term 'extirpation' (column A).

(b) According to the figure, state **two** consequences of the introduction of domesticated animals.

(c) Explain *each* of the two consequences.

(d) (i) In column A, state three possible human processes, X, Y and Z affecting the grasses and herbs compartment.

(ii) Which of the three processes X, Y and Z would lead to an increase in the population of small burrowing animals?

(iii) Explain your answer to **(d)** (ii).

(e) (i) The Prairies of North America represent one major area of temperate grassland. Name **two** other major areas.

(ii) Suggest **two** reasons why temperate grassland sometimes occurs in areas with forest climates.

(University of London, June 1985)

Understanding the question This is a straightforward question, but is not as simple as it first appears. It cannot be answered purely from analysis of the diagram but requires good knowledge of the temperate grassland ecosystems and their world distribution.

Answer plan The structure of your answer has been provided by the series of questions asked about the figure. You should write brief answers to each, in complete sentences.

Part (a) Extirpation means the removal or total destruction of the existing carnivore population.

Part (b) You need to analyse the diagram to identify arrowed lines which indicate the consequences of the introduction of domesticated herbivores, e.g. increased habitat for locusts and grasshoppers.

Part (c) You need to explain that originally, selective grazing by wild animals such as the antelope largely controlled the composition of the plant community. These animals were more likely to eat suitable rapidly growing plants which were conspicuous because of their size, so a plant which grew steadily had a better chance of survival. The soil retained high levels of nutrients and organic matter and in times of drought this helped to retain moisture. When the turf mat was broken, the organic matter helped to prevent erosion. When domesticated animals were introduced, the palatable plants were grazed out and this left a lot of open soil.

The open soil favoured the increase of locusts, grasshoppers etc. The openness of the soil also makes the leaching of nutrients more likely – hence the danger of overgrazing.

Part (d) (i) X, Y and Z could be: ploughing, reseeding and weed control.

(ii) and *(iii)* Reseeding would increase the food supply for some species, and hence the food supply for carnivorous species would also increase.

Part (e) (i) Any two temperate grasslands, e.g. the Russian Steppes, the Argentinian Pampas.

(ii) You could choose two reasons which reflect the fact that during historical time man has converted forested areas into productive arable land or to other desirable purposes. Two reasons might be:

1 Forest clearance to establish an arable economy.

2 Forest clearance to make way for ornamental parklands for wealthy landowners.

13.7 FURTHER READING

Bradshaw, M., *Earth. The Living Planet* (Hodder and Stoughton, 1977)
O'Hare, G., *Soils, Vegetation and Ecosystems* (Longman, 1988)
Pimentel, D. and M., *Food, Energy and Society* (Edward Arnold, 1979)
Simmons, I. G., *Biogeographical Processes* (Unwin Hyman, 1981)
Tivy, J. and O'Hare, G., *Human Impact on the Ecosystem* (Oliver and Boyd, 1981)

14 Growth and distribution of population

14.1 ASSUMED PREVIOUS KNOWLEDGE

If you have studied world population growth for GCSE you will know many basic facts. You may have looked at the problems resulting from this growth with special reference to Third World countries and cities (e.g. shanty towns). Your detailed regional studies will have given you examples of patterns of population distribution e.g. West Africa, Chile, Egypt. You may also have carried out field studies or OS map exercises concerned with the distribution of population.

If you have not studied this topic, you should look at the following sections in Lines, C. J. and Bolwell, L. H., *Revise Geography*, Letts, 1987 – the growth of world population, urbanization and the growth of shanty towns.

14.2 ESSENTIAL KNOWLEDGE

14.2.1 Definitions

Read Unit 26, *West Africa: population issues* for these definitions.

Optimum population	**Death rate**
Underpopulation	**Net reproduction rate**
Overpopulation	**Population density**
Birth rate	**Demographic transition theory**

Read Unit 15 *Movement of population* for a definition of:
Migration

14.2.2 General features

Fig. 14.1 shows how rapidly world population has grown since 1750. Before that time the total population was fairly stable but over the last two hundred years the rate of growth has become increasingly rapid.

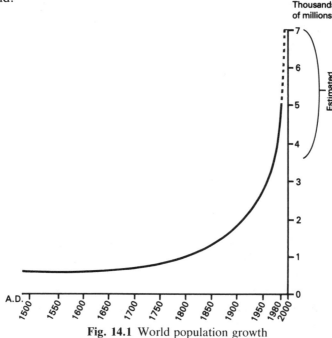

Fig. 14.1 World population growth

Reasons for this growth include:

(a) The creation of modern economic systems as a result of the Agricultural and Industrial Revolutions (see the demographic transition model on page 182).

(b) The vast increase in medical knowledge has increased the proportion of babies born live (out of the total number of births) and decreased death rates.

(c) Modern technology and communications have enabled us to tackle and overcome the worst effects of floods, famine and other natural hazards.

World population is very unevenly distributed. One-fifth of the world's population is in China. The people of China and India together make up one-third of the total population of the world.

There are two main clusters of people in the world – one in the western hemisphere between 40° and 60° north (Europe and North America) and one in the eastern hemisphere between 20° and 40° north.

Three-quarters of the population of the world now live in developing countries and it is in these countries that population growth is the most rapid.

The features of world population distribution do not correlate neatly with physical conditions. Eighty per cent of the world population occupy 10 per cent of the earth's surface. It is possible to say that most people live in warm, humid, lowland areas (90 per cent of humanity lives at altitudes less than 450 metres above sea level). But it is only possible to make very broad generalizations. This is because the environment is a complex of potentialities of which different cultures and societies at different levels of technological advance make changing use.

The distribution of population is the expression of all the factors that affect human societies. These factors may be divided into three broad categories.

For example:
Biological – sex, age, race, morbidity (prevalence and types of diseases)
Social factors – place of residence, occupation, socio-economic class, place of birth, religion, nationality
Dynamic factors – birth rates, death rates, migrations out of and into a given area.

Any explanation of a pattern of distribution has to be historical in nature. It is only through the operation of processes over time and under conditions which prevailed in the past that present day features came into being e.g. the distribution of people and cities in northern England at present is basically the result of the importance of coal as the raw material which enabled rapid industrialization in the eighteenth and nineteenth centuries.

Economic factors generally have more direct effect on distribution patterns than do characteristics of the physical environment. This is partly because the nature of an economy will determine the extent to which a group of people will control the physical features of the region in which they live. As resources are exploited, markets develop and technological changes occur, the distribution of population can alter dramatically. For example, prior to industrialization the distribution of population in Chile reflected the farming opportunities provided by the physical factors of relief and climate. The introduction of mining in the Atacama has resulted in roughly nine times as many people living in one of the driest deserts of the world than live in the cool temperate highland region of south Chile.

Another reason is that a particular economic system may lead to a distribution of population and densities regardless of variations in the physical landscape. For example, in the USA 70 per cent of the people live in a vast urban area which has little direct relationship to the physical nature of the land occupied.

14.2.3 Age structure of the population

The age structure of a population is the proportion of people who fall into particular age categories. Each category usually spans five years. This data is usually represented by *population pyramids*.

Fig. 14.2 is two pyramids which show contrasting patterns of population structure found in the world today. The pyramid for the developing country (A) shows a large infant population but as one moves up the pyramid each consecutive age category gets smaller. So only a small proportion of the total population is more than 50 years of age. In contrast, the population pyramid for a country like the United Kingdom (B) is very different. It is not a pyramid in the geometrical sense for each age category is very like the others until the age of about 60.

The shape of the pyramid is important because population growth and life in a society are affected by the proportion of people in different categories. For example if there are many old or many young people in a country the number of workers may be comparatively small i.e. there is a *high dependency rate*. In a society with many young children there will be a need for resources to be used for education, health care etc. and for consumer goods such as children's clothes, toys and books. Where there are many old people in the population resources are required to provide pensions, nursing homes etc.

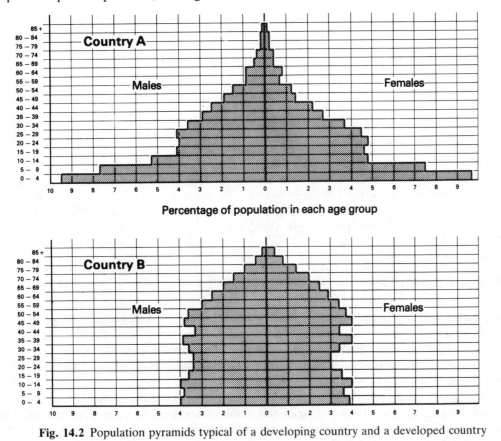

Fig. 14.2 Population pyramids typical of a developing country and a developed country

14.2.4 Models of population growth patterns

The stabilization model (the 'S' curve model) This model assumes that population growth will continue for some time but that then there is an eventual reduction of the growth rate. The graph of such growth shows an 'S' curve which has three main parts. The base represents a period of relatively slow population growth, a steeply inclined portion represents rapid growth and the third portion represents the period in which the rate of population growth declines.

The reduction in growth leads to a stabilization of population at a size which can be supported by the environmental system. Stabilization is achieved when birth rates and death rates are in equilibrium (balanced).

The rapid growth and rapid decline model (the 'J' curve model – Fig. 14.3) This model also assumes that the population will continue to grow. Instead of stabilizing within the capacity of the environment however the model envisages the population increasing until it 'overshoots' the environment's capacity to carry the population. This results in a catastrophic decline in population. This is a 'J' curve growth pattern. It is argued that as the world's non-renewable resources are used up the productive base of agriculture, industry and services will collapse. Food shortages and environmental degradation will lead to a major rise in death rates and a rapid population decline.

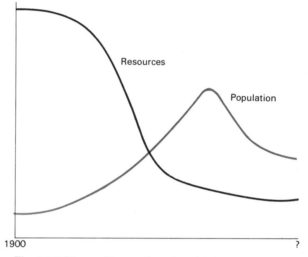

Fig. 14.3 The rapid growth and rapid decline model

14.3   GENERAL CONCEPTS
See Unit 26 *West Africa: population issues* for the following:

Optimum population

Overpopulation

Underpopulation

Demographic transition

Population growth is the sum of population changes due to natural increase and population changes due to migration. Both these variables may have positive or negative effects upon growth. Natural increase is the difference between the birth rate and the death rate. This may also be positive or negative. For example, if the death rate exceeds the birth rate the total population (ignoring the effects of migration) will fall so the natural increase then has a negative value.

14.4   DIFFERENT PERSPECTIVES
Malthusian perspective; Demographic transition perspective See Unit 26 *West Africa: population issues*.

Biological perspective The growth of population of living organisms seems to follow generalized patterns. The population of any species seems unable to grow indefinitely. Population growth is eventually checked by factors such as the depletion of essential resources. For human beings food supply and the availability of living space are important limiting factors. So the identification of the prevalent growth pattern is important. If human population is not stabilized there may be a catastrophic decline of world population if it overshoots the carrying capacity of the environment. Campaigns to recycle non-renewable resources, to limit pollution etc. are attempts to maintain the present level of carrying capacity.

Biologists also see *population regulation* in human populations. There appears to be an intrinsic self-regulating mechanism in some species. By this process a natural population size will be reached and maintained unless significant changes take place in the environment. In technologically underdeveloped societies in the past the human population grew very slowly (see the graph on page 94) and many societies were probably stable.

Social perspective In examining the relationship between population and resources a critical issue which arises is that of population control. In some countries such as India and China official campaigns have been mounted to encourage people to practise birth control in order to limit population growth. In contrast, in other countries e.g. the Catholic countries of South America the Church and governments oppose such measures so population growth is unchecked despite the social and economic problems which result. Cultural factors can also affect attitudes and policies e.g. in Nigeria a large family is regarded as a status symbol so people are reluctant to limit family size.

14.5   RELATED TOPICS

The two most closely related units in this book are 26 *West Africa: population issues* and 15 *Movement of population*. The West African chapter deals with a number of important concepts, perspectives and definitions in detail. It also provides a case study of factors affecting the distribution of population. The movement of population chapter discusses migration as an important influence on the growth and distribution of population.

Since population problems are particularly severe in the Third World, Unit 23 *Less developed countries* is also relevant.

14.6   QUESTION ANALYSIS

1 Explain clearly the meaning of overpopulation, underpopulation and optimum population. Select one appropriate example to illustrate overpopulation. Why is it is more difficult to find an example of optimum population?   (*in the style of Southern Universities' Joint Board*)

Understanding the question You should give equal weighting to the three parts of this question because each will be awarded one-third of the marks. The first part is the definition of the three terms. The second part involves writing about one country or area which exemplifies what overpopulation is. Finally you are asked why it is hard to find examples of optimum population – this is probably the most difficult part and it needs more than just one or two sentences written about it. Overall this is a very full question and it would be easy to spend too much time on it. There is probably not enough time to allow you to draw a sketch map of the country or area you choose. It is vital that you keep strictly to the questions asked.

Answer plan Point out first of all that the three conditions of overpopulation, underpopulation and optimum population are concerned with the relationship between population and resources. All three are also linked with the law of diminishing returns.

Then define the terms.

Begin the second section of your answer by naming the example you have chosen e.g. *Egypt*

Establish the physical environmental conditions first. Egypt has an area of 1 million square kilometres but the habitable area is only about 36 000 km^2 – 3 per cent of the total area. The population of 45 million is concentrated in the Nile Delta, along the River Nile and in the Fayum. Beyond the delta the Mediterranean coast is sparsely populated, the Western desert has sedentary groups living at oases. Apart from the oil well settlements the Sinai provides a home for scattered nomadic groups.

Now point out the chief population characteristics:

The 45 million people are increasing at a rate of 2.7 per cent a year. This is one of the highest growth rates in the world. This rate is the result of the reduction of a very high death rate in modern times. Life expectancy is still fairly low by Western standards (males–56, females–58) and more than 40 per cent of the population are under 15 years of age. The population pyramid is very similar to that for under developed countries on page 95. The crude birth rate is very high, 36.9 per 1000 (compare with France–12.8). So a large proportion of the population is too young to do productive work. Settled areas are very densely populated and population pressure is by far the greatest problem the country has to face.

In 1830 Egypt was by far the most advanced country in Africa but the economy has failed to take off. This is because the country has been constantly faced with the problem of trying to maintain basic food supplies in line with the increase in population. Despite significant improvements in farming practice and despite major development projects income per head has not risen and periodically has fallen. Egypt may therefore be said to be overpopulated. If the population had not increased so rapidly, the improvement in farming and the development projects would have provided more food and a higher standard of development.

Egypt's total cultivable land is only 2½ million hectares (6 million acres). Efforts have been made to increase the area of cultivable land e.g. the Gezira scheme and the building of the Aswan Dam. The introduction of year-round irrigation allowed the growing of more than one crop a year. So the cropped area has grown faster than the cultivated area. Irrigation, intensive farming methods, use of modern fertilizers and pesticides have made yields in Egypt amongst the highest in the world. But income per head has hardly changed and between 1913 and 1952 output per head fell. So the law of diminishing returns may be said to have operated. Even major modern projects have failed to change this basic pattern. The completion of the High Dam at Aswan in the 1960s added 1 million hectares of cultivable land to the farmlands of Egypt but this merely produced additional food and income to maintain the pre-existing standard of living because the population had grown so rapidly.

It is difficult to find an example of a country which has an optimum population because:

(a) Optimum population cannot be measured. It is a useful theoretical concept against which to measure the relationship of people to resources in different countries to gauge whether they are seriously under- or overpopulated but no calculation of an optimum size can be made.

(b) The condition of optimum population must be the result of the operation and interaction of a great variety of factors. Each of these factors is dynamic and subject to change e.g. the value of resources, the size of markets, food prices, the area of cultivable land, exchange rates. So if an optimum figure were calculated it would be for one point in time and subject to rapid change. It is highly unlikely that any country would therefore maintain an optimum state.

(c) It is also impossible to establish what is the optimum productivity of a particular industry, area or country. So we are not able to assess what population size would produce the greatest economic welfare. Therefore even though we can claim that Egypt is overpopulated, we are not in a position to work out what the optimum situation would be, given the present state of development.

Finally, emphasize that optimum population is a theoretical concept which is useful to geographers because it focuses attention on the problem of the relationship of population to resources.

2 What factors affect the distribution and density of population (a) in a small area you have studied of not more than 100 square kilometres; (b) in an area of continental size?   (*in the style of Cambridge*)

Understanding the question The question is clearly divided into two parts which will be equally weighted. Each part has two sections – distribution and density. It is a straightforward question which requires good knowledge of two areas of contrasting size and the ability to select the significant factors when they are operating at quite different scales.

Answer plan First distinguish between distribution and density. This would show the examiners that you have read the question carefully and understand its implications. *Distribution* is concerned with location – where people are found. *Density* relates the number of people to the area in which they live.

$$\text{Density} = \frac{\text{number of people}}{\text{unit of area}} \text{ e.g. no. people per km}^2$$

Since the question lists the two parts as (a) and (b) use the same division.

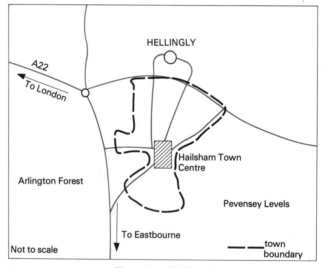

Fig. 14.4. Hailsham

(a) Name your first area, describe its location and its basic geographical features. For example: the area shown in Fig. 14.4 is in East Sussex which focuses on the small market town of Hailsham. The town lies near the A22 nine miles north of Eastbourne. It was originally built on a low sandstone ridge which forms the landward rim of Pevensey levels, a marshy area with heavy clay soils. The distribution of population in the area is the result of both physical and human factors:

Physical The distribution of the rural population reflects the agricultural possibilities offered by the geology and soils. Pevensey levels are grazing areas used for sheep and cattle mainly in summer when farmers rent fields for their stock. There are consequently few farms on the marsh. To the north of Hailsham the lighter, better drained soils provide grazing and arable land. There is therefore a greater density of farms in the north of the area and a nucleated village at Hellingly.

Human Economic factors resulted in the growth and development of Hailsham as a central place. It still has a weekly stock market and serves the surrounding rural area.

The modernization of road transport in the 1930s encouraged the development of ribbon settlement along the main roads except along the by-pass (A22) where planning restrictions forbade residential development. Development is also totally restricted to the west of the A22 where the Forestry Commission has planted Arlington Forest.

The town has expanded appreciably since 1945. This is due to its proximity to Eastbourne, a major retirement centre in south-east England. The retirement function has spilled over into Hailsham and much of the new building is for elderly incomers.

The density of population reflects the factors outlined above but study of the town itself reveals two other factors that influence the pattern.

One is historical – prior to expansion the town of Hailsham was characterized by Victorian terraced housing. Expansion has not involved replacement of the older housing stock. So the central town area and the ribbons along the main roads within the town are areas of relative high density.

Socio-economic factors have influenced the expansion of the town. The new houses are partly council built but mainly privately built. The former Rural District Council had a housing policy which concentrated council housing in the town on a large estate to the south of the town centre. This now forms a high density area. The private housing development was partly geared to the lower end of the market – high density bungalows for people who were retiring but could not afford higher prices in Eastbourne. Although the bungalows are densely built the occupancy rate is low (one or two persons per dwelling) so the density of population is less than in the council estate. More expensive housing has been built for younger professional people who commute to Eastbourne and neighbouring towns. These estates are not as densely built as the other new housing. Despite the fact that they house young families the density of population is not high.

Since Hailsham is not a major service centre there is comparatively little competition for central locations so there are no high blocks of flats. Densities do not therefore reach the levels found in larger towns and cities.

(b) Name your example e.g. Australia.

The distribution of population is best shown by a map (Fig. 14.5). The features shown by the map need to be described briefly:

The distribution is peripheral. The population is found around the edges of the continent. It is found mainly outside the Tropics. The eastern coastlands are more heavily populated than those of the west.

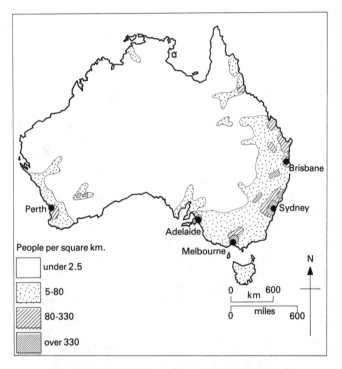

Fig. 14.5 Distribution of population in Australia

Again emphasize that this pattern is the result of the interaction of physical and human factors:

Physical The overwhelmingly important factor is that of climate. Because of its latitude most of central Australia is a tropical desert which discourages settlement. Large areas are unpopulated. The indigenous people are small isolated bands of Aborigines. Typically the desert stretches to the western coastlands but on its eastern flank it is bordered by a temperate grassland region. Settlement is therefore more extensive and denser in the east.

Human The northern coastlands are largely empty as a result of human rather than physical factors. Australia has operated a 'Whites only' immigration policy which has meant that Asian immigration has been largely precluded. White settlers have been unable and unwilling to develop the humid monsoonal northern belt which might have supported a dense Asian population.

Economic factors have played an important role in determining the distribution of population. For example the settlement of the interior areas of western and north-eastern Australia is the result of the exploitation of mineral resources. The peripheral pattern is partly a reflection of former colonial ties with Britain when ports were the main points of exchange for exported food (wheat) and industrial raw materials (wool) and the import of manufactured goods. A further economic influence is revealed by the gradient zoning of densities of population around the major cities.

The settlers in Australia originated mainly in western and central Europe. The majority are urban dwellers and are attracted to the cities. A striking feature of the population geography of Australia is the dominance of the major cities in and around which the densities of settlement are as high as in countries with larger total populations.

The distribution of population and the densities shown by the map also reflect the comparatively small total population of the country. Many areas are sparsely populated because there are still relatively few people to settle such a vast area.

3 'A densely populated area is not necessarily overpopulated, nor a sparsely populated area under-populated.' With reference to local and world examples explain this statement.

Understanding the question You must understand the concepts of overpopulation and underpopulation. You must also be able to draw upon a number of relevant examples at two distinct scales – local (which you could define as your home region e.g. north-east England) and world examples.

The question asks you to discuss the relationship between density of population and the two concepts of overpopulation and underpopulation. Density is a simple relationship of numbers to area. The other two concepts are much more complex because they involve judgements about whether the number of people in an area, given their technological level, capital etc. appear to be achieving the maximum income and standard of living from their use of that land.

This is an ideas question. You are not asked to question the quotation but to explain it. The number of marks you receive depends on the quality of the points you make and how well you back them up with examples.

Answer plan Begin by pointing out that density of population relates people to the land on which they live whereas underpopulation and overpopulation relate people to the resources and standards of living they enjoy in a particular area.

You should then provide brief definitions of the three terms.

The second section should be the main part of your answer. Do not write separate sections on densely populated and sparsely populated areas. Deal with the two together emphasizing factors which influence both.

The standard of living and the per capita income is not merely a function of the number of people living in an area. It is also the result of the character of the economy, the technological level and skills of the people, the capital and resources available and economic and political factors outside the control of the inhabitants themselves. So we can find poor and wealthy sparsely populated areas and poor and wealthy densely populated areas at both local and world scale.

At world scale this point can be supported by reference to two very densely populated regions – Bangladesh and Western Europe. The intensive farming of a country like Holland is based on high technology and high capital investment. This produces high yields which can be sold to wealthy industrial regions and cities. The high density of population does not therefore mean that Holland is overpopulated. Since farming there is now less labour intensive a decrease in rural population could reduce yields and farm income. Contrast this with Bangladesh where there is every indication of overpopulation.

Local scale examples are harder to find but the intensively farmed Fenlands of eastern England are able to maintain a denser and more prosperous farming than equally fertile areas where the small landholdings have become uneconomic and a rise in income depends upon the consolidation of holdings and the movement of some people from the land e.g. along the Welsh borderland.

The same arguments can be advanced for densely populated urban areas. You could for example contrast London or Los Angeles with Calcutta or other Asian cities. Locally you could compare inner city Belfast or other city areas which experience major employment and social problems with densely populated Mayfair or Chelsea.

Similarly you cannot assume that sparsely populated areas are underpopulated e.g. in rural Wales farm subsidies and the work of the Welsh Development Board have tended to stabilize population so there is a balance between population and resources. The emigration from Ireland and the Scottish Highlands in the nineteenth century was partly the result of overpopulation.

At a world scale you could point out that agricultural development and settlement in the interior of Australia has been limited by the lack of people and so Australia could be said to be underpopulated. On the other hand although the Sahel is sparsely populated it can no longer carry the numbers it has (see Unit 26 *West Africa: population issues*). In some regions of the world the contrast between modern economic activities and settlements on the one hand and traditional ways of life on the other make concepts of over- and under- population meaningless e.g. the oil industry in Middle Eastern countries, recent developments in Amazonia.

Finally, make the important point that density is an *average* figure and as such may hide significant differences within a country or region. But it is these variations which may be the key to judgements about whether a particular area is under or overpopulated. So density of population is in itself not a good guide.

14.7 FURTHER READING

Carr, M., *Patterns, Process and Change in Human Geography* (Macmillan, 1985)
Jarrett, H. R., *Tropical Geography* (Macdonald and Evans, 1977)
Lowry, J. H., *World Population and Food Supply* (Hodder and Stoughton, 1989)
Thomas, I., *Population Growth* (Macmillan, 1984)
Woods, R., *Population Analysis in Geography* (Longman, 1979)

15 Movement of population

15.1 ASSUMED PREVIOUS KNOWLEDGE

Most GCSE syllabuses pay little attention to this aspect of human geography so it may well be a new topic. You may however have looked at the process of urbanization in relation to Third World studies. You may also have done a detailed regional study of an area in which the effects of migration have been very important. This background material will give you useful examples which can be related to the more theoretical and analytical A level work.

15.2 ESSENTIAL KNOWLEDGE

15.2.1 Definitions

Migration cannot be defined simply. It is generally defined as a permanent or semi-permanent movement on the part of an individual or group of people. If this definition is accepted migration involves two things, (a) going to a new place to live, and (b) staying there for a certain amount of time.

This cannot be a hard and fast definition e.g. university and polytechnic students satisfy both these conditions but are not regarded as migrants.

There are varied ideas about how long a person must move away from home to become a migrant. As far as international migration is concerned the United Nations defines *permanent* migration as removal from a former place of residence for one year. On the other hand, one in five of the gastarbeiters (guest workers) in Germany lived there for more than seven years but were still regarded as *temporary* migrants.

Within Britain the census definition of migration is movement from one local administrative area to another within one to five years prior to the census.

15.2.2 Classification of migrations

In order to define more clearly and to understand the different types of movement involved in migration, different classifications have been developed.

Scale classification is based upon the distance travelled and the nature of the movements e.g.

international	(Jamaica to Brixton)
inter–regional	(NE England to the SW peninsula)
inter–urban	(Belfast to London)
rural–urban	(Scottish Highlands to Edinburgh)
intra–urban	(Southall to Barnet)

Movements may also be classified according to the *purpose* of the movement e.g.

economic	(to a new job in Canada)
forced	(Ugandan Asians)
leisure/retirement	(New York to the sun belt of the SW USA)
religion and culture	(Jews to Israel)

Time classification is based on the length of time spent in a new location. The major subdivisions are into recurrent (repeated) and non-recurrent movements. *Migration* is a movement away from the place of residence followed by a return to the point of origin (e.g. circulatory movements of African migrant workers to S. African mines; British contract workers in the Middle East). *Emigration* is a change of habitat with no return. *Nomadism* is a constant change of movement with cyclical paths (Bedouins).

15.2.3 Models of migration

The 'Laws' of migration (1885) The earliest model of migration was Ravenstein's 'Laws' of migration which list persistent regularities which characterize the areas of origins, destinations and migrants themselves:
(a) The majority of migrants go only a short distance.
(b) Migration proceeds step by step.
(c) Migrants going long distance generally prefer to go to one of the great centres of commerce or industry.

(d) Each current of migration encourages a compensatory counter current.

(e) Townspeople are less migratory than people from rural areas.

(f) Females are more migratory than males within the kingdom of their birth, but males more frequently venture beyond.

(g) Most migrants are adults; families rarely migrate out of their countries of birth.

(h) Large towns grow more by migration than by natural increase.

(i) Migration increases in volume as industries and commerce develop and transport improves.

(j) The major direction of migration is from the agricultural areas to the centre of industry and commerce.

(k) The major causes of migration are economic.

The gravity model It is a fundamental fact that most migrations take place over relatively short distances. There are relatively few long range movers. This model is based on the premise that migration is some function of distance:

$$I = fD$$

$$I = \text{migration interaction}$$
$$D = \text{distance}$$

Migration occurs according to the degree of attraction of a region or location. The volume of migration depends upon the population of the two localities involved and the distance between them i.e.

$$M_{ij} = \frac{P_i P_j}{d_{ij}}$$

$$M_{ij} = \text{migration between } i \text{ and } j$$
$$\begin{matrix} P_i \\ P_j \end{matrix} = \text{population of the 2 centres}$$
$$d_{ij} = \text{distance between them}$$

The weakness of the model is that although the function of distance is an important factor, there are other variables which also influence the movements e.g. other competing locations may also attract the migrants (intervening opportunities). And some of the greatest migrations recorded have occurred for reasons unrelated to the gravity model.

Intervening opportunities model (Stouffer) This model attempts to translate the distance factor into social as well as economic terms. Stouffer argued that the number of migrants over a distance was *directly* related to the number of 'opportunities' at that distance and *inversely related* to the number of intervening opportunities which exist:

$$I_{ij} = \frac{(P_i P_j)}{(O_{ij})(C_{ij})}$$

$$I \text{ and } P \text{ as above}$$
$$O_{ij} = \text{number of intervening opportunities}$$
$$C_{ij} = \text{number of competing migrants}$$

This model is more elaborate than the simple gravity model but it is still descriptive. It does not explain the causes of migration nor take into account other significant variables.

Multivariate analysis models (e.g. Olsson) Olsson showed that there were five variables with significant relationship with the function of distance:

(a) The level of income at place of out-migration

(b) The size of place of in-migration

(c) The size of place of out-migration

(d) The level of unemployment at place of out-migration

(e) The level of unemployment at place of in-migration

A systems approach to migration Some writers explain the process of migration in terms of systems theory. A potential migrant receives stimuli from his environment, to which he may or may not respond. There are two subsystems – the urban control subsystem and the rural sub-system. The urban control subsystem includes factors such as the organization of employment and the city administration. The rural subsystem includes family, community beliefs and inheritance laws. As movement occurs adjustments take place and the information flows become important. The importance of this model is that it demonstrates that a variety of inter-related factors lead to population movements.

15.2.4 General features of migration

Migration is one of the processes leading to the redistribution of world population e.g. in the nineteenth century Europe had a net loss of 40 million people through migration.

Migration may be viewed as an adjustment to economic inequalities The key motivation may be the better job opportunities that may exist in the receiving place. These inequalities may be different regional employment rates within a country (e.g. Northern Ireland compared with south-east England) or international. They may have a 'push' effect (movement from an over-populated country) or a 'pull' effect (the attraction of Third World cities to the rural poor).

There are regularities in the patterns of movement At present, on a world scale, rural-urban movement (urbanization) is dominant, but now the population of the cities is declining. The movements are *age and sex selective* – it is predominantly young men who migrate. This is partly because, in wage-earning economies, most of the jobs that exist are for male workers and partly because in most societies, men are still the chief wage earners. Another influencing factor is that in many rural societies, the young men have greater independence than the young women, and so are freer to move

Traditionally, immigrants take up jobs which are the least attractive to the indigenous population, e.g. in Britain in the 1960s immigrants took shift work in textile mills, jobs on buses and trains, hospital work etc. These are usually the lower paid jobs and are concentrated in the cities. A poor migrant who then finds a low-paid job is not able to afford to move his family and rent a home for them until he has worked long enough to save the money needed. This is another reason why, in the first instance, men migrate alone. In some countries the wives stay at home to maintain the traditional way of life and to farm the land, e.g. in Kenya, Kikuyu wives stay in the rural village while their husbands migrate to Nairobi looking for well-paid work.

Movement from the less well-off regions is controlled or even prohibited by the more advanced regions which do not welcome large scale immigration of people of lower socio-economic status. The USA operates a quota system to control immigration, Australia previously had a 'White Australia' policy, New Zealand and Canada favour skilled workers; our immigration laws limit immigration from the new Commonwealth.

15.3 GENERAL CONCEPTS

Differential population pressure (Thompson) This is the idea that migrants tend to move from a place of high economic and social pressure to places with lower pressure i.e. people tend to follow the line of least resistance. For example, white people went to the White Highlands of Kenya and to Southern Rhodesia (now Zimbabwe) because it was easier to become a member of a social and economic elite than it was in Britain.

'Push/Pull' factors Migration is conceived of as being the result of two conflicting forces. The 'push' factors encourage people to leave overpopulated and economically depressed regions e.g. the West Indian migration to England, English migration to Canada. The 'pull' effect is the attraction offered by a new location e.g. the magnetism of California in the USA; the cities of the Third World.

Gastarbeiter (guest workers) Advanced countries which are short of labour make use of migrant labour without having to give the immigrants full civic rights as citizens. For example, there are southern Mediterranean workers in Germany and Switzerland. For racialist reasons South Africa has a permit scheme which allows African workers from neighbouring countries to work for a period in South Africa without civic rights. Some workers have stayed in a country for years and have settled with their families. In times of economic depression however work permits can be withdrawn and 'temporary workers' may not have the same right to social benefits, unemployment pay etc. as citizens.

Immigrant This is a loosely used word which often has overtones of prejudice or even racial discrimination. For example, although more white people than black or Asian enter Britain in a year the word now is used for people from new Commonwealth countries (India, Pakistan etc.). Before World War I for a time it was used almost exclusively to describe Jewish people.

15.4 DIFFERENT PERSPECTIVES

The demographic perspective Demographers see migration as a balancing factor in the basic population equation because it provides a dynamic element over and above the factor of natural increase (birth and death rates). The demographer is interested in the effects on the age and sex structure of the sending and receiving areas and in the imbalance of flow between these regions.

The economic perspective This perspective is symbolized by the gravity, intervening opportunities and multivariate analysis models above. Distance is regarded as a frictional cost with man making purely rational decisions.

The behavioural perspective This approach is concerned with why moves are made or not made – with the decision making of the migrants themselves. Pred for example believes that people move according to the quality of information available to them and the use they are able to make of that information. Migrants respond to social and economic factors with varying degrees of rationality. People are not totally rational in their decisions as 'economic man' would be. They act in a *bounded rational way* according to such factors as the level of information, the stage they have reached in the life cycle and the opportunities they perceive in the new location.

15.5 RELATED TOPICS

The two units with which this section is most closely related are *18 Urban land use* and *14 Growth and distribution of population*

15.6 QUESTION ANALYSIS

1 Explain each of the following migration characteristics:
(a) proportionally higher incidence of men to women in rural or urban migration
(b) urban to rural movement of elderly people
(c) movement of people away from the inner areas of a city.   (*in the style of London*)

Understanding the question This question has no hidden pitfalls so long as you concentrate on *explanations*. The three parts will earn equal marks. It is important that you handle the answer at a good theoretical level showing that you understand the basic concepts of migration theory. If you decide to quote examples they should be brief, or the answer will take too long to write.

Answer plan To answer part (a) first discuss the view of migration as an adjustment to economic inequalities. The reasons for the high number of migrant males are discussed in Section 15.2.4.

Thompson's concept of differential population pressure is relevant to part (b) (see Section 15.3). Elderly people move to areas they consider to be more conducive to their way of life. Retired people are no longer tied by their work and so are freer to choose where they live (if they can afford it). *Bid rent theory* is important (see Unit 17 *Central Place Theory*). It is cheaper to move out of expensive city areas to cheaper country areas. Information factors are also very important. In country areas where there are large proportions of elderly people, the facilities are geared to their needs, thus encouraging further migration. Social factors also influence people. Retirement to the country is a mark of social status.

The 'push/pull' concept is a good framework within which to answer part (c). Push factors include:
(a) the decline of job opportunities in the inner cities – many industries have moved to the suburbs to make use of motorways etc.
(b) the overall expense of living in the city.
(c) other disadvantages of the inner city which make people want to move out – congestion, poor housing, social problems etc (see Unit 18 *Urban land use*).
Pull factors include
(a) better housing in suburbia, council estates, new towns etc.
(b) better job opportunities elsewhere.
(c) post-war planning to decrease the population of the inner cities.
(d) greater mobility as car ownership has grown.
(e) increased social mobility – social status is indicated by where people live.

2 Give an account of:
(a) the causes of the movement of population to towns and cities in the Third World
(b) the problems which have arisen as the result of this type of migration.
     (*in the style of the Joint Matriculation Board*)

Understanding the question This is a straightforward question. It is important to realize that the problems arise in regions which migrants leave as well as in the cities to which they go. So a statement about shanty towns in not enough. As the second part of the question has two sides to it, it will probably be worth more marks – the likely division is (a) – 10 marks, (b) – 15 marks.

Answer plan In your introduction you should point out that the causes of migration are complex. Economic, social and political factors are important while the actual movements are the result of decisions made by individuals. It is possible however to identify some basic causes. In the underdeveloped world the most important appears to be the unequal economic opportunities which exist in the countries of the Third World.

(a) *The causes of migration* List the pull effects of the cities and the push effects of the countryside (e.g. sharecropping offers a very low return, overpopulation, inefficient farming methods have led to decreasing productivity of the land). Point out that natural hazards may also push migrants from the countryside – e.g. drought (Sahel, Ethiopia) and flood (Bangladesh). Migrants know that emergency supplies of food and medicines are most available in the cities.

Other points to mention are the fact that large industrial cities may also need large numbers of unskilled workers e.g. underdeveloped Lesotho and Botswana provide migrant workers for S. African cities. Information factors are important – information flows from the cities to the countryside helps people decide to migrate and how to take the necessary steps. In many developing countries (e.g. Kenya) people who live in cities have higher status than those who remain in the villages.

(b) *Resulting problems* In the rural areas, these include imbalance of population (the villages are left with the very old and the very young), destruction of balance of economic systems of the rural areas (the male workers are no longer available, the money sent back from the cities helps put up prices etc.) and the increase of family debts – migrants borrow money for the journey (e.g. Pakistanis migrating to England). Also, traditional society and family patterns are broken up and farming may become even less productive as the experienced farmers migrate.

In the cities there is overcrowding – in the Indian cities over 50 per cent of the families have only one room. Shanty towns grow up and there are health hazards created by overcrowding, lack of pure water, and lack of sanitation. The cities are faced with increased financial burdens as they are faced with the problems of the shanty towns. The poor conditions of the migrant communities can lead to social and political unrest – riots etc.

In conclusion it would be appropriate to point out that the movement to the cities is now the most important feature of all Third World countries and that it is causing serious problems for them.

3

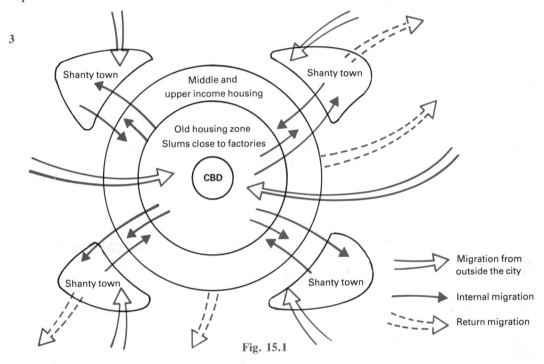

Fig. 15.1

(a) (i) Compare a Third World city you have studied with the model in Fig. 15.1, highlighting the similarities and differences noted.
(ii) For the city chosen in (a) (i) above, explain the patterns of land use and migration.
(b) Describe the problems caused by rapid urban population growth in Third World countries and discuss the ways in which the problems are being tackled.

(Scottish Certificate of Education, Higher Grade, May 1986)

Understanding the question This is a complex question which brings together your theoretical knowledge of land use in a Third World city, an actual example of such a city that you have studied, knowledge of migration patterns in the Third World and the causes of the development of shanty towns, together with policies aimed at tackling the problems they symbolize.

There are three distinct parts to the question. Parts (a) (ii) and (b) require the most analytical work and would probably be allocated more marks than section (a) (i).

Answering the question *Part (a) (i)* The precise form of your answer to this section will vary according to the example you have chosen, but points to think about may be:
Similarities: residential segregation; the existence of a CBD; the existence of shanty towns beyond the city limits; migration into the city – the inner city and shanty towns; return migration to the countryside.
Differences: location of CBD – because of their colonial histories, the location of many important Third World cities is coastal and the CBD is on or near the sea, not in the centre of the city; presence of former colonial residential areas, e.g. military and civilian cantonments; segregation is not based purely on economic factors – caste or religious differences may be the chief determinants of segregation; in some cities, shanty towns occupy space within the city boundaries, e.g. the bustees of Calcutta; most large cities have modern industrial zones on the city edges.
Part (a) (ii) Outline the pattern of land use in your chosen city very briefly – a swiftly drawn sketch map would be very useful.
Explain: the range of land use zones in your city; the reasons for the pattern of distribution of these zones – historic, economic, social, site factors etc.
Outline very briefly the nature of the migratory movements to and from your chosen city. You might use a model like the push-pull model to help explain the existence and extent of these movements (page 103).

Part (b) The problems are listed on page 105 (Unit 15.6 q.2 part (b)).

You need not confine yourself to the city you chose in (a), when answering section (b). The ways in which the problems are being tackled vary from one country to another and it is a good idea to quote particular countries when writing about a particular policy, e.g. for population control, you could quote Indonesia's 'Two is enough' publicity campaign.

Important methods are:

1 Attempts to encourage rural dwellers to stay in the countryside, through rural development plans, extension of the Green Revolution, the establishment of small industries, improving health services etc.

2 Controlling population growth, by adopting population control programmes to reduce the birth rate.

3 Since the most rapid growth occurs in the primate cities, attempts have been made to spread industrial development to peripheral regions, which will disperse in-migration and make the effects of urban growth more manageable.

4 The problems of the shanty towns are being tackled by means of international aid programmes, national development policies, and by attempts to alleviate the worst conditions by providing basic facilities and concrete bases for homes upon which squatters may build healthier and better shacks which they can afford.

5 Health programmes are introduced to reduce infant mortality and disease. In solving the worst health problems of the inner cities and shanty towns, however, these programmes increase the number of live births and raise life expectancy. So the problems are intensified.

15.7   Further reading

Carr, M., *Patterns, Process and Change in Human Geography* (Macmillan, 1985)
Woods, R., *Population Analysis in Geography* (Longman, 1979)

16 Rural settlement patterns

16.1   Assumed previous knowledge

You will probably have studied rural settlement pattern in two different ways for GCSE – through the study of OS maps and in the field. The background gained from such work is invaluable for A level work. If you did a detailed study of particular map extracts, your local area or another locality in which you did field work, you should find the knowledge you have to be very useful in providing examples to support this section. In studying some regions of the world in detail you may also have worked on case studies or sample studies of particular localities or villages. This material is also very useful in providing you with actual examples to back up your ideas.

16.2   Essential information

16.2.1 Definitions

Site The actual land upon which the settlement is built. The initial site will continue to influence the plan of the settlement even when the settlement has outgrown it. The significance of factors of siting change over time e.g. factors such as the availability of spring water or a defensive location are no longer key factors to rural settlements in Britain.

Morphology The form of the settlement, e.g. street villages have a linear form with the houses, farms and other buildings strung along a road. In contrast a green village will originally have had a more compact appearance with the buildings clustered around a village green.

Nucleated settlements are those in which farms and other buildings are grouped together as a nucleus from which the village has evolved. Both street villages and green villages are forms of nucleated settlement.

Dispersed settlements are those in which individual farms are scattered over the cultivable land.

16.2.2 Classifications of settlements

We classify rural settlements in order to understand them and to try to identify the processes by which they have achieved their present forms and distribution.

(a) **Physical classification** In GCSE studies you may have labelled villages according to initial site factors e.g. *gap settlement, spring line village, bridging point* etc. This is a physical classification. Although these labels relate villages to specific physical features they do not help us understand the nature of the settlements as they now are. If you do examine the physical features of sites it is important that you ask what the people who originated the settlements believed to be the chief priorities in choosing a site. For example, when establishing villages in south-east England the Saxons would have had a shopping list of items which made up an ideal site for a village such as: easy to defend, ready water supply, materials to build houses and barns, grazing land for animals, arable land for crops. At different times and at different places in the south-east the order of priority of these factors would have varied.

(b) **Morphological classifications** Villages may also be classified according to their form. The form of the village reflects vital characteristics such as the basic type of farming practised, the people who founded the settlement, the conditions under which it was originally established.

(c) **Evolutionary classifications** This approach is concerned with the classification of settlements according to their origins. Residual or relic features are identified in the present landscape and the distribution and forms of these remains are related to the economic and social systems which operated at the time of origin e.g. Section 16.2.4 is a form of evolutionary classification.

(d) **Social classification** Villages may be classified in terms of their social make-up e.g. classification according to the social groups living in the village, classification based upon patterns of land ownership. The concepts of 'open' and 'closed' villages (Section 16.2.7) is a way of identifying social characteristics according to social criteria.

16.2.4 The evolution of rural settlement in England and Wales

The main phases of rural settlement were:

Celtic settlement This formed isolated clusters of housing usually located on easily defended sites e.g. hill tops. They are found mainly in the north and west of Britain and most of the sites still evident have been abandoned.

Roman settlement Remains of Roman villas are found in lowland Britain e.g. Bignor, Sussex; Chedworth, Glos. The villas were largely self-supporting country estates. Many of them had their own small industries such as pottery, cloth making and milling. The estates provided nearby towns with farm produce which could be transported along the new road system.

Saxon settlement Many street and green villages in England are of Saxon origin. Many Saxon villages still exist, mainly in south and central England. They too were linked by new routeways which followed the grain of the land and cut across the Roman road system.

Scandinavian settlement (Viking, Norse) These settlements were set up by invaders in the ninth and tenth centuries. They are found in eastern and north-western parts of England and in Ireland. At first they built defensive forts but these evolved into villages and towns such as Stamford and Gainsborough in Lincolnshire.

Norman settlement This was a very important period of expanding rural settlement. The growth of population led to the need to cultivate more land. More villages were therefore established. The typical village in cultivable areas was the nucleated village farming large open fields.

Late medieval settlement This was a negative period when many villages disappeared. The population was decimated by the Black Death (1348). Depopulation led to the desertion of many village sites. There were fewer people to work the land and the badly hit towns needed less food so arable land was converted to pasture and sheep greatly outnumbered people in England.

Since the Middle Ages the development of trade has at times brought great prosperity to rural areas e.g. the sixteenth century saw the building of 'wool' churches in the Cotswolds and East Anglia. The need for capital to invest in industry has led to the concentration of land ownership into fewer hands. The common lands and open spaces have also been enclosed with a resultant secondary dispersion of people from the original nucleated villages.

16.2.5 Factors leading to the nucleation of settlement

Nucleation is a settlement form which is often related to the ways in which the land is farmed and owned. It is encouraged by:
(a) a co-operative system of working the land (open field)
(b) defence (hill top, inside a meander)
(c) water supply considerations (spring line)
(d) need for dry sites in marshy areas (Fenland)

(e) scarcity of building materials – settlements concentrate where they can be obtained e.g. near a source of brick clay

(f) planned villages established by the land owner

16.2.6 Factors leading to a dispersed settlement pattern

(a) dependence on livestock farming (Scotland)

(b) specialist intensive farming (market gardening)

(c) Celtic influence (Wales)

(d) very low densities of population (W. Highlands)

(e) dissolution of large estates (of monasteries during Reformation)

(f) secondary movement away from nucleated villages (as a result of enclosure)

(g) planned dispersal (on to new holdings in Sicily)

16.2.7 Open and closed villages

This is a basis for classifying villages according to patterns of land ownership.

Closed village This was a village in which all aspects of life were dominated by the landowner or his squire. The land is usually divided into a few large farms. The farmers were tenant farmers with high social status in the village. Traditionally the farmers employed many workers. The size of the village has been controlled by limiting the number of estate cottages. Farming has totally dominated the economy, there were few village industries and few tradespeople. On the OS map the closed village may be identified by such features as: a large country house, a model estate village, landscaped parkland, a few large farms, plenty of woodland to provide shooting.

Open village This is a village in which land ownership was shared by a number of landlords. The farms are therefore sometimes not very large. Small landowners often diversified their interests by developing industries and trades. The open village often became the service centre for nearby closed villages. Because there was less social control the population was not restricted. Population densities are therefore higher than in closed villages. On a map, instead of a great house you are likely to find more than one manor. Shops, workshops and small industries may also be indicated. Fields are usually not very large and have irregular shapes. There is no evidence of domination e.g. estate lodges, parkland etc.

16.3   GENERAL CONCEPTS

The choice of settlement sites and the ways in which they develop reflect the relationship between physical and human factors (e.g. level of technology) These can be examined through the study of residual (relic) settlement features.

Sequent occupance Britain has been settled by successive waves of invaders. Each band of new settlers exercised choice in where they would settle and the types of settlements they would establish. Sequent occupance implies that each successive wave of new settlers entered existing settlements and in so doing added to their form and character; thus the landscape was constantly adapted and changed.

Palimpsest A landscape may be regarded as being made up of a series of layers of different settlement patterns which have in turn been established on top of relics of previous times. For example, Saxon settlements were established in a landscape which already contained relic settlements from Celtic and Roman times with distinctive forms and patterns of distribution. This is an alternative concept to that of sequent occupance.

16.4   DIFFERENT PERSPECTIVES

There are three main perspectives from which the study of rural settlement patterns may be approached by the geographer:

(a) the physical determinist approach (this is now outdated) It assumed that in rural areas where agriculture is the prime economic activity, the physical environment largely determined where people lived and the types of settlements in which they were found (nucleated or dispersed).

(b) the evolutionary approach This approach focuses on the question of how the forms and distribution of rural settlement have changed over time and so evolved into the distribution and morphological patterns we find today. In this approach geographers examine the residual features which reflect the criteria by which people in the past operated in selecting particular sites and establishing particular forms of settlement. The origin and development of settlement are therefore central studies in this approach.

(c) the sociological perspective This is concerned with identifying the characters of rural settlements according to their social form. The approach is especially concerned with systems of land ownership which in turn influenced the social groupings and divisions within the settlements.

16.5 RELATED TOPICS

This is a distinctive section with few links with other topics in geography. It does have important links with central place theory. The theoretical landscapes of Christaller and Lösch offer economic explanations of the evolution of rural as well as urban settlements.

16.6 QUESTION ANALYSIS

1 What are the factors encouraging **either** nucleation **or** dispersion in rural settlement patterns.

(in the style of London)

Understanding the question This question is straightforward as long as you realise that it is an 'either/or' question. The argument you present will be more convincing if you back up your points with actual examples.

You should also make a statement about what you take 'rural settlement' to mean. If, for example, you intend to include market and other country towns as forms of nucleated rural settlement you should say so in your introduction.

Answer plan Begin by stating quite clearly which aspect of settlement pattern you have chosen to write about e.g. nucleated settlements.

You should then briefly describe what is meant by nucleation. The main part of your answer should be devoted to discussing the factors which have encouraged nucleation.

In this answer plan it has been assumed that we are concerned with village forms so market towns and other country towns have been excluded. If you wish to see how towns may be fitted in, see Section 16.6.3(b).

Main points to include are:

(a) Nucleation was partly the result of the type of farming practised. Co-operative farming, like the open field system in Norman times, was most efficiently run from central nucleated settlements. This pattern can still be seen in Laxton, Nottinghamshire.

(b) Other physical factors also encouraged the grouping of farms and associated buildings. In the chalk and limestone areas there was little surface drainage. The availability of a permanent water supply was a strong influence on the location of settlements. In areas of extensive marshland suitable dry sites for villages were equally important. In some areas such as the chalklands of south-east England there was a paucity of building material. The existence of suitable building stone in a particular locality (e.g. Greensand) encouraged the grouping of settlement nearby.

(c) An important cause of settlement nucleation was the need for defence. Sites on top of hills and inside a river meander were often favoured. A recent phase of nucleation occurred in Malaysia in the 1950s when the British built fortified villages to defend rural peasants from communist guerillas.

(d) The development of modern transport encouraged the development of modern nucleation in rural areas e.g. Stoke Bruerne, Northants is an example of a canal village; Craven Arms, Shropshire owes its modern development to the railway.

(e) Nucleation may also be the result of the action of landowners. In the nineteenth century the need to provide better houses for workers resulted in the establishment of planned or model villages in the countryside.

(f) Political planning has also played its part e.g. the reclamation of the Zuider Zee resulted in the building of nucleated settlements on the new farmland.

Your concluding section could point out that nucleation is therefore the result of the interaction of a number of influences, both human and physical. Different peoples at different times and with different levels of technology have decided that nucleated forms of settlement offered the most desirable pattern of settlement given the choice of site they had.

If you chose to write about *dispersed* forms of settlement you would need to include:

(a) Farm economy – livestock farming favours dispersal.
Specialized labour-intensive farming also favours dispersed forms
(b) Very low densities of population
(c) Dissolution of large estates (e.g. after the Reformation)
(d) Secondary dispersion from nucleated settlements (e.g. after the 18th century enclosure of common land)
(e) Political planning decisions e.g. the agricultural development of Sicily has meant a more intensive use of labour and workers have moved out of the traditional agrotowns to dispersed cottages on the land they work.

2 With reference to specific examples you have studied from maps or when working in the field, compare the factors which have resulted in (a) nucleated (b) dispersed patterns of settlement.

(in the style of Associated Examining Board)

Understanding the question This question requires detailed answers. You are expected to refer to actual small scale examples. To prove that you are writing from knowledge of actual examples it would be a good idea to include a map of your examples if they are drawn largely from a single region. Since the question is divided into two distinct parts treat it as such.

Answer plan In your introduction specify the area or areas from which you intend to draw your examples.

Example: the answer will be related to two contrasting areas in Wales – the Aberystwyth district in West Wales (part of the upland core of Wales) and the St Fagans area of the southern lowland plain, the Vale of Glamorgan. Together they illustrate the main factors which have produced dispersed and nucleated settlements in Wales. See Fig. 16.1 and Fig. 16.2.

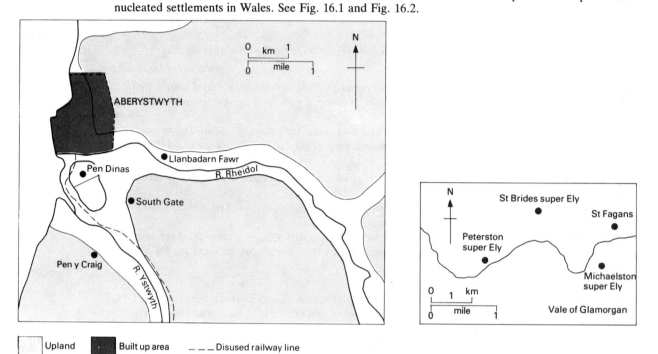

Fig. 16.1 Aberystwyth area, West Wales   **Fig. 16.2** St Fagans area, South Glamorgan

(a) *Nucleated rural settlement* In chronological order the main influences which created nucleated settlements in Wales were:

1 the influence of the location of monastic cells of the early Celtic Saints. These cells became the nucleus of small clustered settlements. Immediately to the east of Aberystwyth (see sketch map) the village of Llanbadarn Fawr is such a settlement.

2 the Anglo-Norman invasion of the lowlands and main river valleys. The main southern route was along the Vale of Glamorgan. Castles and towns were established at major defensive points e.g. Cardiff and Neath. In the intervening areas, where the land is suitable for arable farming, the medieval open plan farming system was established. This created nucleated farm villages with anglicized names. Examples are shown on the St Fagans map.

Nucleation also occurred at minor defensive sites. Although the chief castle was eventually built at Aberystwyth on the west coast an earlier motte and bailey was first built further south and a nucleated settlement, Pen y Craig, has developed at that point.

3 The development of modern transport also encouraged nucleated settlement. Toll roads created accessible locations from new hamlets and villages. South Gate to the south-east of Aberystwyth developed around a tollgate. The building of a railway line south from Aberystwyth to Carmarthen also stimulated the grouping of settlements around stations and halts.

4 Mining (lead) developed in the Rheidol valley in the nineteenth century. The mines were small but the miners settled around them e.g. at Goginan which is located to the east of the area shown on the sketch map.

(b) *Dispersed rural settlement* Factors encouraging dispersed settlement were:

1 The upland farm economy. Most of upland Wales is used for livestock farming. This economic system favours the dispersal of settlement. The uplands to the east of Aberystwyth are characterized by dispersed single farmsteads.

2 The economic system was reinforced by cultural factors. The Welsh inheritance system (gavelkind) led to the sharing of land amongst all the children when the landowner died. So gradually the size of landholdings decreased and the landscape became characterized by small fields and scattered homesteads.

3 During the eighteenth and nineteenth centuries there was a land hunger in Wales. Landless peasants encroached on the common lands. They claimed the ancient right of building *Tai un nos* whereby if a house could be built overnight and smoke rose from the chimney in the morning the squatter then owned the land.

He could also claim the right to own the land around the house to a distance determined by how far he could throw his axe in different directions. Such squats are not easy to identify today and are not shown on the map.

The question asks you to *compare* so a vital final section is to bring the two sets of influences together. There are two parts to this process. First you should bring out the similarities – the nucleation of settlement in some areas and at some periods in history, and the dispersion of settlement at other periods and in other locations are both the result of economic factors – farm economy, effects of transport development, and mining activities; and social and cultural factors – Celtic inheritance patterns, the Celtic church influence, Norman French influence, Welsh common law. Then emphasize that since Wales is essentially an upland country in which the Celtic influences have remained very strong both the physical and human factors have led to the dispersed forms of settlement being more significant than nucleated forms in the rural landscape.

3 (a) Show how physical factors have influenced the distribution and morphology of rural settlements in Britain.
(b) Use the Christaller model to suggest why some rural settlements have grown to be larger than others.
(in the style of the Joint Matriculation Board)

Understanding the question The question has two distinct parts which should be given equal weighting in your answer. You are not asked to outline the Christaller model in detail nor to point out its weaknesses or modifications by Lösch. The examiner wants to see if you are capable of outlining an explanation based upon Christaller's work for variations in settlement size. Remember too that the first part of the question asks you to write about distribution and morphology *only*. If you fail to focus sharply on these two aspects of rural settlement your answer is likely to be too long.

Answer plan

(a) *Distribution and morphology* This part of the answer should be introduced with a statement that at particular times in history the choice and development of sites for settlements have been the result of the interaction of human and physical factors. You should also establish at the start that the basic division of rural settlement form is into nucleated and dispersed settlement.

The physical factors which have influenced the distribution and forms of settlement include: the agricultural possibilities offered by the environment – influencing factors are climate, altitude, aspect and soil.

In rural areas the density of population is also largely a function of the ability of the land to support a particular size of population. Low densities inevitably lead to dispersed settlement (e.g. the Highlands of Scotland).

In certain locations at different periods certain physical features have strongly influenced the nature of settlement: e.g. in times of invasion easily defended sites were of prime attraction (hill top sites). In areas where permanent water supply sources were not widespread location near a spring or well was important. In waterlogged and marshy areas dry sites were at a premium. In mountain areas isolated farm sites which were sheltered and had a southern aspect were important.

Physical conditions also had a less direct effect upon the distribution and form of settlements e.g. the development of transportation routes was greatly influenced by the nature of the terrain. The lines of transportation themselves then attracted nucleated settlements.

An important point to make finally in this section is that although physical factors have important influences upon the distribution and form of settlement it cannot be assumed that there is a simple (determinist) relationship, as social, cultural and technological factors influence decisions about exploiting the environment.

(b) *Christaller's model* To find out about this model, read Unit 17 *Central place theory*. In this part of the question you must relate the model to rural land settlement. In his study of south-west Germany, Christaller recognized seven levels of a hierarchy of central places. As far as rural settlements are concerned, he suggested the emergence of market hamlets, township centres and country seats.

In offering an explanation in economic terms of the size and spacing of settlements, Christaller suggested that rural areas will be characterized by central places which function at different levels of the hierarchy. Rural populations will generally support the types of settlement at the lower end of the hierarchy. The size of the sphere of influence of any settlement is dependent upon its place in the hierarchy which reflects its ability to compete successfully with other service centres. The more successful and significant a central place becomes, the more likely it is to grow in size as it attracts new services and serves an increasing market. Economically this is an important factor since different goods and services require different threshold populations to purchase or support them.

16.7 FURTHER READING

Carter, H., *Urban and Rural Settlements* (Longman, 1988)
Chisholm, M., *Rural Settlement and Land Use* (Hutchinson, 1970)
Daniel, P. and Hopkinson, M., *The Geography of Settlement* (Oliver and Boyd, 1981)

17 Central place theory

17.1 ASSUMED PREVIOUS KNOWLEDGE

If you followed a 'concept geography' approach for GCSE you may be familiar with basic definitions and principles of central place theory. You will, in any case, almost certainly have studied your local area and/or another part of the country in the field. Your urban field work will probably have included examination of the functions of towns and their spheres of influence. This is a useful basis on which to build and it should provide you with relevant examples to illustrate your answers.

17.2 ESSENTIAL KNOWLEDGE

17.2.1 Definitions

Central place A settlement which provides one or more services for people living outside it.

Hierarchy Organization into successive orders or grades (see Section 17.3 *General concepts*)

Functions Lower order functions include the kinds of services provided by a small general store or corner shop; higher order functions or services are provided by a departmental store.

Threshold population The minimum number of people needed to support a function.

Range of services or goods The maximum distance over which people will travel to purchase goods or obtain a service offered by a central place. At some distance from the centre the increasing inconvenience of travel (measured by time, cost or trouble) will outweigh the value of obtaining the goods in that central place.

External economies Some central place functions or services are interdependent. Banks, for example, need to be near their customers to discuss business; shops need to be near sources of supply. If a service is located near the functions with which it has close contact it will make savings which are called external economies.

Isotropic surface A flat, featureless plain with uniform population density and with no variation of wealth and income amongst the inhabitants. This is the hypothetical landscape upon which Christaller developed his model.

17.2.2 Christaller's Theory

A certain amount of productive land supports an urban centre. The centre exists because essential services have to be performed for the surrounding area. These services are *central functions*, the places which perform them are *central places*

Ideally each central place will have a circular service area. Circles do not fit together well. The closest regular geometrical figure to the circle which will completely fill an area is a hexagon. So Christaller envisaged a hexagonal pattern of central places and service areas.

In order to explain variations in settlement size and importance Christaller postulated the existence of an isotropic surface upon which small nucleated settlements were originally evenly distributed. If the whole area is covered with market areas *hexagonal networks* grow up with each village receiving trade equal to 3 times the trade produced by its own population. The total trade value is known as the *k-value*

Table 17.1 Hierarchy of settlements

Settlement form	Distance apart (km)	Population	Tributary area size (km²)	Population
Market hamlet (Markort)	7	800	45	2 700
Township centre (Amtsort)	12	1 500	135	8 100
County seat (Kreidstadt)	21	3 500	400	24 000
District seat (Bezirksstadt)	36	9 000	1 200	75 000
Small state capital (Gaudstadt)	62	27 000	3 600	225 000
Provincial head capital (Provinzhaupstadt)	108	90 000	10 800	675 000
Regional capital city	186	300 000	32 400	2 025 000

Christaller recognized seven typical size settlements (Table 17.1) and stated that the number

k = 3   k = 4   k = 7

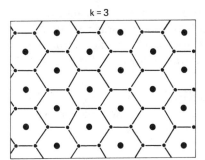 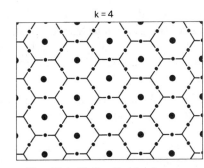 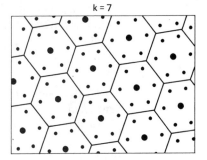

● Village   • Hamlet   **Fig. 17.1** Christaller landscape

of central places followed a norm from the largest to the smallest in the following order $1:2:6:18\ldots$. The larger the central place the larger is its trade area. So each larger class of settlement in the table was spaced on a hexagon of the next order size. The distances between similar centres in the table increase by $\sqrt{3}$ over the smaller preceding class.

Determining k-values Once a k-value is established in an area it is fixed (uniform) for all levels of the hierarchy. The landscape which evolved would be one of three patterns (Fig. 17.1).
(a) $k = 3$ landscape. This is one in which the principal influence is the *marketing principle*. All areas are served from a minimum set of central places. The reason why $k = 3$ is because the settlement's trade area is composed of its own population plus an area which creates twice as much trade as its own population i.e. one-third of the surrounding trade areas; so $k = 1 + (\frac{1}{3} \times 6) = 3$.
(b) $k = 4$ landscape. This landscape is based on the *transport principle*. As many places as possible lie on main transport routes connecting the highest order centres. $k = 1 + (\frac{1}{2} \times 6) = 4$.
(c) $k = 7$ landscape. All the complementary regions are clearly separated. The trade area of a central place is composed of its own and its tributary settlements' population. This pattern is based on the *administrative principle*

17.2.3 The Lösch Model

The Christaller model is very rigid. It maintains that settlements in particular regions are related to each other exactly according to fixed k-values. Lösch produced a more sophisticated model which made central place theory more flexible.

Lösch called the pattern of hexagons *lattices*. By superimposing lattices based upon different principles on each other he produced a pattern of *sectors*. Six sectors contained more higher order centres than others. These he termed *city rich sectors*. Six others were sparsely populated – the *city poor sectors*. So Lösch introduced a variable k hierarchy. However, in adopting this approach it became harder to support the concept of a clear-cut hierarchy of settlements as shown in Table 17.1.

17.3 GENERAL CONCEPTS

Central place theory is concerned with the principles which determine the number, size and spacing of settlements. It provides a framework by which settlement systems all over the world may be studied.

Hierarchy The organization into ranks and orders of towns and cities (urban hierarchy) and functions (hierarchy of functions) is a basic concept in central place theory. In the hierarchy of functions there are lower order and higher order functions (see Section 17.2.1 *Definitions*). In the United Kingdom whenever a high order function occurs in a central place, the full range of lower order functions is usually present. However, in other countries, such as France and the USA, large out-of-town specialist stores are quite common.

Spatial competition This is the process which determines which central places will attract new functions and how the patterns made by centres which provide a particular function will evolve. Centres which provide the same functions compete for customers. Customers are distributed in space so central places compete for space. As people become more mobile, and as some centres become more attractive because of the range of services offered, the space served by a particular function or central place may change.

17.4 DIFFERENT PERSPECTIVES

One vital contribution the Christaller model made to our understanding of settlements was its emphasis on the concept of centrality. Because it is so inflexible, however, it is not completely satisfactory. Factors which disrupt the Christaller model include:

(a) Transport centres Many towns developed as stopping points on roads, railways and.water routes. Along major routes central places are frequently strung out at short distances with the tributary areas stretching out at right angles to the line of the route. The traffic stimulated demand for services, for industries such as repair workshops and warehouses and for housing. Example: Swindon in the railway age.

(b) Centres located in relation to physical resources Towns which exist essentially as industrial centres may be located in relation to resources such as iron ore and coal e.g. Scunthorpe. The physical resources of sea, beaches and landscape similarly locate resort towns clustered in a particular area e.g. south coast resorts. Weber's location theory (see Unit 21 *Location of industry*) may offer a better explanation of their distribution than does central place theory.

(c) Planned towns Towns developed as a result of administrative decisions e.g. the new towns around London do not conform to the Christaller model. Since planning rather than market forces is crucial as far as urban development is concerned in the Eastern bloc, the model may be less applicable in communist countries than in Western countries.

The rank-size rule

Christaller developed a deductive theory which included the concept of a clearly bounded urban hierarchy. By working from empirical evidence, however, Zipf arrived at a different view of the ordering of settlements and their relationship to each other. Zipf proposed the rank-size rule.

$$Pn = P_1(n)^{-1}$$

Pn is the nth town in a series 1,2,3, . . . n in which the towns are arranged in descending order of size. So P_1 is the largest town. The formula used is

If P_1 has a population of 1 000 000 then P_2 has a population of 1 000 000 $(2)^{-1}$ or

$$\frac{1\,000\,000}{2} = 500\,000$$

If rank and size are graphed on log-logarithmic paper for an area then the plotted points would produce a smooth curve. This contradicts the steps which the Christaller concept of hierarchy envisages. This apparent conflict arises in part from the fact that whereas Christaller was concerned with centrality, Zipf measured population. The rank-size rule essentially applies only to large areas e.g. the USA.

17.5 RELATED TOPICS

The units on the location of industry, transport and transport networks, urban and rural land use, rural settlement patterns and the distribution of population are related to this topic.

17.6 QUESTION ANALYSIS

1 'Central place theory alone is an inadequate basis for understanding the urban geography of a country.' How far do you agree with this statement? *(in the style of Oxford)*

Understanding the question Make sure that you are able to outline the main features of Central Place Theory (Christaller and Lösch). This is not enough though. You are expected to assess how completely these models explain urban geography. The question suggests that the theory is not enough but there is no reason why, if you think the evidence points that way, you should not disagree with the quotation. A key phrase in the quotation is *urban geography*. Central place theory is about the size, spacing and distribution of settlements. You are expected to consider whether this is a complete definition of urban geography.

Answer plan First make clear what you understand by central place theory and emphasize its concern with size, spacing and distribution. Make this section brief.

It is important that you then discuss what is meant by 'urban geography'. In addition to the areas covered by central place theory geographers are also concerned with the evolution and morphology of towns. They are also interested in the towns and cities which are the result of national and regional planning decisions e.g. the new towns. So you may argue that central place theory relates to only part of the urban geography of a country.

One of the criticisms made of the Christaller theory is that it is good in analysing sparsely populated rural areas but is not as applicable to areas which are industrial or in which towns perform a major function which is related to a particular physical resource. This is a relevant point because it suggests that in many countries central place theory does not offer a comprehensive basis for understanding the urban geography.

Another significant point is that one of the basic concepts of the model – the urban hierarchy, is disputed by empirical evidence collected on the basis of a large country – Zipf's Rank-size rule. This points to a degree of inadequacy in the theory.

There is also evidence that trade areas are not as clear cut as the theory suggests. Writers such as Huff argue that decisions by individuals as to where they should shop and work leads to considerable over-flowing of trade between areas. The decisions can depend on age, the type of goods required, the status and income of the individual, the views of the groups to which he belongs. Such decisions can seriously affect the competitiveness of the different centres and the size and spacing of those settlements.

Increasing mobility of the population has also worked against the basic Christaller model. In many countries the smaller centres are declining and the larger centres becoming more dominant as people ignore the local services and make for the major towns.

So you can argue that the theory offers only a partial understanding of the urban geography. On the other hand it can also be argued that in concentrating on the basic economic forces that operate the theory helps us to understand not only the size and spacing but also the way in which some centres have grown and evolved as dominant service centres. So it does help us understand the growth and evolution and the morphology of towns.

The concluding paragraph should stress that in concentrating on processes it provides us with a frame-work for understanding apparently unrelated features of the urban geography of a country. To that extent it provides a valuable basis for work in urban geography.

2 (a) How and why do the spheres of influence of central places vary in size in the Christaller model?
(b) Discuss the factors which might influence the size and shape of spheres of influence of *each* of the following in Britain:
(i) a market town in a rural area; (ii) a supermarket on the edge of a major town; (iii) a ladies' hairdressing salon.   (*in the style of Scottish Higher*)

Understanding the question In a question like this it is reasonable to assume that about 40 per cent of the marks will be allocated to part (a) and the other marks shared equally between the three parts (5/25 for each). The theoretical part (a) must be dealt with precisely. It is concerned with spheres of influence. You are not expected to outline the whole of central place theory. Part (b) is concerned with the factors which influence the size and shape of spheres of influence at different scales of service provision. There will be an overlap of factors but there are also influences which specifically operate at a particular level.

Answer plan It would be helpful to establish right at the start that the sphere of influence is synonymous with Christaller's concept of a trade area. His initial assumption was that small nucleated settlements were evenly distributed on an isotropic surface. Each nucleated settlement was a central place as it provided one or more services for people living outside it. The isotropic surface is completely covered by trade areas which are hexagonal in shape. Each one focuses on a single settlement. Initially therefore there is no variation in the size of the spheres of influence.

Since the urban spheres relate to trade, the pattern of evolution which is relevant is that based on the marketing principle (k = 3). The development of spheres of influence of different sizes is the result of the evolution of an urban hierarchy of central places. The table on page 130 would be useful here.

This hierarchy results from the operation of economic processes which result from the fact that central place functions occur at different levels and require different threshold populations. The relationship of settlements at different levels in the hierarchy is determined by the k-value. So are the trade areas. Each larger order of settlement has a larger trade area which is spaced on a hexagon of a larger order. A diagram of the k = 3 landscape should be included here.

Since part (b) says 'in Britain' it is important that you relate your answer to British conditions and, if appropriate, mention British examples.
The market town. Factors which might influence the size and shape might include: the *level in the regional hierarchy* at which the town lies and the *density of population* in the tributary area – towns in more densely populated areas will tend to have smaller trade areas.

If the market town is *near to a larger competing centre* – the sphere of influence will be asymmetrical with a smaller radius on the side near to the larger centre. *Transport patterns* are important e.g. a motorway network passing near a rural market town may siphon off the trade of the small town to a larger town or city. This will lead to a decrease in size and eventually in shape of the rural market centre's sphere of influence. The shape and size may also be affected by *topography*. In North Wales for instance the size and shape of spheres of influence are affected by the influence of relief upon accessibility and convenience. Another point is *nearness to county boundaries* – twin towns will tend to develop on either side of the boundary and both towns then have truncated spheres of influence. Classic examples before local government reorganization in Wales were Narbeth (Pembs) and Whitland (Carms).

You should also consider *the socio-economic character* of the tributary population. Market towns in the south-east for example may be centres serving wealthy commuting or retired groups. If the centre contains the types of shops and higher order services which attract these groups the size of its sphere of influence may be significantly larger than might be expected for such a town.
The supermarket on the edge of the major town. The *amount of competition* it faces from town centre supermarkets is an important factor. This would produce an asymmetrical trade area restricted on the side near the town. The *attractiveness of other shops nearby* is another influence. If, for example, it is part of a new shopping centre it is likely to have a wider pull.

Consider also *ease of accessibility* e.g. nearness to a ring road or by-pass and *the range of goods* the supermarket offers. Some cut-price stores concentrate entirely on low-order goods, others with 'home and wear' departments etc. have a range of higher order goods. Higher order goods require a larger threshold population and the sphere of influence is likely to be larger. *Store image* is also important e.g. the success of Sainsbury supermarkets in the southern half of the country. *The frequency with which the supermarkets belonging to the same firm occur within a region* has a bearing on the sphere of influence e.g. in Sussex the sphere of influence of the Sainsbury supermarket in Crawley was reduced in area when another Sainsbury's was built eight miles to the east in East Grinstead. *The socio-economic character of the population* in the surrounding area will also influence the extent of the sphere of influence e.g. in a region with a high proportion of lower socio-economic groups a supermarket with a strong cut-price image may be expected to have an extensive sphere of influence. Although, to use a supermarket on the edge of town, most people would require a car.

The ladies' hairdressing salon. If the salon is a 'run of the mill' establishment key factors which would produce a very limited sphere of influence would be the *range of goods* – this is a lower order function – and the fact that *demand* for such a service is said to be very *elastic in economic terms* i.e. easy substitution is possible and distance is an important frictional cost. So the number of other salons in the locality is significant. Another vital factor is the *physical context* in which the salon is found. If it is part of a shopping precinct, shoppers are likely to find the trip to the shop more attractive because it can be part of a wider shopping trip. With a lower order service of this nature *accessibility and convenience* are important, and so, unless the salon aims at a high-cost, exclusive trade is the *price of the service* compared with prices charged by competing salons.

If it is an exclusive salon its sphere of influence is wider because the *threshold population* required to support it i.e. the wealthy higher socio-economic group is spread over a wider area. If the salon has a *good reputation* customers are prepared to travel longer distances to patronize it.

3 Calculate the k-value of the landscape shown in the diagram below and draw the network of roads you would expect to serve a higher order centre. State *four* human factors that are likely to influence the spacing of central places. (Accompanying this question is a diagram showing a network of central places on a hexagonal frame.)

(in the style of London)

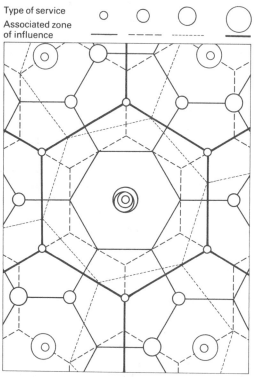

Fig. 17.2 k-value landscape

Understanding the question Most of the marks for this question are awarded for the written part – only about one-third are awarded for the practical section. It is therefore important not to spend too long on the practical part. To complete it quickly you should be able to recognize the k = 3, 4 and 7 landscapes shown in the diagram above. The final part of the question is testing your awareness of important human factors which modify the logical but rigid framework of Christaller's theoretical landscapes.

Answer plan (written part) Begin with a statement that this theoretical model is rigid. As human factors have a significant influence in determining the spacing of settlements the landscapes produced are not as logical and uniform as those developed by Christaller. The following four significant factors have been dealt with above.

(a) The influence of routeways which create transportation centres.

(b) The exploitation of physical resources which creates point production towns at key locations.

(c) Population factors – Zipf's rank size rule model.

(d) Planning decisions relating to the location and size of settlements.

Another vital consideration is the density of population in a region. Central place functions depend upon threshold population sizes to support them. In a sparsely populated area central places tend to be more widely spread.

17.7 FURTHER READING

Burtenshaw, D., *Cities and Towns* (Unwin Hyman, 1983)
Carter, H., *Urban and Rural Settlements* (Longman, 1988)
Daniel, P. and Hopkinson, M., *The Geography of Settlement* (Oliver and Boyd, 1981)

18 Urban land use

18.1 ASSUMED PREVIOUS KNOWLEDGE

You will probably have studied a local town or city in the field as part of your work for GCSE.

18.2 ESSENTIAL KNOWLEDGE

18.2.1 Definitions

Functional areas or zones As towns grow the different functions they perform tend to become spatially segregated i.e. to occupy different areas in the town. Housing, commercial activities, industry – each becomes the chief function of a different part or zone of the town. The development of these zones is the result of various processes such as suburbanization of industry or social segregation. Within these broadly defined functional zones finer, more detailed contrasts in land use are found. These are the results of factors such as the age of the buildings, the density of building, land ownerships and variations in commercial activities, for example, between shops and warehouses.

Morphology This is the layout of the buildings together with the functions. Functional zones are also called morphological zones.

Urban structure The spatial relationships between the functional zones. Models have been developed which describe the structure of towns and cities. The models attempt to identify characteristics common to all towns and so help us to understand the processes which produce the structure.

18.2.2 Models of urban structure and growth

The concentric zone model (Burgess) (Fig. 18.1) Burgess argued that because different types of functions have to compete with each other for limited space in the city certain functions become dominant in certain areas. So functional zones develop. These zones are arranged concentrically around the city centre. Lower status residents live near the centre of the city, higher status groups on the outer edges. Within these zones and across their boundaries there may also be *natural areas*. These are distinctive areas where a particular ethnic group has concentrated. The concentric zones may also be broken up as a result of such factors as the occurrence of high ground which, for instance, might become an attractive residential area for high status people although it is near the city centre.

The pattern is also affected by processes which operate as the town grows. For example, ethnic minorities may be absorbed into the population as a whole and move away from a ghetto. Members of other groups may improve their status and move away from the inner city zones. This process is called *invasion and succession*.

The Burgess model is regarded as a weak structural model because:

(a) It was developed nearly 60 years ago and great changes have occurred in the nature of cities which make the model less applicable.

(b) It was based on the study of Chicago and other American cities. It was therefore relevant to a particular historical and cultural context but not to other parts of the world – it is said to lack universality.

(c) The model suggests that there are sharp boundaries between the functional zones. In reality these abrupt changes do not occur.

(d) It fails to recognise that in pre-industrial cities of countries such as India the high status groups are likely to live near the centres of the cities and the poorest families are found on the edges of the built-up areas (e.g. shanty towns).

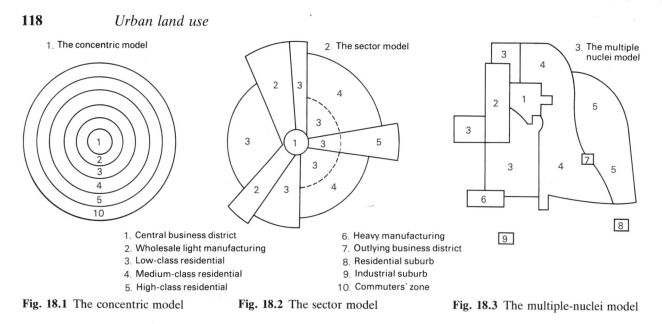

1. The concentric model

2. The sector model

3. The multiple nuclei model

1. Central business district
2. Wholesale light manufacturing
3. Low-class residential
4. Medium-class residential
5. High-class residential

6. Heavy manufacturing
7. Outlying business district
8. Residential suburb
9. Industrial suburb
10. Commuters' zone

Fig. 18.1 The concentric model   **Fig. 18.2** The sector model   **Fig. 18.3** The multiple-nuclei model

The sector model (Hoyt) (Fig. 18.2) The sector model complements the Burgess model because it considers an additional factor – direction as well as distance. It is based on the idea that accessibility to the town or city centre influences the location of functional zones. So the zones are shown in the model as sectors with wedges of residential areas developing outwards from the centre. High-grade residential areas are located in the most convenient locations and have a strong influence on the pattern of urban growth because other zones have to fit around them.

The multiple-nuclei model (Harris and Ullman) (Fig. 18.3) Instead of envisaging functional zones which develop outwards from a single centre, Harris and Ullman took the view that the functional zones would develop around a number of nuclei of which the Central Business District was only one. Other nuclei might be a suburban shopping centre or old villages which have been absorbed into the growing town. So the town or city is a series of distinctive cells. The number of nuclei depends upon the size of the city.

It is argued that towns and cities developed in this way because:

(a) Certain activities need to occupy specific locations e.g. large industrial complexes need cheap land.

(b) Some activities group together for mutual advantage e.g. shops located in a block of shops get more customers than an isolated store.

(c) Some functions are incompatible, for example, the most expensive residential area is not located alongside heavy industry.

(d) Different functions have differing abilities to pay rents and rates. So only some functions can afford to locate near the city centre.

Bid rent theory (Ratcliff) This theory explains the location of functions within a town or city in terms of economic factors – it envisages land-use as being determined by the relative efficiencies of using land in different ways in particular locations.

Efficiency in use is measured by rent-paying ability. Competition for different locations

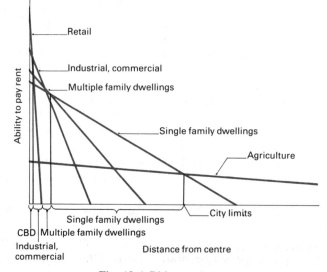

Fig. 18.4 Bid rent theory

within the city by different functions produces the most efficient pattern of land use. So the structure of the city is determined by the financial evaluation of the importance of convenience. So rents are seen as a payment for saving the costs of transport.

In the graph (Fig. 18.4) each category of land user's ability to pay rent is plotted against distance from the town centre or CBD. Retail shops need to be in the centre to have maximum accessibility for all the city population. Since accessibility decreases outwards from the centre so does willingness to pay high rents. Solicitors, accountants etc. also like being near the centre but cannot afford the highest rents. So their offices are found on the edges of the CBD. On the graph therefore the slope for commercial activities is not as steep as for retail shops. Multi-storey blocks of flats produce more income than single family houses and so the flats can compete for land nearer the city centre although they cannot compete with shops and offices. Agricultural land is the least competitive and has the least steep rent/distance slope, so it is found outside the city limits.

So in this model the urban land market is seen as a land value surface. The market centre is the point of highest site value. As rent declines with distance the value of the land falls. As the land value falls the land use changes. This simple pattern is modified by other factors e.g. main transportation routes have higher land values than surrounding areas; where these routes intersect secondary commercial centres develop.

18.3 GENERAL CONCEPTS

Evolution and growth The population of most British cities has grown over the centuries. The cities have therefore grown outwards from the centre. If old villages or towns have been absorbed into the growing city they form zones of old houses e.g. in Norwich. So the most valuable sites – in the city centre – are occupied by the oldest properties.

Urban renewal is a policy designed to regenerate inner city areas. It is a response to public awareness that the scale of urban problems has now increased to a point where it threatens the functioning of our major cities.

It is one of the solutions to urban decay which have been developed since 1945. It is complementary to the programmes for building new towns and overspill estates in which people from the cities were rehoused. Urban renewal is aimed at keeping people and jobs in the inner city areas.

Urban renewal may occur as *spontaneous renewal* or as *comprehensive renewal*. Spontaneous renewal occurs when demand for land in the city centre exceeds supply. Renewal is worthwhile because the new development will bring in higher rents. Much of the redevelopment of the centre of London in the City and along the river is of this type. So is the gentrification of older housing areas in London – fairly well-off people buy older houses as good investments and improve them. Comprehensive renewal may be undertaken by local authorities or by private companies. Because large sites are needed property is often subject to compulsory purchase. Demolition is quicker than rebuilding so large cleared areas may be left as areas of *urban blight*. Comprehensive renewal provides an opportunity for well-planned rebuilding e.g. the new city centre of Sheffield. It can also lead to the break-up of well-established communities.

18.4 DIFFERENT PERSPECTIVES

Burgess and Hoyt both developed simple models which were essentially descriptive. They were based on empirical research, i.e. on practical studies and are therefore inductive – the general patterns they show have been worked out from particular instances. When attempts were made to develop a deductive theory (reasoning from cause to effect) the bid rent theory was the result.

All the models are based on principles of market economies. They are less applicable therefore to pre-industrial cities of the Third World or to the planned cities of the Eastern bloc.

18.5 RELATED TOPICS

This section has close links with Unit 21 *Location of industry* and Unit 17 *Central place theory*.

18.6 QUESTION ANALYSIS

1 Discuss the roles of rents and land values in the creation of internal urban differentiation.

(in the style of Cambridge)

Understanding the question This question requires good knowledge of bid rent theory and an understanding of the limitations of this model. You must bring out the links between rent and land values in cities. It is also important that you define what is meant by *internal urban differentiation*. This may be taken as the creation of distinctive land use zones, the functional zoning. An important part of the question is the need to discuss the importance of the rent and land value factor in creating these zones.

Answer plan Preface the first section with a brief definition of internal urban differentiation. It is vital that you then develop the argument which links rent and land values. Robert Haig (1920) stated that rent in the city is a *charge for accessibility* – the saving of transport costs. Land users who can pay the highest rents occupy the most convenient or accessible parts of the settlement. Ratcliff, in his development of the argument stated that the land for which there is most competition yields the highest rents and is the point of maximum land values in the city. Away from the centre rent declines with distance and values fall.

Having linked rents and land values you should now relate the argument to land uses.

The successful bidder for a location is the one who can use the site to obtain the greatest return. Over time, therefore, the most efficient land use pattern emerges and this creates the functional zoning. The diagram on page 118 should be included here.

Land use is divided into broad categories, each category varying in its ability to pay rents i.e. to compete successfully for particular locations. Develop this point with reference to the diagrams above. You could also add a paragraph showing how this works out in a town or city you have studied in the field. Bring the section together by emphasizing that as rents decline from the centre of the city outwards, land values fall and land uses change. This creates the internal differentiation.

In the third section of your answer discuss the significance of this model. It presents an economic explanation for urban differentiation within cities but it considers one variable only – distance from the centre. This simple relationship is complicated by other factors.

(a) *Economic factors* The location of lines of communication stimulates higher values. They complicate the simple gradient pattern of decrease from the centre. They complicate land use by, for example, the creation of subsidiary shopping centres. The rent curves shown in the diagram are also affected by considerations such as the quantity of land available. Developers may choose less accessible locations in order to buy large sites.

(b) *Social and cultural factors* Some wealthy families choose to live in the city centre; other equally wealthy families choose to live on the edges of the city. So the ability to pay rent is not purely an economic decision. Similarly, those who buy and sell may not be motivated solely by the idea of maximum profit (look at the satisficer concept in Unit 21 *Location of Industry*).

(c) Alonso has suggested that land use pattern in a city is the result of equilibrium solutions to problems of the best location of industry, business or home. So, rent and land values only partly explain patterns.

In the final section point out that the importance of market forces in western countries is so great that the broad details of functional zoning are the result of conditions of rent and land values. Other factors produce some variation in the broad pattern. This has however been seriously affected by town planning decisions. For example, the presence of high rise office blocks in Croydon was the result of planning restrictions on office blocks in central London rather than the competition for land in Croydon. Compulsory purchase of inner city land and decisions to build large council estates have also cut across the economically produced functional patterns.

In Eastern Bloc countries the domination of central planning and the absence of a capitalist economy means that bid rent theory does not have much significance in decisions about land use. In the Third World countries cultural factors, e.g. the caste system in India may provide a historic framework within which modern urban differentiation occurs.

2 What do you understand by *urban renewal*? Discuss with examples its importance in understanding the geography of British cities.   (*in the style of Oxford*)

Understanding the question This is a harder question than it looks at first sight. It would be a mistake to rush into it because you know what urban renewal is – only one-third of the marks will be given for the first part.

The question requires a definition of urban renewal. Since it is one of two parts to the question a brief one-sentence definition is not enough. The question also requires you to discuss what you think should be included in the phrase *the geography of British cities*. This would include the size and spacing of cities, location, evolution and growth, functions, morphology, the economic and social processes which create the spatial patterns, the powers of institutions and authorities to make key decisions. Reference to actual examples is vital. A range of reasonably detailed references to more than one British city will impress the examiner.

Answer plan You should begin with a definition of the concept of urban renewal (see Section 18.3).

In your second section detail the ways in which the process of renewal helps understanding of different aspects of the geography of towns.

(a) *Size and spacing* The establishment of new towns after 1945 was the main attempt to solve our city problems. It had various effects on the cities – causing decline of their populations and facilities etc. So urban renewal policies were developed to solve the problems within the cities to stop further decline.

(b) *Evolution and growth* (see Section 18.3).

(c) *Functions* Cities are dynamic features of the landscape. They change over time as do the needs of their inhabitants. So buildings become obsolete (e.g. the warehouses along the Thames), houses no longer meet the standards that people expect (e.g. indoor baths and toilets) and suitability for a particular land use also changes (e.g. many industries now prefer sites along ring roads to city centre locations).

(d) *Morphology* The Burgess model places low status people near the city centre; the bid rent theory states that city centre land use reflects the ability to pay rent. Both patterns are disrupted by comprehensive renewal programmes. Urban renewal programmes therefore make us look at the processes which create the spatial patterns of functional zoning within cities.

In the third section you should look at these processes:

(a) *Economic processes* – when renewal occurs the developer wishes to obtain maximum profits – property which gives the greatest returns. So low-income families are forced to move e.g. this has happened in North Kensington and Southwark.

(b) *Social processes* Some people see the movement out of the low-income people as socially unacceptable. So a few areas have been rehabilitated, e.g. the Arlington area of Norwich contains old houses which have been renovated and the environment has been improved by the provision of off-street parking and open spaces.

(c) *Decision-making processes* Each urban renewal programme is the result of decisions taken by many different people and groups on both local and national level – landowners, builders, planners, local authorities, government. At the local level they provide a *decision-making filter* whereby national programmes are adapted to suit local conditions.

The government and local authorities also have important effects. For example the Government bill of 1969 changed the emphasis from new and expanding towns to rehabilitation of city areas.

Finally it could be pointed out that the factors and processes influencing policies and decisions about urban renewal are the factors and processes which have to be examined to understand the geography of British cities as a whole.

3 You may be given diagrams of the concentric, sector and multiple nuclei models shown in Fig. 18.1.
(a) Suggest why, in urban areas, distinctive land use zones develop.
(b) With reference to a specific town or city which you have studied discuss how far the models are a valid description of the land use pattern. *(in the style of the Joint Matriculation Board)*

Understanding the question The question requires knowledge of the basic characteristics of all three models and the principles upon which they are based. It also requires knowledge of the types and distribution of functional zones in *one* town or city you have studied. You must deal with *all three* models. It is also vital that you do not just describe the urban structure of your example but apply each of the models to see how well they fit i.e. how fully they explain the pattern of functional zones of your town.

The question is in two parts with marks shared 13/12 between the two. In the second part you can keep the length of your answer to a reasonable scale by using a sketch map of the functional zones to show the chief characteristics of your example.

Answer plan In your introduction point out that distinctive functional features emerge as cities evolve because the different functions have to compete with each other for limited space. This competition results in certain functions or groups of activities being concentrated in certain areas. So we can recognize broad categories of land use in the urban structure. The different models of this pattern are the result of emphasizing the effects of different factors or processes.

Your second section should be.made up of brief descriptions of the basic features of each model and their underlying principles.

Then draw a diagram or sketch map of the town or city you have chosen as your example and accompany it with a very brief description of what the figure shows. Develop the important section of your answer in which you consider the extent to which each of the models offers a valid description of your town and the processes at work in it. Ask yourself the following questions.

(a) Is there one point of origin from which it has grown? Do land use zones change in a regular way as you move out from the city centre? How are the residential areas patterned? These questions will help you decide how well the Burgess model fits.

(b) Are the lines of transportation the chief factors determining zoning in your town? Can you identify sectors? Where are the high status residential areas in relation to transport routes? How good a fit is the Hoyt sector model?

(c) Does your map show a series of nuclei? Did the town grow from more than one point? Are incompatible functions widely separated? Does the pattern reflect the ability to pay rents? (See Fig. 18.2.) So is the Ullman and Harris model the most relevant model as far as your town is concerned?

When you have answered this final question bring together the strands of the argument by pointing out the weaknesses of each of the models in terms of explaining the urban structure of your town. Then come to a conclusion as to which model relates most closely to the fundamental factors underlying the pattern shown by your sketch map.

18.7 FURTHER READING

Briggs, K., *Introducing Urban Structure* (Hodder and Stoughton, 1977)
Daniel, P. and Hopkinson, M., *The Geography of Settlement* (Oliver and Boyd, 1981)
Everson, J. A. and Fitzgerald, B. P., *Settlement Patterns* (Longman, 1969)
Everson, J. A. and Fitzgerald, B. P., *Inside the City* (Longman, 1972)
Gordon, G. and Dick, W., *Urban Geography* (Holmes McDougall, 1988)
Pounds, N., *Success in Economic Geography* (John Murray, 1981)

19 Cities and their problems

19.1 ASSUMED PREVIOUS KNOWLEDGE

You will have studied a local town or city in the field. This practical knowledge must be related to the theoretical models which you study as part of your A level course. Problems and issues will also have been raised as you studied topics such as urban land use, traffic in towns, migration and underdevelopment. In order to achieve a suitably analytical approach, it is important that you draw on knowledge you have gained about topics such as the location of industry (unit 21), urban land use (unit 18), central place theory (unit 17), less developed countries (unit 23) and movement of population (unit 15).

19.2 ESSENTIAL KNOWLEDGE

19.2.1 Definitions

Check the definitions in the units listed above. In particular make sure that you understand the definitions of *urban structure* (18.2.1), *GNP* (23.2.1), *urbanization* (23.5), *migration* (15.2.1), *central place* (20.2.1), *functional areas* and *urban structure* (17.2.1). Other definitions which are relevant to this topic are:

Inner city The area near the centre of the city which is densely populated and which contains dilapidated housing, often in multiple occupation. The inner city is characterized by a declining industrial and economic base, higher rates of unemployment than the city as a whole, and recent loss of population. It is often a reception area for immigrants.

Residential segregration The segregation of groups according to socio-economic status, religious beliefs or ethnic characteristics, into distinct sections of the residential areas of cities. A clearly identifiable area in which members of a single cultural or ethnic group are concentrated may also be known as a **ghetto**.

19.2.2 Models of urban structure and growth relevant to this unit

This topic draws upon knowledge from a number of related units. Models of economic development (23.2.2), for example, are relevant to analysis of Third World cities. Urban land use models and bid rent theory (18.2.2), migration (15.2.3) and central place theory (unit 17) will all help you place city problems and issues in a sound theoretical context.

Push-pull model This is an explanatory model of population movement into cities in which migration is seen to be the result of two sets of forces whose effects are complementary. The 'push' factors are those which encourage people to leave rural areas and include low wages, lack of work, natural disasters. The 'pull' factors are the economic and social attractions (real or imagined) exerted by towns and cities – better job prospects, higher wages, leisure facilities etc.

19.2.3 Problems relevant to cities throughout the world

Problems arising from physical factors, e.g. (a) decay and obsolescence of the oldest part of the city – the inner city, old industrial areas, derelict docklands; (b) inadequate road systems and consequent traffic problems; (c) provision of public services – adequate supplies of pure water, effective waste disposal; (d) uncontrolled expansion onto unsuitable sites – squatter settlements; (e) effects of natural disasters – floods (Dacca, Bangladesh), earthquakes (San Francisco, Mexico City), drought (Timbuktu, Tigre (Ethiopia)).

Economic problems, e.g. (a) decline of former inner city industries and consequent unemployment; (b) decline of traditional manufacturing industries; (c) suburbanization of industry and consequent land use conflicts on city edges; (d) industrial growth and pollution.

Problems arising from governmental problems and planning processes, e.g. (a) administrative fragmentation of the city area – decisions may be difficult to make because so many authorities are involved in the administration of the built-up area; (b) tension and conflict between planners and entrepreneurs who demand uncontrolled economic development; (c) problems of social justice – decision-making in cities may merely reflect exisiting patterns of power and wealth; (d) difficulty of establishing agreed criteria for change – varying groups may have genuinely conflicting interests which cause social tensions. For instance, in the redevelopment of the London docklands, the dispute over whether housing should be for locals or for 'yuppies'.

Social problems, e.g. (a) contrast between rich and poor can lead to political unrest; (b) the contrast between suburban growth and inner city decay; (c) immigration and problems relating to multicultural populations; (d) housing and employment; (e) squatter settlements (shanty towns).

Problems arising from population shifts, e.g. (a) aged and deprived elements of the indigenous population left behind in the inner city when upwardly mobile and young move away; (b) emergence of ghettos and racial tension; (c) squatter settlements.

Urban problems are generally complex, so although the groupings listed above provide a framework for study, important problems do not necessarily fit into a single section. For example, the problems of the inner city are the result of the ageing of the oldest parts of the city; of economic processes such as the decline of traditional industries and the establishment of new ones away from the city centre; of planning decisions which took people and jobs out of cities such as London to planned new towns; and of racial tension as the centres become multicultural. Similarly, the problem of illegal squatter settlements, which is common to many Third World cities, is a complex one, caused by the interplay of a number of factors.

Cities in both the developed and less-developed countries face many contemporary problems. The cities of the less-developed world, however, are faced with problems which differ in scale and intensity from those found in the wealthy industrial countries. This is because:

(a) The process of urbanization is occurring very rapidly and at a vast scale in the Third World.
(b) The Third World countries have neither the wealth nor a sufficiently large reservoir of skilled labour with which to tackle serious problems swiftly and efficiently.
(c) Lack of an advanced technology and a modern economic infrastructure means that Third World countries are less able than more advanced countries to respond swiftly to sudden crises. For instance, during the Ethiopian famine it was easier for advanced countries to get food to Ethiopia than it was to distribute that food within the country to where it was most needed.
(d) Because of their lack of wealth and international power, many Third World countries are compelled to react to urban problems in ways which meet with the approval of rich creditor nations and powerful multinational companies.

19.3 GENERAL CONCEPTS

Key ideas relating to urban problems are:

Urbanization Throughout the world, people are gravitating from rural areas to cities. As far as the volume and speed of movement is concerned, the process is especially significant in the cities of the Third World.

Movement to the suburbs Movement to the edges of the cities is more than a reflection of upward social mobility. For example, many cities tackled their housing problems in the inner city areas by building new estates on the city edges. Shopping and leisure facilities and industrial estates have also moved to outer areas, where lower land prices and locational convenience by new motorways have been tremendous attractions.

Urban renewal See unit 18.3.

Environmental considerations Concern about environmental quality may be reflected in:
(a) decisions to restrain growth e.g. by the establishment of a green belt;
(b) regulations to restrict change, e.g. the designation of urban conservation areas which are specially protected because of their special historical or architectural qualities.

19.4 DIFFERENT PERSPECTIVES

Many of the models developed for the study of cities and their problems were developed with particular reference to modern industrial cities in advanced countries. Third World cities have often developed in a different social and cultural context. For instance, the largest and most powerful cities of the developed world achieved their status as a result of the development of modern industry. The growth of many Third World cities, in contrast, was not based upon industrialization.

There is a very different perspective in the socialist countries of Eastern Europe, the USSR, China, and other countries of the communist block. For example, free market economic processes such as competition for land on a bid rent basis does not apply in these countries. Their economies are 'command economies', with basic decisions about urban land use and the functions of cities being made centrally on the basis of agreed national priorities. Until recently, Eastern Europe gave relatively little priority to the development of service industries to meet consumer demands. Housing patterns were determined, in theory, by family needs and not simply the ability to pay, so residential patterns in socialist cities differ significantly from those in capitalist cities.

19.5   RELATED TOPICS

These have been identified in 19.1 above.

19.6   QUESTION ANALYSIS

1 Figure 19.1 is a model of urban development in a Western city in the nineteenth century.

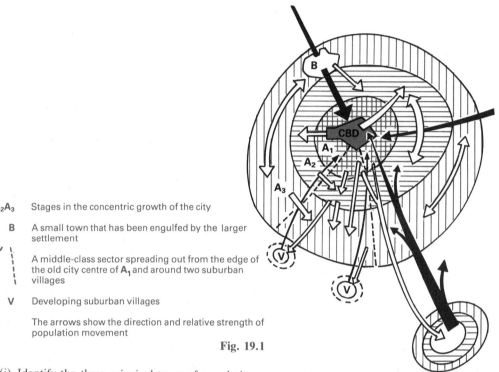

A₁A₂A₃   Stages in the concentric growth of the city

B   A small town that has been engulfed by the larger settlement

A middle-class sector spreading out from the edge of the old city centre of **A₁** and around two suburban villages

V   Developing suburban villages

The arrows show the direction and relative strength of population movement

Fig. 19.1

(a) (i) Identify the three principal types of population movement.
(ii) Suggest reasons for each type of movement.
(b) (i) How might the pattern of population movements have changed in the twentieth century?
(ii) What might have been the effects of these changes on the structure of the city?
(c) Contrast concentric zones A₁ and A₂ in terms of their present-day residential character, both:
(i) inside the middle-class sector;
(ii) outside the middle-class sector.

(COSSEC, specimen paper for AS level)

Understanding the question The question tests your understanding of how Western cities have changed in the last 100 years or so. To answer the question you must first study the diagram carefully.

Answer plan The structure of the question means that you have little to do in planning your answer.
Part (a) (i) Principal in this context is best defined as the movements which involve the greatest volume of people. the three main movements are rural-to-urban migration, inter-urban (urban-to-urban) migration, and intra-urban (within the city) migration (see unit 19.2.2).
Part (a) (ii) **Rural-urban** The growth of industry and commerce in the cities attracted workers from rural areas.
Inter-urban Some migrants moved from small towns to larger cities because of the greater range of jobs available in the larger centres. There was also a counter flow as people from the cities moved to smaller towns to set up new activities or subsidiary industries, e.g. to set up a local newspaper.
Intra-urban As residents of the city become socially upwardly mobile, they were likely to move to new residential areas. The building of new housing estates and early suburbs encouraged further movement. The expansion of the CBD also displaced residents, who sought homes elsewhere in the city.
Part (b) (i) Greatly increased movement out to suburbs and movement out of the city to a rural commuting belt; significant movement away from the city centre as a result of new housing areas developing on the city edges; the decline of inner city industries. In-migration is now on an international scale and most cities are increasingly multi-cultural.
Part (b) (ii) Decline of the inner city; extension of the city limits; absorption of neighbouring towns into the city; development of modern industrial zones in outer city; creation of a green belt to restrict uncontrolled physical growth of the city.
Part (c) (i) Within the middle-class sector, in zone A₁ large family houses disappear. The buildings change in function and become solicitors' and accountants' offices, etc. Housing is replaced by business and commercial buildings as the CBD expands towards the middle class sector.

In zone A$_2$ the area is likely to be subject to gentrification as some of the middle class move back towards the city centre as land values and house prices rise sharply in the favoured residential areas. There is some invasion of the areas the middle class moved out of, by lower income families who can afford these older houses but not good quality modern suburban houses. Houses are sub-divided into flats, apartments, student hostels etc.

Part (c) (ii) Outside the middle-class sector, A$_1$ is now part of the twilight zone. Local industry has declined, the indigenous population has declined, and the area has become one of multicultural immigration.

Many former buildings have been removed to make way for comprehensive redevelopment – council estates or prestige schemes, such as in the London dockland area.

Zone A$_2$ is being invaded by the upwardly mobile immigrant community. As younger people move to newer and better residential areas, the elderly are left behind.

2 The map below shows the net migration of black population in the USA, between 1960 and 1970.

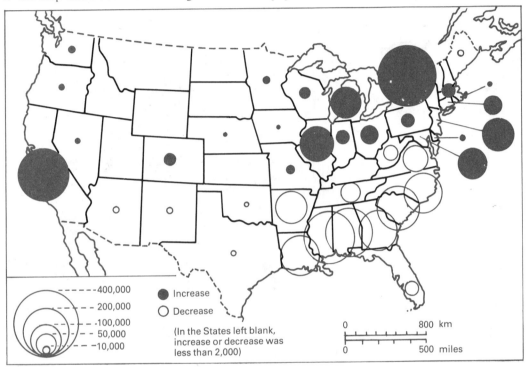

Fig. 19.2 Black population, net migration 1960–70

(a) Describe and comment on the regional variations in distribution shown.
(b) Discuss the economic and social implications of the migrant patterns described in (a).

(*WJEC, June 1985*)

Understanding the question This question falls into two equally important parts. An important point relating to (a) is that you are asked to consider *regional* variations, and not details of individual states.

Answer plan *Part (a)* There are three distinctive features as far as regional variation is concerned:
 (i) the traditional 'South' with Florida is a region of out-migration, with the deep south states experiencing the greatest movement;
 (ii) New England and the north-eastern states form the chief region of in-migration;
 (iii) California is distinctive in the rest of the USA in being a major centre of in-migration.
There is little or no movement in the remainder of the country.
    Your comments could include:
 (i) The black population is moving in large numbers from its traditional home in the South, where new technology has destroyed the intensive labour character of traditional agricultural activities.
 (ii) They are moving away from the states in which colour prejudice is deeply ingrained, and perceive more equal opportunities elsewhere.
 (iii) Despite the relative economic decline of New England and the north-eastern states, the region retains its long-standing magnetism for black people from the South.
 (iv) California is the new 'land of opportunity', and like the white population of the USA, the blacks are moving to what is now the wealthiest and most powerful state.
 (v) The small outward movement from the south-western states except California may reflect the fact that the influx of Spanish-speaking immigrants (Hispanics) in that area is making it difficult for unskilled blacks to find the low-paid jobs they traditionally did.
    Part (b) Basic points to make are:
 (i) The black problem is not a purely southern problem, but now has a national significance which has resulted in considerable improvement in opportunities for blacks in the last 20 years.

(ii) Many of the blacks who move to the north-east move to cheap housing areas in the inner cities. A counter movement of whites to the suburbs has increased the segregation between the groups and deprived the inner cities of the local tax income needed to improve the inner city.

(iii) Movement away from the South has reduced unemployment and deprivation in areas where traditional 'black' jobs have now disappeared. So the social and economic problems of these areas are being exported to other parts of the USA.

(iv) California now has the major social and economic problem of absorbing black immigration at the same time as it is failing to deal effectively with the illegal immigration of Hispanics across the Mexican border.

3 With reference to the model Asian city shown in Fig. 19.3 describe and account for the morphological changes that would be likely to follow its designation as the capital of a newly independent country.

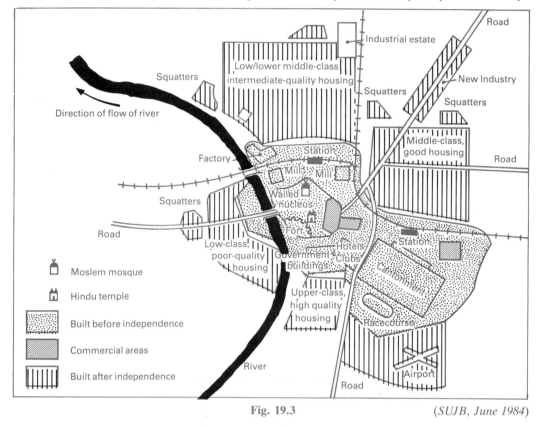

Fig. 19.3 (*SUJB, June 1984*)

Understanding the question The questions asks you to link your knowledge of urban land-use patterns in Third World cities with your understanding of the likely effects of a major change in function upon such patterns. You need to:

(a) draw on your knowledge of what has happened to capital cities of newly independent countries in Asia;

(b) select land use changes which are likely to occur as a result of the city becoming a capital;

(c) explain why becoming a capital has direct effects upon the spatial land-use pattern.

Answer plan The chief morphological changes are likely to be:

(a) The establishment of an area of government buildings – probably in historic buildings which may have housed the colonial administration. Since the new nation has wider governmental functions than a colony, the government buildings zone will either expand, or a new national administrative centre will be built.

(b) The upper-class housing area is likely to see replacement of senior colonial officers by members of the government. This zone will also have to expand, or a second zone will have to be established, to house the embassies of foreign powers.

(c) Newly independent countries acquire their own civil service and minor officials. These will occupy middle-class residential areas of varying quality according to income and status. The white-collar groups may replace ex-patriates who have returned to their own country, but again existing areas are likely to expand outwards, or new residential areas may emerge in suitable locations on the edges of the city.

(d) The existence of the airfield symbolizes the need for a newly independent country to establish its status in the world. Establishing a national airline and linking into prestigious international air links is likely to result in the redevelopment of the airport and extension of the airport complex.

(e) Most newly independent countries have seen industrial development as a means of increasing national wealth. The capital city in particular has to share in this process and, because of its facilities, it may be more attractive to foreigners. Consequently it may well be the major focus of post-independence industrialization. Existing modern industrial zones are likely to expand and new ones to appear at advantageous locations such as near the airport or along main roads.

(f) All cities act as magnets for in-migration, but capital cities offer additional attractions as they are likely to have more prestigious new jobs, better health and entertainment facilities, and better food and consumer goods supplies than less important and remoter urban centres. So there is a rapid growth of cheap housing and of illegal squatter settlements.

(g) Increased international links which result from independence, industrialization etc. means that there may be a substantial need for modern international-class hotels. A new hotel zone is likely to appear which is conveniently located in relation to the airport and centre of government.

(h) All these changes place tremendous strains upon the existing infrastructure – the road network in particular. A major new feature of the city may therefore be a redesigned road system.

Note: These changes are consistent with the operation of a relatively free-market, competitive economy. If the newly independent country becomes communist, the morphological changes will reflect the priorities established by the national government for planned development. Generally this means that high priority is given to the establishment of industry and to the provision of housing for the workers. At the same time, however, prestigious developments are seen as important status symbols which indicate the benefits of communism.

19.7 FURTHER READING

Bradnock, R. W., *Urbanization in India* (Murray, 1984)
Hall, P. J., *World Cities* (Weidenfeld and Nicolson, 1984)
Knox, P., *Urban Social Geography* (Longman, 1982)
Lawless, P., *Britain's Inner Cities* (Harper & Row, 1981)
Scargill, D. I., *The Form of Cities* (Bell & Hyman, 1979)

20 Agricultural land use

20.1 ASSUMED PREVIOUS KNOWLEDGE

Your GCSE geography course would not have covered agricultural land use in any great detail. In some schools land use maps are examined and the different categories of use are identified and discussed. These studies are sometimes connected with a local project in which one of the exercises involves the making of a local land use map. Land use is confined to straightforward descriptions of the natural and man-made landscapes to be found in such areas as granitic uplands, the Fens, the Vale of Kent and the North York Moors. In regional geography you will have studied the distribution patterns of some of the major farming types such as dairying, market gardening and cereal cultivation. You will also have studied the distribution of some types of farming such as wheat cultivation throughout the world. If you have not taken GCSE geography you should look at the relevant chapters of such books as Marsden, W. E. and V. M., *Britain* (Oliver and Boyd, 1987).

20.2 ESSENTIAL INFORMATION

20.2.1 Definitions

A level studies of agricultural land use are concerned with the processes which help to determine present-day patterns of farming. There is a particular emphasis on the part played by economic factors and the influence exercised by governments in deciding what is grown or produced on farmland. Some of the terms used by economists may be unfamiliar to you.

Locational rent The difference between the total revenue received by a farmer for a crop grown on a unit of land and the total cost of production and transport of that crop. Locational rent is not the same as the rent a farmer may be charged for a unit of land by the owner.

Intensity of agricultural production The greater the input of labour and capital on a unit of land, the greater the intensity of agricultural production.

Diminishing returns This economic law states that a certain point in production additional units of input will yield proportionately smaller units of output and the additional cost incurred will be greater than the additional revenue received. This can be represented as a graph (Fig. 20.1), where O – X represents the input and O – Y the net returns i.e. the returns which remain after the farmer has paid his production and transport costs. At first the curve rises but with increases in input a point is reached at Z beyond which the net return to the farmer declines and finally assumes negative values.

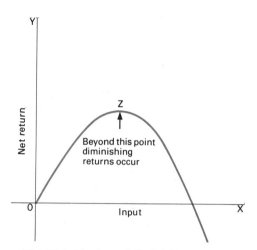

Fig. 20.1 The law of diminishing returns

Marginal farming A farmer whose total revenue only just covers his total costs is a marginal producer. If the farm is located in an area where total costs of cultivating the land just balance the total revenue it is said to be at the margin of cultivation. Beyond this margin farming would not be worthwhile. A margin also exists between growing different crops on the same land if costs and revenue differ for each crop grown. The net income for wheat on a parcel of land may be very little or nothing but there may be a large net income for using the same land for the rearing of cattle. In this case the land is marginal for wheat cultivation and the farmer is likely to transfer his capital and labour to the rearing of cattle.

20.2.2 Assumptions made by Von Thünen

Von Thünen, who farmed near Rostock in Eastern Europe in the 1820s, tried to establish why farmers behaved in a particular way in his locality. The result was a book called *The Isolated State* (1826), in which he attempted to explain how and why agricultural land use varied with distance from a market. Underlying his theories were a number of assumptions.
(a) An *isolated state*, that is land surrounded by an uncultivated wilderness.
(b) One central city as the sole market for the products of the land surrounding it.
(c) A uniform plain surrounding the city where fertility, climate and other physical factors do not vary.
(d) The plain is inhabited by farmers who supply the city.
(e) The farmers aim to maximize their profits and have full knowledge of the needs of the market.
(f) Transport is by horse and cart and the cost of transport is directly proportional to distance.

20.2.3 The Von Thünen models

Von Thünen introduced two models. The first was concerned with the intensity of production while the second examined the location of crops in relation to the market.

The intensity of production model is best explained by an example. Two farmers, Mr Green and Mr Brown cultivate the same crop, they have identical inputs and yields but Brown is located 20 kilometres from the market whereas Green is only 2 kilometres away. Assuming that the market price for the crop is £50 per tonne and the transport cost is £1 per tonne/kilometre, the farmers' locational rent can be calculated as in Table 20.1.

The transport cost is higher for Brown and this substantially increases his total costs. Green has a higher locational rent even though both farmers had the same production costs and yields.

If both farmers decide to increase the intensity of their production by doubling their production costs i.e. by using more labour and/or capital, yields will increase but the law of diminishing returns may apply and the increase in yields may be only 50 per cent compared with the 100 per cent increase in production costs. The two farmers' locational rents can be calculated as in Table 20.2.

By intensifying his production Farmer Brown is worse off than previously when his cultivation was more extensive i.e. inputs were lower. His returns will therefore be greater if he adopts his previous more extensive method of cultivation. By contrast, Farmer Green is better off after intensifying his production.

This example shows that other things being equal, the intensity of production of a particular crop will decline with distance from the market (Fig. 20.2).

Table 20.1 Locational rent (a)

	Farmer Green	*Farmer Brown*
(a) Distance from market	2 km	20 km
(b) Cost of production	£2000	£2000
(c) Yield	100 tonnes	100 tonnes
(d) Transport cost	£1 per tonne/km	£1 per tonne/km
(e) Total transport cost (a×c×d)	£200	£2000
(f) Market price	£50 per tonne	£50 per tonne
(g) Total cost (b+e)	£2200	£4000
(h) Total revenue (c×f)	£5000	£5000
Locational rent (h−g)	£2800	£1000

Table 20.2 Locational rent (b)

	Green	*Brown*
(a) Distance from market	2 km	20 km
(b) Cost of production	£4000	£4000
(c) Yield	150 tonnes	150 tonnes
(d) Transport cost	£1 per tonne/km	£1 per tonne/km
(e) Total transport cost (a×c×d)	£300	£3000
(f) Market price	£50 per tonne	£50 per tonne
(g) Total cost (b+e)	£4300	£7000
(h) Total revenue (c×f)	£7500	£7500
Locational rent (h−g)	£3200	£500

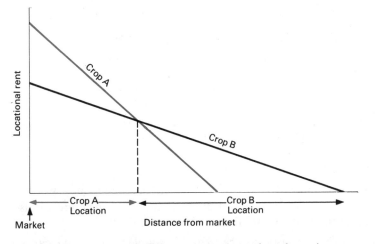

Fig. 20.2 Two crops with different transport and market prices

Von Thünen's second model In this model Von Thünen looked at the location of more than one crop in relation to the market. The location of different crops is determined by production costs, yields per hectare, transport costs and market prices. The crop with the highest locational rent will be grown since the return will be at its greatest and the farmer will maximize his profits. One example is shown above (Fig. 20.2).

If two crops A and B have the same production costs and yields but A has higher transport costs and a higher market price than B, A will be grown closer to the market than B.

The explanation for this is as follows. A has high transport costs so the locational rent for A will fall more sharply with distance than will the locational rent for B. As the market price is higher for A than for B the locational rent of A at the market will be higher than for B and A will be grown closer to the market than B.

In reality production costs, yields, transport costs and the market price vary between farm products. If the farmer does not grow the crop with the highest locational rent he will not maximize his profits and may find the farm is running at a loss.

The following formula will enable you to calculate the locational rent of a crop.

$$LR = Y (m - c - td)$$

LR – Locational rent per unit of land   c – Production cost per unit of product
Y – Yield per unit of land   t – Transport cost per unit of product
m – Market price per unit of product   d – Distance from the market

20.2.4 Spatial application of Von Thünen's models

Von Thünen combined his model of intensity of production with that for spatial variations in land use and applied them to his 'isolated state' using the assumptions listed in Section 20.2.2. Fig. 20.3 shows the theoretical pattern which would result. Nearest the city would be concentrated the production of vegetables and fresh milk because the products are perishable and the fertility of the land could be maintained by manure from the cattle and the city. Further away from the city timber would be cut. It is bulky and so has high transport costs and a high locational rent.

In the next three zones rye would be grown in varying degrees of intensity with the yield decreasing as the crop became more extensive.

Beyond these zones there would be livestock farming for products such as meat, butter and cheese.

Von Thünen also considered modified versions of this model. With a navigable river to speed up transport and reduce its cost the pattern of land use would tend to focus on the river. By introducing a smaller city with its own land use pattern the model became more complex but closer to reality.

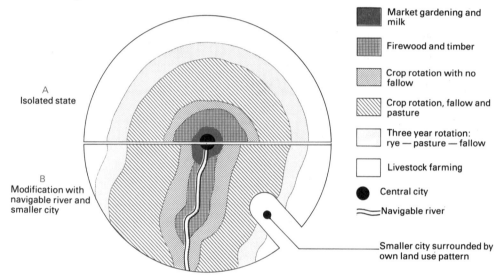

A
Isolated state

B
Modification with navigable river and smaller city

- Market gardening and milk
- Firewood and timber
- Crop rotation with no fallow
- Crop rotation, fallow and pasture
- Three year rotation: rye — pasture — fallow
- Livestock farming
- Central city
- Navigable river
- Smaller city surrounded by own land use pattern

Fig. 20.3 Von Thünen's agricultural zones

20.2.5 Limitation's to the Von Thünen model

The technological and communication improvements since Von Thünen's time have brought about changes which make the original assumptions out-of-date. At the same time there are also objections to the model which, with the changes, are listed below.

(a) There have been extensive changes in transport since 1826. Perishable goods can be carried long distances.

(b) Storage capacity, including refrigerated stores, has resulted in the possibility of keeping goods such as apples for months before they are sold.

(c) Pricing policies may encourage production away from the market. The Milk Marketing Board use differential transport charges to encourage more remote areas to produce milk.

(d) Cities no longer supply large quantities of manure and cheap labour for neighbouring farms. Labour has been largely replaced by machinery.

(e) The marketing of agricultural produce has changed drastically. Farmers sell much produce to the food-processing industry. Furthermore, governments have set up marketing agencies which help to control production by subsidizing the farmer.

(f) Soil fertility can still be a dominant factor in the production of a crop. Although there are three sugar-beet processing plants in Norfolk, sugar-beet is not grown extensively near them. Instead it is grown on the loams of north-east Norfolk and the silts of the southern Fens.

(g) Decisions made by farmers are not based on complete information. The farmer achieves what to him appears to be a satisfactory level of returns. This is a satisficer solution (see Section 18.3), which is dependent on two factors – the level of knowledge of the farmer and the level of uncertainty or risk in the production.

(h) Economies of scale tend to extend the area under one crop.

(i) R. Sinclair has made the suggestion, based on field evidence in the Mid-west of the USA, that the Von Thünen zonations should be inverted so that the intensity of agricultural activity increases with increasing distance from the city. He argues that cities are expanding rapidly and because of anticipated expansion very little, or no investment is made on land close to them. He suggests that the inner zone should be labelled *land speculation,* and that moving from the centre the other zones should be *vacant grazing, field crops and grazing, dairying and field crops, specialist field grains* and *livestock.*

20.2.6 The present-day significance of Von Thünen's model

A number of geographers and economists have tested Von Thünen's agricultural zone theory in the field and support his basic concepts. Michael Chisholm in *Rural Settlement and Land Use* (Hutchinson, 1966) cites a number of examples from different parts of the world where zoning takes place around villages. He shows that distance, irrespective of natural fertility, exercises a strong control over intensity of cultivation and that fertility is likely to be highest near a village where manure is available. Other studies have identified zones along the coast of New South Wales and around Hamburg.

J. R. Peet argues in 'The spatial expansion of commercial agriculture in the nineteenth century: a Von Thünen interpretation', *Economic Geography 45* (1969) that the developed world of Western Europe and the north-eastern United States forms a world city with zones of decreasing intensity surrounding the highly developed area. S. Van Valkenburg and C. C. Held in *Regional Geography of Europe* (Wiley, 1952) show that the average yield of eight crops in Europe follows a concentric pattern, declining away from the central market (Fig. 20.4).

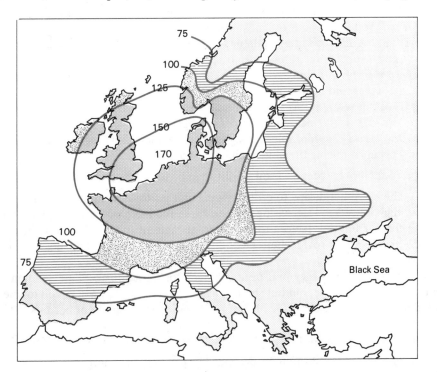

Fig. 20.4 Zones of decreasing agricultural intensity

Various studies of farming in the less developed world suggest that conditions may be similar to those in the Rostock area investigated by Von Thünen (see R. J. Horvath, 'Von Thünen's isolated state and the area around Addis Ababa', *Annals of the Association of American Geographers,* 59, 1969).

Von Thünen's analysis is evidently still significant. He postulated a normative pattern of land use, that is one which may be reasonably expected given a number of stated premises.

20.2.7 Government influences on agricultural land use

Governments can indirectly influence what the farmer grows by tariffs, import quotas and other forms of import control. These controls are aimed to protect high cost producers from low cost imports. For example, before Britain joined the European Economic Community there was a tariff of 0.83p per lb on imported lamb.

The government's deficiency payment system which operated in the 1950s and 1960s, together with guaranteed prices and markets for certain farm products such as fat cattle, fat pigs, milk, eggs, potatoes and cereals, meant that farmers at the margin of production for these commodities remained profitable and land which would otherwise have remained unproductive was utilized.

The EEC has a complex Common Agricultural Policy which includes guaranteed prices and subsidies to member countries. However, membership of the EEC means that trade barriers with other member countries must disappear. In Britain this means for example, that apple producers are no longer being shielded from the highly efficient French growers. French orchards produce 16 tonnes of apples per acre compared with only five tonnes from British orchards. This may reduce the area in Britain used for the growing of apple trees.

Other forms of government influence in the production of specific crops have included the creation of *soil banks* in the United States and the encouragement of milk production in Britain. In the 1950s and 1960s the Federal Government, concerned by the increase in crop surpluses, particularly of corn, barley and oats, introduced the acreage reserve scheme which enabled farmers to be paid for placing land previously used to grow crops in surplus supply in the 'bank'. The scheme did not achieve its purpose because farmers deposited their poorest land and continued to grow crops more intensively on the more fertile land, increasing the surpluses still further. In Britain the government encouraged milk production by setting up the Milk Marketing Board and maintaining a national pricing policy which enables small farms to remain in business and encourages dairy farming in remote areas where transport costs are high.

20.2.8 Physical factors influencing agricultural land use

The emphasis in this unit has been on economic and governmental factors as determinants of agricultural land use but it is important to remember that physical factors also play a significant part. A full account can be found in Symons, L., *Agricultural Geography* (Bell, 1978) and only a summary of the factors which are involved will be given here. There are three main types of physical factors, soil, relief and, the most significant, climate.

Climate (a) The amount of water available for plants and animals is highly significant. Requirements vary and evaporation rates must be considered as well as the amount and seasonal distribution of the precipitation.

(b) There are threshold temperatures (5°–6°C for wheat), below which the crop will not germinate. Average temperature requirements in the growing season vary from crop to crop and limiting factors such as the incidence of frost are significant.

(c) Winds can cause considerable damage to crops and where conditions are suitable soil erosion may occur, reducing the land available for agriculture.

Soils An account of soil types and soil fertility can be found on pages 32–39.

Relief (a) There are handicaps to crop growing and pastoral farming at high altitudes just as there are at high latitudes. Decreasing temperatures and increasing rainfall, humidity and wind speeds are further limiting factors to be found at high altitudes. However, in tropical areas increased altitude may provide better conditions for agriculture than nearby low-lying and coastal regions.

(b) Slopes provide advantages and disadvantages for agriculture. Slope gradients may limit cultivation and soils may be thin but slopes facing towards the sun where soil temperatures are increased by the sun's angle may be ideal for cultivation if the gradient permits.

20.3   GENERAL CONCEPTS

Economies of scale are important in present-day production There are savings in costs at a certain size of enterprise; these are known as scale economies and occur in larger units of production. The net effect of economies of scale in farming is a tendency towards regional specialization of some activities.

Man is a satisficer rather than an optimizer This concept, fundamental to behaviourist thinking, conflicts with the concept of economic man. It considers that instead of trying to maximize their profits, people strive to achieve a satisfactory level and establish a pattern and routine of activity which provide them with this level. They do not strive for more. This is the satisfactory situation and man is the satisficer. This state of equilibrium persists in farming until a new technological advance is adopted. This disrupts the routine of activity of the farmer who now tries to achieve a new level.

A general 'decay' in agrarian economic activity increases in proportion to the distance from the city (which is the market for the products of its hinterland). This is the distance-decay concept. In the Von Thünen model the intensity of agricultural activity decreases with increasing distance from the city. Yields fall and the capital and labour input decreases with increasing distance. This distance-decay concept can also be applied to a number of other aspects of social geography including urban shopping habits and land use values within cities.

20.4 DIFFERENT PERSPECTIVES

The ideas of Von Thünen and recent supporters of his theories have been described in Sections 20.2.2 to 20.2.6. The inversion of the distance-intensity relationship by R. Sinclair in 'Von Thünen and urban sprawl', *AAAG 57* (1967), is worth careful examination, although its universal application can be questioned (see Section 20.2.5 (a)).

Many geographers now challenge the concept of *economic man* and have developed theories which recognize that farmers have imperfect knowledge and are not always guided by the need to maximize profits. Once a satisfactory income has been achieved a farmer might be interested in more leisure time and less profit.

One problem being explored   *the basis upon which farmers take decisions*. This introduces such concepts as the perception of farmers – how far they are aware of such environmental influences as storm hazards to their crops, or the risk of drought. Another aspect of farmers' decision-making skills is *their awareness of new ideas and innovations,* such as improved seeds and new machinery. The study of the diffusion of innovation shows that distance is significant in the spread of new ideas. You will find a detailed summary of these behavioural elements which influence agricultural land use in Tidswell, V., *Pattern and Process in Human Geography* (UTP, 1976).

20.5 RELATED TOPICS

The farm, like a factory, is a risk-taking business and a number of economic laws such as the laws of supply and demand and of diminishing returns apply to agriculture just as much as they do to industry. Von Thünen's attempt to explain the location of different types of farming is similar to the theories relating to the distribution of manufacturing industry put forward by Weber and others.

One aspect of agricultural geography which is closely related to land use theory is the identification of agricultural regions. Attempts to do this by O. E. Baker and others are described in Symonds, L., *Agricultural Geography* (Bell, 1967).

When you are studying agriculture in underdeveloped countries you may find that you can apply Von Thünen's model in some areas. This is because the transport facilities and relative isolation may bear a close resemblance to the conditions which Von Thünen observed around Rostock.

20.6 QUESTION ANALYSIS

1 How far do you consider that Von Thünen's assumptions about agricultural land use and distance from the market can be applied to present-day rural land use patterns?

(in the style of London)

Understanding the question To answer this question you must fully understand what Von Thünen's assumptions were and remember the details of his agricultural zones model. The model is obviously out-of-date in such details as the zone of firewood close to the city centre, but the broader concept that land use is related to distance from the market is still relevant. In order to answer the question you must know examples which show Von Thünen's model is close to reality in some areas today. These examples are to be found in case studies carried out by geographers in different parts of the world (see Section 20.2.4). You should also be aware of the limitations of the model.

Answer plan Start with a short introduction explaining what Von Thünen set out to accomplish in his book *The Isolated State*. The underlying assumptions made by Von Thünen should be listed and explained (see Section 20.2.3). Draw a diagram similar to Fig. 20.1 to explain Von Thünen's concentric rings. Do not become involved in the minor modifications which he introduced.

A paragraph should then follow pointing out that technological changes in energy production, transport, refrigeration and communications have modified the patterns seen by Von Thünen 160 years ago. Nevertheless the model can be applied today and examples should be given which include:
(a) The concept of viewing Western Europe and the north-east United States as central cities with concentric zones of farming on a continental scale (see Section 20.2.6).
(b) Case studies carried out in the less-developed lands of India, Brazil, Africa and elsewhere.

A final paragraph should state how far you consider Von Thünen's assumptions can be applied today as well as some of the limitations of the model (see Section 20.2.5). Point out that it is a model of only partial equilibrium, it does not take into account non-economic factors and it does not consider differences in the scale of the central city.

2 'The significance of climate and soil differences in determining the pattern of British farming today is far less than the effects of market and government influences.' Discuss.

(Oxford and Cambridge, July 1977)

Understanding the question This question consists of a statement which may be true, partly true, or without any foundation whatsoever. It is reasonable to presume that the statement is at least partly true. What your answer must make clear is how far you agree with the statement. The question does not categorically state that the market and government influences determine the pattern of British farming, but that climate and soil differences have *far less* effect. Your answer should, therefore, refer to examples where physical factors continue to determine the farming pattern. It would be wrong to assume that the examiner expects you to answer the question by repeating all you know about the Von Thünen model. The model can be referred to when explaining market influence but it should not become a major part of your answer.

Answer plan Introduce your answer by accepting that the market and government influences play an important part in determining the pattern of British farming but emphasize that other factors such as climate and soil are also important, particularly for certain crops.

Write a section on market influences, indicating that Von Thünen's theory can be applied on a limited scale to some perishable market garden products such as tomatoes and lettuces. Similarly crops for freezing, such as peas, must be grown close to their market, i.e. the processing plant which buys them from the farmer.

A section on government intervention should follow which explains that the government can influence production both directly by price fixing and acreage or output controls, as well as indirectly by import controls and improving marketing channels. Give examples of the work done by such bodies as the Milk Marketing Board (see Tarrant, J. R., *Agricultural Geography* (David and Charles, 1974). The question uses the words *government influences* but it is worth mentioning that the government may have limited powers when confronted by European Economic Community regulations to which it must subscribe.

A further section is needed giving examples of when climate and soils can determine the crops grown, for example barley cultivation in eastern and south-east England, celery in the Fens, cauliflowers in the south and west and sugar-beet in north-east Norfolk.

A final paragraph should point out that on a national scale climate and soils are still important, but where crops are interchangeable, for example barley or wheat, wheat or sugar-beet, then the economic factors of the market and government influence may determine what the farmer grows.

3 Discuss the factors which have given rise to the pattern of rural land use in an area of less than 250 square kilometres which you have studied. Illustrate the land use pattern with a labelled sketch map.

(in the style of the Joint Matriculation Board)

Understanding the question First you must appreciate the size of the area involved. It is less than 10 km by 25 km (6 miles by 15 miles). The 1 : 50 000 sheets of the Ordnance Survey cover an area 40 km by 40 km (1600 km²) or about seven times the area to be considered.

The question requires a labelled sketch map. This is best presented as a detailed land use map of the area using suitable categories such as those used by the Second Land Use Survey organized by A. Coleman (see Coleman, A. and Maggs, K. R. A., *Land Use Survey Handbook* (Isle of Thanet Geographical Ass., 1961). The sketch map should be reasonably detailed with an explanatory key. If there are marked soil changes in the area covered or distinctive physical features such as hill slopes which influence the land use pattern, they should also be shown on the map. Since you will refer to the map in your answer it should be set in a grid so that locations can be quickly found. You can invent your own grid if you do not know the official grid numbers. Apart from the map the question requires a detailed account of the factors which have given rise to the rural land use pattern which the map displays.

Answer plan Draw an annotated sketch map of the area you are discussing. Do not forget to add a scale line, compass direction and a key. Give the map a title so that the region can be clearly identified by the examiner, for example *Land use to the south-west of Evesham, Worcestershire*.

Describe the main distribution pattern shown on the map and then proceed to account for the pattern, bearing in mind that some, if not all of the following factors may be responsible for the distribution you have described.

(a) Physical factors – climate, soils and relief.
(b) Economic factors – distance from market, cost of labour, government influence.
(c) Behavioural factors – perception and decisions of farmers.

As far as possible attempt to examine each crop or farm product shown on the map in turn and explain which factor or factors have helped to decide the amount and distribution of that product in the area which you are concerned with. To do this adequately will probably involve describing recent changes in land use and placing the area in the wider context of the region in which it is located.

20.7 Further reading

Bayliss-Smith, T. P., *The Ecology of Agricultural Systems* (Cambridge, 1982)
Bradford, M. G. and Kent, W. A., *Human Geography. Theories and their applications* (OUP, 1977)
Coppock, J. T., *Agriculture in Developed Countries* (Macmillan, 1985)
Horsfall, D., *Agriculture* (Blackwell, 1983)
Symons, L., *Agricultural Geography* (Bell, 1978)
Tarrant, J. R., *Agricultural Geography* (David and Charles, 1974)
Tidswell, V., *Pattern and Process in Human Geography* (UTP, 1976)

21 Location of industry

21.1 ASSUMED PREVIOUS KNOWLEDGE

The study of industrial location for GCSE consists of a brief examination of the factors which influence location such as raw materials, power supply, labour, the market, transport and the attitude of the government. These general factors are seen to determine whether individual factories are located near their raw materials, the market or at some intermediate point. Specific industries, e.g. steel, textiles, oil refining, are used as examples, usually in connection with areas studied for regional geography.

21.2 ESSENTIAL KNOWLEDGE

21.2.1 Definitions and formulae

A level studies of industrial location are concerned with classifying industries, analysing their distribution patterns to measure such things as industrial concentration, and examining the attempts which have been made to develop models of industrial location.

Material index This measures the loss of weight during processing by comparing the weight of the raw material with the weight of the finished product.

$$\text{Material index} = \frac{\text{weight of localized raw material inputs}}{\text{weight of finished product}}$$

The more the index exceeds 1 the greater the significance of the cost of moving the raw material to the factory.

Location quotient This measures the degree of concentration of an industry in a particular area. It is obtained by using the formula

$$LQ = \frac{\dfrac{\text{number of people in industry A in area X}}{\text{number of people employed in all manufacturing in area X}}}{\dfrac{\text{number employed nationally in industry A}}{\text{number employed nationally in all manufacturing industry}}}$$

If 2000 people are employed in industry A in area X out of a total local workforce of 40 000 then 5 per cent of the local workforce is employed in industry A.

If 100 000 are employed nationally in industry A and 1 000 000 are employed nationally in all manufacturing industry then 10 per cent of the national workforce is employed in industry A but in area X the proportion is only 5 per cent.

$$LQ = \frac{5}{10} = 0.5$$

An LQ of more than 1.0 reveals that the region has more than its share of a particular industry. Conversely, a value of less than 1 indicates that it has less than its share.

Industrial linkage The operational contacts which exist between separate industrial firms. These contacts are strongest in firms which are pursuing the same kind of process or participating in a sequence of operations.

Some firms perform one stage in a series of operations to make a particular product, this is known as *vertical linkage*. For example, in the non-ferrous metal industry one firm refines the metal, another shapes it, another machines it and so on until the product is finished.

Horizontal linkage is common in the automobile industry where many components of a car, such as the battery and tyres are made by specialist firms and then assembled at the automobile plant.

Diagonal linkage occurs when a firm makes a product, or provides a service which is part of a chain of processes, but the firm is not supplying one plant as in vertical linkage. Instead a variety of separate plants are supplied. An example is a firm which makes plastic mouldings required by a number of other firms in the district.

Firms may obtain benefits from local services and such things as a local pool of specialist labour. These firms are not necessarily linked functionally, they have in common certain services or skills which may not be available in other areas, for example the cutlery industry of Sheffield which has *common roots* in the district.

Industrial inertia Some industries continue to survive in an area where the cost benefits they once enjoyed no longer exist. An example is the continuation of textile machinery manufacture in New England, even though most textile mills are now located elsewhere.

21.2.2 Advantages and disadvantages of industrial concentration

Advantages These can be summed up as similar industries having similar needs, for example: (a) a local pool of skilled labour; (b) local specialist trade associations; (c) availability of local services such as cleaning and maintenance; (d) local financial services and expertise which understands local requirements; (e) local research and educational facilities; (f) a specialist quarter where valuable links with other firms can be established; (g) components bought in bulk may be cheaper because the supplier is also supplying other local firms. This factor and (f) are known as external economies.

Diseconomies of concentration Although firms may find costs are lower if they are located close to similar firms, there are also diseconomies, for example: (a) the prices of factors of production may be increased by intense local demand; (b) labour may be strongly unionized; (c) services and amenities may have costs which are excessive; (d) transport congestion.

21.2.3 The costs of production

Costs of production can be summarized as follows.

Labour costs These vary from place to place; their supply and productivity can also vary.

Entrepreneurship The skills of the entrepreneur are more likely to be available in large cities. Managers may also have locational preferences based on such things as their personal life styles.

Capital Costs of building vary; small firms cannot obtain capital easily outside their own area.

Energy Some firms require vast quantities of energy e.g. aluminium producers. However, the national grid makes supplies of electricity widespread in the UK.

Raw materials Improved technology may reduce costs and less raw material may be required. The cost of extracting ore from the ground is partly determined by the amount of waste which is involved.

Transport costs There are two types, line haul charges and terminal charges. Various rates are imposed e.g. mileage rate, blanket rate with stepped charges, 'postage stamp' rate i.e. same charge over any distance.

Land costs Local variations in land costs can be considerable.

21.2.4 Weber's model of industrial location

Initial premises (1909)

(a) Homogeneous area in terms of climate and topography.
(b) Conditions of perfect competition with large numbers of buyers and sellers.
(c) Some raw materials such as water and sand are ubiquitous, others are localized.
(d) Labour is available at fixed locations.
(e) Transport costs are dependent on weight and distance.
(f) Markets occur at specified fixed points.
(g) Man is an 'economic' animal. People tend to seek locations at which lowest costs are incurred. At such locations the highest profits will be achieved.

Weber's model illustrated by a locational triangle It is possible to illustrate some aspects of Weber's model by using a locational triangle (Fig. 21.1).

Assume two raw materials RM1 and RM2. 1 tonne of RM1 combines with 3 tonnes of RM2 to make a product weighing 2 tonnes which is consumed at A. In the diagram each corner of the triangle exerts a force proportional to the weight attached to it. The optimum location for the firm will be at OL which is nearer to RM2 than RM1 because it is cheaper to transport raw material from RM1 than from RM2. OL will be nearer to RM1 and RM2 than to A because of the loss in weight before the product is sold at A.

Isodapanes These are lines joining places with equal total transport costs. An isodapane is shown in Fig. 21.2; the concept was introduced by Weber. In Fig. 21.2, A represents the market and RM the raw material source for an industry with one raw material. It costs twice as much to transport the raw material from RM as it costs to transport the finished product from A. If the contour interval is the same for both sets of costs, lines drawn around RM will be closer together than those drawn around A.

At the intersecting points CDEFG of these lines total transport costs are the same e.g. $C = 2 + 5 = 7$; $D = 3 + 4 = 7$; $E = 4 + 3 = 7$. If these points are joined an isodapane has been

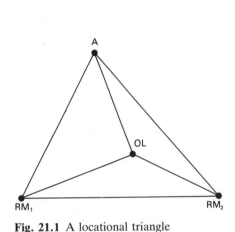

Fig. 21.1 A locational triangle

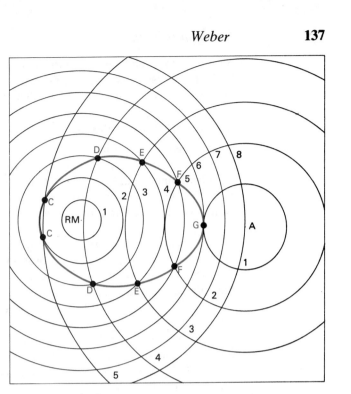

Fig. 21.2 An isodapane

formed, that is a line joining places where the total transport costs for raw material and product are the same.

Labour costs Weber also recognized that the least transport cost location could be modified by a pool of cheap labour. This would be particularly true for industries where labour cost ratios were high when compared with the costs of the combined weights of material inputs and product outputs. In these circumstances the location of an industry would be pulled towards the pool of cheap labour, provided the savings from using cheaper labour were greater than the extra transport costs incurred in marketing the finished product.

Weber devised an index of labour costs; this is the average cost of labour needed to produce one unit weight of output. The higher the index, the more likely is the industry to move away from the least transport cost location if a pool of cheap labour becomes available elsewhere.

Agglomeration and deglomeration Weber also stated that the least transport cost location might be rejected in favour of a location where there were cost savings resulting from the spatial association of industries. This grouping of industries in a specific area is known as agglomeration (see above).

Although there are economies arising from industries concentrating in one area there are also diseconomies. In recent years congestion in large industrial cities and the high price of land has encouraged many industries to leave the cities and find locations in less congested areas. This is known as deglomeration.

21.2.5 Weaknesses in Weber's model

(a) Perfect competition is an unrealistic concept It assumes that demand is constant irrespective of distance from the plant. However, increased transport costs will increase prices as distance from the plant increases. When this happens demand will decrease accordingly.

(b) The model does not allow for possible spatial changes in the supply of raw materials or demand for the finished product The supply of raw materials is rarely from a fixed point, a number of alternative sources of supply are available to the manufacturers. Weber also located the market at a fixed point but in reality the market for a finished product is scattered. Furthermore demand is not constant but varies from place to place.

(c) Transport costs are not directly proportional to distance, instead they tend to be stepped, rising suddenly at certain points Moreover, transport costs make up a relatively small part of total costs of production for modern industry.

(d) Labour is not fixed but mobile Weber's assumption that labour is immobile has been weakened by the growth of transport facilities and the movement of the unemployed to find work elsewhere in times of depression. Mobility is however limited by such things as the need to learn new skills, family ties and lack of funds to move. These limitations tend to support Weber's assumption.

(e) 'Economic man' does not exist Many decisions are taken on a personal rather than a rational basis. Businessmen will choose satisfactory locations which enable them to operate at a profit, not necessarily the maximum profit. They are satisficers not optimizers.

21.2.6 Market area analysis

Weber's assumption that demand (i.e. the market), was centred on one point is unreal since demand, in practice, is spread over a wide area. A German economist, August Lösch, introduced the market area concept, that is the optimum marketing area for firms in competing industries in a given locality. He suggested that large volumes of sales could enable the manufacturer to obtain profits which would be sufficiently large to offset possible high transport costs.

For his model he assumed an isotropic plain, that is a uniform land surface with an evenly distributed population of farm households each demanding identical goods. A number of producers of, for example, beer, located in this region would set up breweries which would serve the population for a distance around the plant with the price increasing away from the brewery. The market area for each brewery would be a circle with demand greatest at the centre and diminishing with distance as transport costs increase the price (Fig. 21.3). A series of these trade areas will develop (Fig. 21.4A). Beyond these trade areas would be potential markets with no breweries which would encourage new producers to enter the market until the circular trade areas touch each other leaving small unserved areas in between (Fig. 21.4B). The most efficient shape for the market area is a hexagon (Fig. 21.4C), as this shape will give each brewer a monopoly over an area and leave no part of the region without a brewery.

If other products are introduced into the model each will have a market area of a different size depending on the importance of transport costs and the significance of economies of scale. The different marketing areas will produce a system of networks which will form an economic region or landscape.

Lösch went on to show that by rotating these networks around a common producing centre there will be sectors where production is concentrated containing a wide range of activities, and sectors where production will be more dispersed. These ideas link closely with the work on central place theory discussed in Unit 17.

Lösch went on to modify his model by introducing situations from the real world. In his theory, however, he ignored the behavioural aspects of locational choice. Instead he used the 'economic man' concept which is unrealistic. He also ignored the situation when competing producers locate close to one another. The reasons why this may happen are discussed below.

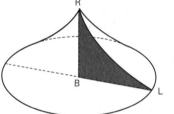

B Brewery
BR Demand at B
L Limit of demand
BLR Total volume of sales

Fig. 21.3 Demand in a market area: a three dimensional demand cone

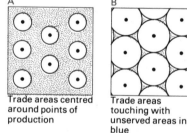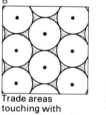

Trade areas centred around points of production

Trade areas touching with unserved areas in blue

Hexagons represent the most efficient trade areas

Fig. 21.4 Trade areas

21.2.7 Locational interdependence

Like the market area analysis approach, locational interdependence is concerned with the impact of demand upon location. Cost factors are ignored but entrepreneurial decisions are introduced to provide a behavioural aspect to the model.

The best known model was put forward by H. Hotelling in 1929. He assumed that there were two competing suppliers each having the same production costs and each capable of supplying the entire market with identical products. Demand is assumed to be totally inelastic (not affected by price changes), and the only variable cost is transport which varies in cost directly with distance.

If this situation is described by taking the example of two ice-cream salesmen on a narrow strip of beach, the location of the ice-cream stalls will significantly affect each salesman's profits. Each will have a monopoly over a certain market area, which in this case is linear, and at some point the two market areas will meet, just as the edges of the hexagons meet in Fig. 21.4C. A number of alternative positions are possible along the beach for the location of the two ice-cream stalls. They could, for example, be close together, or at either end of the beach or at the mid-points of each half of the beach. The optimal solution given the assumptions listed above is for each seller to be at the mid-point of his market. At this point transport costs for customers are minimized and sales are maximized (see D281 II 3–4, *Economic Geography – Industrial Location Theory*, pp 35–39, The Open University Press, 1972).

In reality the situation is much more complex. For example, there are likely to be more than two sellers and the population will not be evenly distributed. Centres of population will attract sellers and buyers. This results in further concentration of population and industries until diseconomies set in which will result in deglomeration with new clusters of population and industries arising in other areas (see Section 21.2.2).

21.3 GENERAL CONCEPTS

Some useful concepts are described in Section 21.2.

One of the main features of manufacturing industry is regional concentration There are certain advantages to be gained by firms which are in close proximity to each other.

The fully comprehensive location model has been partially replaced by an appreciation that:

Final decisions on location are made by businessmen who have imperfect knowledge of the cost-benefits involved or who have personal reasons for their choice.

National planning policy may also play a part in the location of industry.

Social needs may determine industrial locations in a free society and governments have sometimes intervened to redistribute industry from wealthy regions to those in need.

21.4 DIFFERENT PERSPECTIVES

Like Von Thünen with his rural land use, Weber framed his model in an isolated state with transport and labour costs to determine the location of an individual firm. Many of Weber's assumptions are unreal but his ideas highlight the importance of transfer costs and the possible different orientations of industries to materials, labour and markets.

Since Weber's time there has been a greater emphasis on location under conditions where there is not perfect competition. Lösch, for example, attempted to identify the optimum market area for firms in competing industries.

It is worth remembering that the satisficer principle gives a more realistic approach to industrial location.

Finally, do not forget that one-third of the world's population lives in countries which are organized as planned economies under Communist governments. By centralized planning Marxist societies eliminate the market forces which operate in capitalist countries. However, even in centrally planned economies it is necessary for decision makers to draw up lists of priorities. Although central planning on a Communist scale does not exist in capitalist countries there is a growing tendency for governments to intervene in industrial location, encouraging developments in some regions and discouraging them in others.

21.5 RELATED TOPICS

There are some similarities between the basic premises on which Von Thünen based his model of rural land use and the work by Weber on industrial location. There are clear parallels between the forces which influence decision making in industry with those in agriculture, but manufacturing is more complex in that the scale may be of international size. Furthermore, the product of one firm, such as a factory making car components, is the raw material of another, i.e. car assembly. This interdependence of units in industry has no parallel in agriculture.

In your studies of central place theory you will find close connections between the hierarchies developed by Christaller and the distribution of market-orientated industries described by Lösch. Large cities also attract large-scale industries and a greater range of industries than smaller towns. Just as some urban centres can support a range of chain stores, so cities of different sizes are likely to attract specific types of industries. There are exceptions to this rule and specialization may depend on natural resources, but the basic concept remains.

Studies of regional problems in Britain or other developed industrial countries which you may undertake during your A level course are also closely connected with studies of industrial location, unemployment and evidence of industrial inertia.

21.6 QUESTION ANALYSIS

1 The structure of manufacturing employment in Great Britain has changed considerably since the Second World War.

(a) Table 21.1 shows the employment in the textile industry and the paper, printing and publishing industry for the period 1921–81. On a single sheet of semi-logarithmic graph paper, draw two separate line graphs, one to show the changes in employment in the textile industry and the other to show the changes in employment in the paper, printing and publishing industry for the period 1921–81.

(b) Table 21.2 shows the regional employment in the textile industry, the paper, printing and publishing industry and all manufacturing industry for 1961 and 1981. Complete Table 21.3, which shows the location quotients for textiles and paper, printing and publishing, by calculating the six remaining location quotients.

(c) (i) Contrast the rate of change of employment in the textile industry with that in the paper, printing and publishing industry, as shown by the line graphs you have drawn.

(ii) Describe the major variations in employment in the textile industry and the paper, printing and publishing industry as shown by the completed table of location quotient values (Table 21.3).

(d) Indicate the usefulness and limitations of location quotients in assessing changes in the regional distribution of industry with time.

Table 21.1 Great Britain: employment in two major occupations (totals in thousands)

Industry	1921	1931	1941	1951	1961	1971	1981
Textiles	1308	1116	No data: war year	986	790	591	323
Paper, printing and publishing	378	436		515	605	612	510

Table 21.2 Great Britain: manufacturing employment by region (totals in thousands)

Year	Industry	South East	East Anglia	South West	West Mids	East Mids	Yorks and Humb	North West	North	Wales	Scotland	Great Britain
1961	Textiles	39.0	3.2	13.2	37.5	126.7	191.5	246.6	17.6	16.4	98.2	789.9
	Paper, printing and publishing	298.2	11.4	35.0	32.0	27.6	36.0	82.4	15.4	10.1	56.8	604.9
	All manufacturing	2422.7	139.1	377.7	1154.9	670.9	822.9	1338.7	439.1	300.1	712.3	8378.4
1981	Textiles	19.6	2.6	8.5	16.6	85.4	65.0	64.2	11.0	8.5	41.7	323.1
	Paper, printing and publishing	225.4	19.7	32.6	30.7	30.6	33.2	63.7	21.3	13.3	39.2	509.7
	All manufacturing	1683.5	186.0	395.7	800.7	533.4	578.9	799.8	339.4	238.2	502.0	6057.6

Table 21.3 Location quotients (Q) for textiles and paper, printing and publishing

Industry	Year	South East	East Anglia	South West	West Mids	East Mids	Yorks and Humb	North West	North	Wales	Scotland
Textiles	1961	0.17	0.24	0.37	0.34	2.00	2.47	1.95	0.43	0.58	1.46
	1981	0.22	0.26	0.40	0.39		2.11		0.61	0.67	
Paper, printing and publishing	1961		1.14	1.28	0.38	0.57	0.61	0.85	0.49		1.10
	1981	1.59	1.26		0.46	0.68	0.68	0.95	0.75	0.66	0.93

The location quotient (Q) may be calculated by using the formula $Q = \dfrac{X_1/Y_1}{X/Y}$

where X_1 = number employed in a given industry in the region
Y_1 = total number of all manufacturing employees in that region
X = number employed in a given industry in Great Britain
Y = total number of all manufacturing employees in Great Britain.

(*Joint Matriculation Board, May 1986*)

Understanding the question If you have never used semi-logarithmic graph paper, or do not fully understand what a location quotient is, do not attempt this question! Your mathematical skills may tempt you to use the location quotient formula which is provided to calculate the six remaining L.Q.s, but in fact this part of the question is only worth six marks. Failure to give appropriate answers to the other parts of the question would result in a maximum of 6 out of 20, which is a fail mark.

Do not be put off by the three tables of statistics. The largest (Table 21.2) is only used to extract the essential statistics required to complete Table 21.3. When using semi-logarithmic graph paper, remember that the logarithmic axis is divided into 'cycles' and you must first look at the range of values needed in order to select the appropriate number of cycles and number the axis. An explanation of how to use logarithmic paper can be found on page 117 of D.I. Smith and P. Stopp, *The River Basin,* Cambridge University Press, 1978.

Answer plan When you have plotted the two line graphs, you will find that they have different shapes. You should understand the significance of these differences. The distinction will help you to answer part **(c)** (i), which is concerned with the rate of change in the two industries. The location quotient calculations are straightforward. For example, using the formula and figures provided, the L.Q. for textiles in the East Midlands in 1981 is:

$$Q = \frac{85.4 \div 533.4}{323.1 \div 6057.6} = \frac{0.160}{0.053} = 3.02$$

When answering part **(c)** (ii), deal with the situation in 1961 and then in 1981 in each industry separately. For example, the L.Q.s for textiles show that in 1961 the East Midlands, Yorkshire and Humberside, the North West and Scotland had more than their share of the industry, i.e. had L.Q.s greater than 1. In 1981 the same areas continued to dominate the industry, but the East Midlands had increased its share considerably, whereas Yorkshire and Humberside and the North West had declined in importance. Scotland showed a slight increase.

The location quotient indicates the degree of concentration of an industry, based on the numbers employed. It does not show whether the industry is expanding or contracting, nor does it tell us whether productivity in one region is greater than in another. To understand more about the structure and significance of an industry, quotients are also needed for such things as the number of factories, value of production and average wages.

21.7 FURTHER READING

Bale, J., *The Location of Manufacturing Industry* (Oliver and Boyd, 1981)
Clark, G., *Industrial Location* (Macmillan, 1983)
Estall, R. C. and Buchanan, R. O., *Industrial Activity and Economic Geography* (Hutchinson, 1973)
Hoare, T., *The Location of Industry in Britain* (Cambridge, 1983)
Paterson, J. H., *Land, Work and Resources* (Arnold, 1972)

22 Transport and transport networks

22.1 ASSUMED PREVIOUS KNOWLEDGE

Few GCSE syllabuses include a section on transport. Those that do usually relate the various types of transport to the countries or regions being studied, or to the importance of transport in the study of industrial development.

The geography of the British Isles is one area where some attention is paid to communication networks and methods of transport. Questions are frequently set on the road and rail networks shown on the Ordnance Survey map extracts which are compulsory questions for many examining boards. Transport questions on other countries appear on the regional papers. They may require details of a route and an explanation of the significance of a particular routeway, or an exploration of the relationship between routeways and industry.

To sum up, transport geography for GCSE is mainly concerned with the distribution of routeways and their significance to the regions through which they pass. Without this background knowledge you will find the A level approach, which is partly theoretical, difficult to understand. If you have not studied transport for GCSE geography, you should read those sections of up-to-date text books which deal with transport in Britain and other countries. In addition read articles in newspapers and magazines which describe new transport development in various parts of the world.

22.2 Essential information

22.2.1 The importance of examples

Much of the A level work on transport and transport networks is concerned with costs, network patterns, the development of networks and traffic densities. The emphasis is on theory and relationships which you will understand much better if you have built up a bank of knowledge about recent transport developments in different parts of the world and the variety of network patterns which have been established in particular countries. This information will be of great value when answering questions since many examining boards ask for specific examples or use expressions such as 'Discuss and illustrate this statement'.

22.2.2 Definitions and network relationships

Network A set of routes which connect junctions and termini.

Topological map or graph Networks which have been simplified as a map or graph are called topological maps or graphs. The junctions and routes are preserved but distances and directions may not be accurate. The London Underground and the British Rail Inter-city maps are examples of topological maps. By simplifying the system they are particularly useful for passengers who wish to solve routing problems quickly.

Vertex A location on a network such as a road junction or a station. The word *node* is sometimes used instead of vertex.

Edge (also known as an *arc* or *link*) A direct route connecting two vertices.

Connectivity One important structural property of a network is the degree to which the vertices are interconnected. The degree of connection between all vertices is defined as the connectivity of the network. It is a particularly valuable concept when one network is compared with another or when changes in the same network over a period of time are being compared.

A developed region with an extensive demand for transport facilities to move goods and people will have a transport network with a higher degree of connectivity than the network to be found in a less developed region where movements are not so intense.

Comparing networks In Fig. 22.1 there are 8 edges and 9 vertices. Because the number of edges is one fewer than the number of vertices the network is described as minimally connected. If one of the edges is removed one part of the network will be disconnected from the rest. The formula for a minimally connected network is

$$e = (v - 1)$$

In Fig. 22.2 the network is more complex. Most of the vertices are connected to more than one other vertex and between most pairs of vertices there is more than one sequence of edges. If one edge is removed from this diagram the network will remain connected. This is not a minimally connected network so the formula given above does not apply.

Measuring connectivity A number of different methods have been developed to measure the connectivity of networks. Of these the beta index is one of the simplest.

The beta (β) index is found by dividing the total number of edges by the total number of vertices.

$$\beta = \frac{e}{v}$$

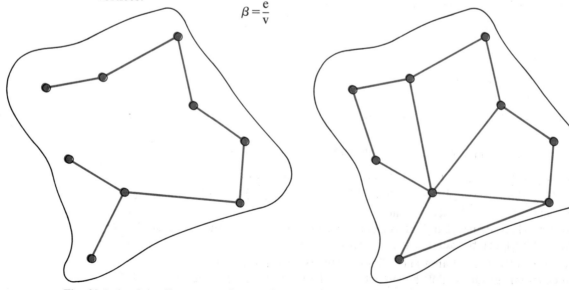

Fig. 22.1 A minimally connected network   **Fig. 22.2** A complex network

The beta index for Fig. 22.2 is $\beta = \dfrac{12}{9} = 1.33$

For a given number of vertices, the more edges there are that connect, the greater the connectivity and the higher the beta index.

The beta index has little value when complex networks are being considered. The alpha (α) index is more useful when comparing networks. To understand it you must first understand the following term:

Circuit The additional linkages added to Fig. 22.1 to increase its connectivity and produce Fig. 22.2 have created circuitry. A circuit is defined as a closed path in which the initial vertex of the linkage sequence coincides with the terminal vertex. Another way of putting this is to describe a circuit as a path starting and finishing at the same point and traversing some or all of the network by the shortest route. In Fig. 22.2 circuitry exists because additional or alternative paths between vertices have been established. The number of circuits is determined by the number of additional edges added to a minimally connected network.

In Fig. 22.2 there are 12 edges. In Fig. 22.1 which had minimal linkage there were 8 edges. The number of additional edges in Fig. 22.2 is therefore $12 - 8 = 4$. So there are 4 circuits in Fig. 22.2 and if you remember the definition of what constitutes a circuit they are quite easy to locate on the network map.

Expressed as a formula the number of circuits is

$$(v - 1)$$

i.e. the number of edges needed for a minimally connected network, subtracted from the number of edges actually present – e.

This can be expressed as $e - (v - 1)$ which algebraically is the same as

$$e - v + 1$$

The resulting number is sometimes called the cyclomatic number.

The alpha index compares the observed number of circuits (the numerator) with the maximum possible number of circuits for a given number of vertices (denominator). As a formula this is

$$\alpha = \frac{e - v + 1}{2v - 5}$$

For Figs. 22.1 and 22.2 the alpha values are

Fig. 22.1
$$\alpha = \frac{8 - 9 + 1}{2 \times 9 - 5} = \frac{0}{13} = 0$$

Fig. 22.2
$$\alpha = \frac{12 - 9 + 1}{2 \times 9 - 5} = \frac{4}{13} = .31$$

There is no circuitry in Fig. 22.1 and minimally connected networks have an alpha index value of 0. A network with the maximum circuitry has an index value of 1. The alpha index value is normally expressed as a percentage of the maximum so that in the example given the network circuitry is 31 per cent of the maximum.

22.2.3 Factors which influence network patterns

Transport networks are built for the following purposes:

(a) To make a flow of goods and services possible between existing settlements – the new road network in Brazil is partly being built to connect existing settlements which previously were not connected or relied on inadequate links by water.

(b) To open up an area or a resource – networks were built in many parts of Africa by colonial powers to provide easy access to raw materials.

(c) Strategic reasons – to provide a good network of roads and railways to transport troops swiftly.

In 1963 K. J. Kansky identified five factors which influence network patterns. They are:

Relief Highlands, rivers and marshland are obstacles for road and rail builders. The degree of hindrance can be registered by means of a relief index (see V. Tidswell, *Pattern and Process in Human Geography*, UTP, 1976).

Shape A country like Japan is likely to have a different network pattern from a compact country such as France. Shape can be measured statistically.

Size In a country with a small area and a dense network, one form of transport may replace another on economic grounds. This has happened in Britain; the canals were replaced by the railways, which in turn have declined in the face of road competition.

Population Density is important, but it must be related to the standard of living of the population. Nations with high standards of living will have denser networks than those which are poor.

Degree of economic development Networks will be affected by such things as energy consumption and the level of imports. These will be higher in a rich country.

Additional factors Kansky accepted that other factors, apart from the five already listed, can affect transport networks. Among these are political boundaries and social and historical factors as well as chance decisions.

22.2.4 Development and density of a network

Using West Africa as an example, E. J. Taaffe, R. L. Morrill and P. R. Gould have developed a model which identifies the stages through which a network is presumed to pass (Fig. 22.3).

Initially there are small scattered ports along a coast with no links with each other, although each has a small hinterland (Fig. 22.3a). Then one or two ports begin to grow and links with towns inland speed their growth (Fig. 22.3b). This process continues and intermediate centres appear (Fig. 22.3c). Then the vertices become interconnected (Fig. 22.3d and Fig. 22.3e), leading to a final stage (Fig. 22.3f) in which some links become more important than others as some cities prosper while others remain poor.

The density of a network will depend on: (a) density of population; (b) volume of circulation of goods and people within a region (this is related to the standard of living); (c) availability of capital for investment. Of these three factors the least important is the first.

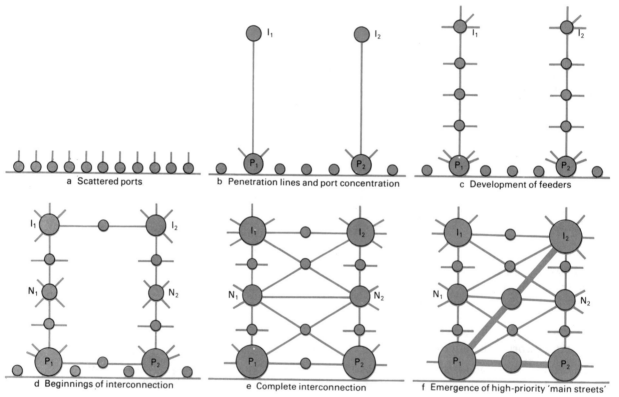

a Scattered ports

b Penetration lines and port concentration

c Development of feeders

d Beginnings of interconnection

e Complete interconnection

f Emergence of high-priority 'main streets'

Fig. 22.3 Ideal sequence of network development

22.2.5 Gravity model

This model gives a simple measure of interaction between places and is based solely on population and distance. It is therefore a limited interpretation of what may happen.

The formula is: Movement between A and B =

$$\frac{\text{Population of A} \times \text{Population of B}}{(\text{Distance})^2}$$

For example, using Fig. 22.4

$$\text{Movement A} - \text{B} = \frac{2000 \times 2000}{(1)^2}$$

$$= \frac{4\,000\,000}{1} = 4\,000\,000$$

$$\text{Movement A} - \text{C} = \frac{2000 \times 2000}{4} = 1\,000\,000$$

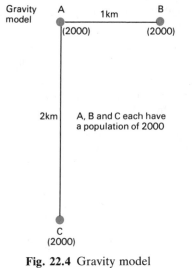

Fig. 22.4 Gravity model

In this example there would be four times as many trips between A and B as between A and C.

22.2.6 Intensity of movement between locations

E. Ullman identified three principles which will help to determine the intensity of movement between locations.

Complementarity For two areas to interact there must be a demand in one and a supply in the other. The main reasons for travelling are work, firm's business, education, shopping and personal, social and entertainment, to reach home and for no specific purpose.

Intervening opportunity This is a complex concept. The basic assumption is that all trips made by individuals will be as short as possible and trips to a particular place would not, therefore, be made if there were *intervening opportunities*, i.e. other opportunities between the individuals' starting point and the place to which they travel. The strength of the attraction of a place will, therefore, be directly related to the number of opportunities in that place and inversely related to the number of opportunities between the originating point and that place.

Transferability The ability to move. This is the over-riding consideration since the cost of moving may be too high for the prospective traveller.

22.2.7 Transport costs

There are two main types of costs incurred by transport operators.

Capital costs *Track costs* are the costs of providing and maintaining a surface over which the transport service can operate. In the case of roads this is normally provided by the country or region. *Interchange costs* are the costs of providing facilities at the beginning and end of journeys. These are sometimes called terminal costs.

Running costs These can be either fixed or variable. *Fixed costs* include the maintenance of the network and vehicles, depreciation and the interest on borrowed capital. These fixed costs would be incurred whether or not the vehicles or other means of transport were being used. *Variable costs* relate to the costs incurred as a result of movement in a transport system. These costs will largely depend on the distance travelled and the number of journeys which take place. Variable costs include the cost of fuel and servicing.

Cost distance This, not distance, is the basic variable in transport. It is the distance multiplied by the freight rates per tonne kilometre.

Government intervention Both passenger transport and freight rates have become increasingly subject to this. For political, social, strategic and economic grounds governments intervene both in the provision of services and their pricing.

22.2.8 Relative costs of different forms of transport

The comparative costs of road, rail and water transport are summarized in an idealized form on Fig. 22.5. For the distance O-A road transport is the most competitive. For the distance A-B rail transport is the most competitive while for distances beyond B water transport is most competitive.

Railways There are high capital costs which increase with the distance to be covered by the track. Fixed running costs are also high but variable running costs are low making rail costs more competitive over long distances.

Water transport Costs are high over short distances because terminal costs are high but over long distances costs are lower than for railways or roads. Bulky and heavy cargoes can be carried cheaply in specialist ships such as oil tankers, bulk ore carriers and container ships. The larger the ship the more economical it is to run. Artificial waterways involve a heavy capital outlay and continuous maintenance which makes the fixed running costs high.

Road transport Although movement costs are higher for road than for rail over long distances, over shorter distances road transport is more economical because fixed costs are low and terminal costs are much lower than for rail or water.

Air transport Capital costs are high and so are terminal costs but speed makes air transport attractive for passenger travel and certain types of goods. The great advantage of air transport is that it is three-dimensional and routes as a result are more direct.

Pipelines There is a heavy capital outlay and high fixed costs. Carriage over long distances by water is much cheaper than by pipeline.

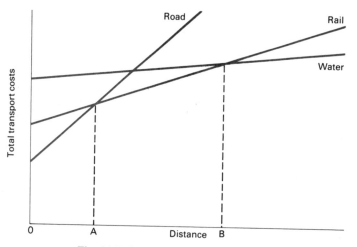

Fig. 22.5 Comparative transport costs

22.3 GENERAL CONCEPTS

Shrinkage of distance This, the result of improvements in technology, is a continuing phenomenon. Expressed positively it has been described as space convergence and this concept is highly significant to geographers in the study of spatial relationships.

Route network This concept introduces a systems approach to the study of transport and provides a more coherent viewpoint than the examination of individual routes in isolation.

Relative costs are very important in determining the nature of a transport network. Route networks will develop where the economic demand is sufficiently strong, despite physical difficulties.

22.4 DIFFERENT PERSPECTIVES

The study of transport networks follows a recent trend in geography away from regional synthesis towards viewing phenomena as part of a system. The links between elements in the system form networks and the study of these networks leads to a greater understanding of relationships within a system.

22.5 RELATED TOPICS

A theme which runs through many aspects of human geography is the problem of distance separating locations, how this space can be traversed and the sociological and economic significance of the links which are formed.

Transport and transfer costs play a very important part in industrial location theory and feature in the Weber model which is described in the unit on *Location of Industry* (pages 135-41). Transport costs and distance are also central to the von Thünen model of rural land use (see pages 128-32).

One of the key concepts underlying central place theory (see pages 112-17), is the range of a good, this is the maximum distance which people are prepared to travel in order to obtain a particular good or service.

The description of the network development model based on the evolution of the transport network in West Africa provides a valuable basis for research into similar networks in the developing world (see Section 22.2.4). Many countries in the Third World have networks super-imposed by colonial powers and this is an important factor to consider in your studies of these areas.

Since networks consist of edges connecting vertices, studies of population distribution must be related to the transport facilities of the region. The motorway network in Britain is a good example of a response to the distribution of centres with high populations.

22.6 QUESTION ANALYSIS

1 (a) Give reasons why transport costs differ in rail, road, water and air transport systems according to the distance travelled.
(b) Use specific examples to explain why the comparative volumes of rail, road, water and air traffic differ from country to country. *(in the style of the Joint Matriculation Board)*

Understanding the question Part (a) asks for reasons why transport costs vary over distance depending on the form of transport used. The question is concerned with cost-distance factors and it is best to exclude such aspects as government pricing policies which can vary from time to time and from one form of transport to another. Do not forget that water transport includes both inland and oceanic systems and that costs are likely to be higher on inland systems where maintenance charges are high.

Part (b) is concerned with the reasons why the density of traffic is much greater in some countries than in others. This part of the question must be answered by giving examples of different densities. It is important that the examples are drawn from countries which display contrasting stages of development.

Answer plan For part (a) write a short introductory paragraph explaining that transport costs consist of fixed and variable costs and that there are considerable differences in the costs of different forms of transport. Then analyse the costs incurred by each of the transport systems, rail, road, water (inland and oceanic) and air, using examples wherever possible (see Section 22.2.7). Describe how the four forms of transport compare over short, medium and long journeys (see Section 22.2.8) and give examples.

Introduce part (b) by describing the factors which determine the volume of traffic from country to country, quoting Ullman's three principles (see Section 22.2.6), and then explain that volume depends on network density which in turn depends on a number of factors (see Section 22.2.4). Give examples for each of the four means of transport (five if you separate inland from oceanic waterways), first in countries where the volume of movement is high and then for countries where the volume of movement is low. These two divisions should emphasize the differences between developed and developing economies.

2 'In transport geography the physical distance involved in transporting goods and passengers is often less important than other factors.' Discuss and illustrate this statement. *(in the style of Oxford)*

Understanding the question The statement must be read carefully. It says that the distance involved *is often* less important than other factors. In other words there are times when the distance involved is as important or more important than other factors. You will need to know of such instances if you intend to answer this question.

The term *transport geography* is very generalized and you will need to consider carefully what the *other factors* are. Before answering this question write down the other factors. They will include economic, political and social factors.

In discussing the statement you must make it clear as to how far you agree with it since you may consider that 'often' should be replaced by 'seldom' or alternatively 'always'.

Although the world 'illustrate' is used it should not be taken literally. You do not have to provide sketches or diagrams as part of your answer, illustrations are required in the form of written examples.

Answer plan Write a brief introduction outlining the various factors which you intend to discuss in addition to physical distance.

Take each of these factors in turn and give examples of their significance in determining route networks, degree of connectivity and type of traffic and preference for a specific form of transport. Give examples from different parts of the world.

Write another section which discusses instances where physical distance is a very important factor in transport geography, for example, in deciding the forms of transport and network pattern in such areas as Northern Canada and Siberia.

Write a concluding paragraph which explains how far you agree with the statement and why you have come to this decision.

3 Identify and discuss the main factors which appear to determine the location of routes and the form of route networks. *(Oxford and Cambridge, July 1980)*

Understanding the question The first part of this question is concerned with the location of routes while the second focuses on the form of route networks. In both cases you are asked to identify and discuss the main factors involved.

Your answer should, therefore, be divided into two parts. In the first part the main factors which determine the location of routes should be listed and then discussed. In the second part a similar structure should be adopted when writing about the forms of route networks. In each case include practical examples of the factors; these should be taken from both highly developed countries and from countries of the developing world.

Answer plan List the main factors which help to determine the location of routes, using the Kansky summary described in Section 22.2.3. In discussing these factors remember that physical constraints may be less important than economic needs and that the size of the population is less important than the standard of living. Give examples for each of the factors mentioned.

For the second part of the question on the form of route networks, use the model based on West Africa to explain how the number of edges and vertices helps to determine the form and is related to the stage of development the country has reached (see Section 22.2.4 and Fig. 22.3).

Describe the forms of different networks and the factors which have influenced these forms, for example, the road network along the US–Canadian border in Manitoba, the climax network in industrial Britain and the limited railway network in heavily populated China.

22.7 FURTHER READING

Bradshaw, M. G. and Kent, W. A., *Human Geography. Theories and their Applications* (OUP, 1977)
Briggs, K., *Introducing Transport Networks* (University of London Press, 1972)
Robinson, R., *Ways to Move – Geography of Networks and Accessibility* (CUP, 1977)
Robinson, H. and Bamford, C. G., *Geography of Transport* (Macdonald and Evans, 1978)
Taaffe, E. J., Morrill, R. J. and Gould, P. R., 'Transport expansion in underdeveloped countries: a comparative analysis', *Geographical Review* (1963) 53, pp 503-529
Tidswell, V., *Pattern and Process in Human Geography* (UTP, 1976)

23 Less-developed countries

23.1 ASSUMED PREVIOUS KNOWLEDGE

Not all GCSE syllabuses deal with Third World topics. Those that do cover (a) developing countries such as India, Nigeria, Brazil, Egypt; (b) problems which Third World countries have to face – overpopulation, poverty, overdependence on primary products etc., (c) examples of development projects which tackle the problems; (d) how richer countries help to finance these projects – through international aid, the World Bank and investment through multinational companies.

23.2 ESSENTIAL INFORMATION

23.2.1 Definitions

Underdevelopment is not easy to define. It is generally agreed that the underdeveloped countries make up most of the continents of Africa, Asia and Central and Southern America. These countries contrast economically with the developed rich, advanced, industrial countries of Western Europe, North America and Japan (Fig. 23.1). Terms such as *less-developed countries (LDCs), underdeveloped countries*, the *Third World* and *developing countries* are interchangeable.

Gross National Product (GNP) When we compare the volume of goods and services which one individual, group or nation receives compared to others, we often use the phrase *standard of living*. This is not easy to measure or express as a figure. One way of comparing countries is to use estimates of income. Most countries calculate their GNP.

> GNP = net value of all goods produced and all services rendered in one year in a particular country (the country's exports are subtracted and the imports added when the final figure is calculated).

Per capita income (PCI) GNP is not much use as an index on its own as it is important to know how many people have to share this income. So an index is calculated which gives the average income per person of the country. This is the PCI.

23.2.2 Some models of economic development

A number of models have been built to describe and explain the process of development. The models are attempts to simplify a very complex situation so that we may understand why some countries have remained poor. The models also help us to examine the inter-relationship of the

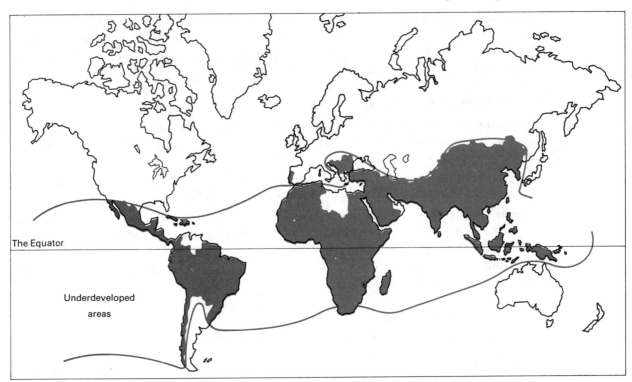

Fig. 23.1 Less-developed areas of the world

factors which affect development and the stages in the process by which some countries have become more advanced. Models are useful in planning for future development and in pointing out ways in which the economic and social gaps between rich and poor countries may be narrowed.

The vicious circle model (Nurske) (Fig. 23.2) This model highlights the problem of lack of capital in underdeveloped countries. It does not explain why processes occur in LDCs and suggests that nothing changes.

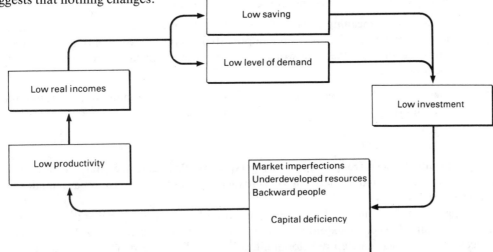

Fig. 23.2 Vicious circle theories of underdevelopment

Rostow's model is concerned with the economic growth of countries as part of a single process of development. A key stage in the model is take-off – when old ideas give way to forces of economic progress. So the problem of underdeveloped countries is how to achieve take-off. The model is criticized because it ignores social factors.

Myrdal's model is also called the cumulative causation model. The model suggests that economic development leads to an increase rather than a decrease in the difference between regions. Two main effects of rapid growth of a thriving region are *spread effects* – surrounding regions benefit from the greater demand for foodstuffs and raw materials, and *backwash effects* – capital and people move to the thriving regions so other regions are worse off.

23.2.3 The bases of development

Industrialization is seen as the main agent of change.

Vast amounts of capital are needed for industrialization. These are obtained: either through *international aid schemes* (World Bank) or by *private investment by foreign firms* (multinationals).

Trade agreements help development – guaranteeing markets and/or prices e.g. EEC arrangements for former British colonies.

23.3.4 Strategies for development

National plans (India, Tanzania) These plans may aim at *balanced* growth (trying to keep a balance between the different parts of the economy) or *unbalanced* growth (rapid development of key sectors such as iron and steel).

Low technology strategy Traditional industries are modernized and expanded, mainly to meet home market demands e.g. the textile industry in village India.

Export stimulation or import substitution Countries have to decide whether to encourage exports of raw materials to obtain currency to buy other goods or to spend less on imports by making their own goods to meet home demands.

Revolution Countries like Cuba have decided to leave the western economic system and to try to achieve development by this means.

23.3 GENERAL CONCEPTS

Underdevelopment is defined essentially in economic terms Third World countries have not undergone modern industrialization. They have obsolete methods of production so their poverty is not entirely due to poor national resources. But such countries are characterized by mass poverty which is not just the consequence of a short-term crisis or national disaster. The poverty could be lessened by the introduction of methods which are already successfully used in other lands.

Development is a progressive transformation of society Countries try to achieve development by means of deliberate planning of large-scale economic and social change. The use of natural resources is co-ordinated and ways of trying to catch up with the wealthier nations are devised. However development is not the same as economic growth. Economic growth may mean *quantitative* change, i.e. the increase of existing means of production; development means *qualitative* change – new forms of economic activity are created.

Development has a social meaning It involves creating the conditions which make it possible for the people of a country to realize their human potential, so that they get enough food, receive sound education which trains them for work, have good job opportunities and the right to take part in independent government.

Economic links between developed and less-developed countries are maintained through international trade, international aid and the investment of capital.

23.4 DIFFERENT PERSPECTIVES

Development may be seen from an economic perspective Problems are analysed in a rational manner, plans are formed as a result of statistical forecasting; decisions are made about resource allocation.

This approach is fraught with difficulties. Conditions beyond the control of the planners may change rapidly and to such an extent as to make the plan less practicable e.g. the rise in world oil prices has seriously affected development policies. When the plan is finalized it may be extremely difficult to implement it because of problems such as civil disorder or the lack of enough skilled people.

Development may be viewed from a social perspective The main concern is then to seek ways to overcome problems of overpopulation, hunger, malnutrition etc.

Underdevelopment in some countries is the direct result of the ways in which wealthy capitalist countries have developed This is the radical Marxist view. Wealthy countries are said to use the resources of the poorer countries to their own advantage and to exploit the Third World.

No single perspective gives a complete picture of the real situations in Third World countries or of the process of development. People and groups with different sets of beliefs and values see economic and social problems in different ways. For example, a member of the government of the People's Republic of China will have very different ideas on how to tackle development problems from the ruler of an oil-rich Arab state. It is important therefore that we are aware of the existence of different viewpoints, explanations and solutions to development problems.

23.5 RELATED TOPICS

You should look at the following sections of this book: *Movement of population; Growth and distribution of population; West Africa: population issues; India: development issues; Brazil: regional strategies.*

Other general topics which are closely related to the topic of underdevelopment are:

World population growth Most less-developed countries face severe population problems. Population control programmes are made ineffective by low educational standards and the effects of social and religious customs.

Urbanization The urbanization of populations is occurring most rapidly in the Third World. The cities act as magnets and as a result many Third World cities have shanty towns on their edges and problems of water supply, health and education are severe.

Agriculture Non-commercial primitive forms of farming such as shifting cultivation, subsistence farming and nomadic herding are found in less-developed countries. Yet the Third World also grows important raw materials and food crops needed by the rich, advanced industrial countries.

23.6 QUESTION ANALYSIS

1 Study the population graphs and pyramids for Mexico and Japan (Figures 23.3 and 23.4).

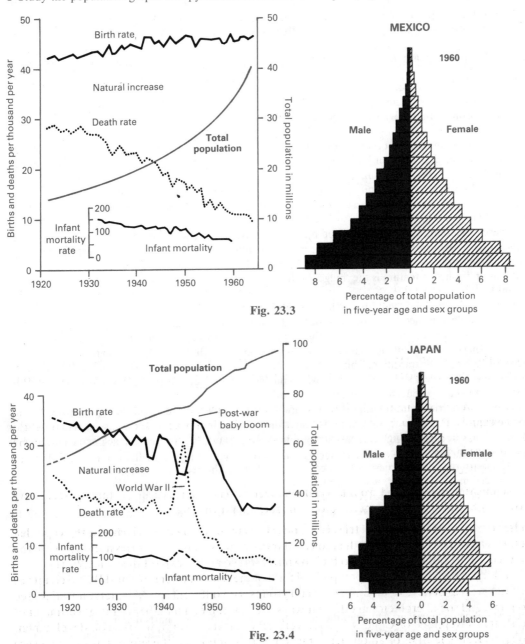

Fig. 23.3

Fig. 23.4

(a) What stage had each country reached by 1960 in the demographic transition development model?
(b) Outline three major demographic differences between the two countries.

(c) With reference to Japan, how do you explain the apparent contradiction of a declining birth rate and an increasing population?

(d) Suggest three factors which might possibly contribute to the falling death rate in Mexico?

(e) (i) For Japan, describe the characteristics of the pyramid for the age group under 30 years.

(ii) Account for these characteristics.

(f) (i) Sketch a population pyramid for a country with an ageing population.

(ii) For a named country, give three reasons for such a population structure.

(University of London, 1985)

Understanding the question This question will require you to draw upon your basic geographical of the meaning of terms such as *natural increase* and *population pyramids*, and upon your understanding of models such as the demographic transition model. It is not essential, however, for you to have detailed knowledge of the geography of Mexico or Japan to answer the question well.

An important point to remember is that *all* the data presented in the diagram are relevant to the question, so all aspects of the graphs and the pyramids need to be examined together.

Answering the question The structure of your answer has been determined by the form of the question. Make sure that you answer each section precisely and present your answers in complete sentences.

Part (a) From your knowledge of the demographic transition model (page 182), you can quickly establish that Mexico is in the *early expanding* phase or *early transitional* phase, while Japan is in the *late expanding* or *late transitional* phase.

Part (b) Use the graphs and pyramids.

(i) Comparison of the pyramids shows that a higher percentage of the population is below 45 years of age in Mexico than in Japan.

(ii) Comparison of the birth rate graphs shows that the overall trend in Mexico is upwards, while in Japan it is falling.

(iii) The natural increase in Mexico is greater than in Japan. Note: There are other relevant points you could list.

Part (c) Three factors are:

(i) The death rate is declining faster than the birth rate, so there is a positive natural increase.

(ii) Apart from during the period of the Second World War, the birth rate has been consistently higher than the death rate.

(iii) The decline in infant mortality – fewer babies are being born, but a higher percentage stay alive.

Part (d) You can draw on your knowledge of population features of developing countries to answer this section. You could list, for example:

1 improved medical and maternity services, which have led to a decline in infant mortality;

2 economic development has raised the standard of living, so general health standards have improved and life expectancy has increased;

3 the use of modern drugs and improved medical practise have reduced the death rate.

Part (e) (*i*) The horizontal bars in the pyramid each represent five years, so you must confine your examination to the lowest six bars. You could point out:

1 the most distinctive feature is the 'bulge' of the 10–15 age range, which represents the post-war baby boom;

2 the 0–10 groups show the effects of a declining birth rate;

3 There is little evidence in the pyramid of the dip in the birth rate during the Second World War.

(ii) The baby boom was the result of members of the armed services returning home, and of people not wishing to have children until the dangers of war were over. After the boom, Japan is in the late expanding phase of demographic transition. This phase is characterized by a decline in the birth rate. The dip in the birth rate was for a very short period, after which the birth rate increased rapidly. Over a five year period, these sudden changes effectively cancelled each other out.

Part (f) (i) Countries with ageing populations are found in the developed world; Britain and Sweden are good examples. Draw a pyramid which illustrates the main demographic features of such countries – low birth rate, low death rate, high life expectancy. Check the pyramid for one of the developed countries you are studying in depth.

(ii) Choose three relevant reasons for a country you know. The reasons might include:

1 advanced medical services, which result in low birth and death rates;

2 a high standard of living, which results in a very high general standard of health and long life expectancy;

3 high levels of education, which result in many families and family sizes being carefully controlled by the parents.

2 With reference to named and located examples:

(a) explain what you understand by the term *underdevelopment*

(b) indicate how such a condition may be recognized

(c) suggest various ways by which the problems of underdevelopment may be tackled and indicate their relative effectiveness.   *(Associated Examining Board, Summer, 1979)*

Understanding the question Although the question is divided into three parts if you read it carefully you will see that (c) has two aspects to it – you have to show a knowledge of different ways in which under-development may be tackled and then judge how effective they are in comparison with others. This gives you an idea of the balance you should aim for in your answer (25 per cent, 25 per cent, 50 per cent for the three parts).

Answer plan The outline of underdevelopment on page 100 largely answers (a). Additional points to be made are:

(a) There is no precise agreement as to which countries should be labelled as 'underdeveloped'. Some countries have certain features of underdevelopment and other features similar to advanced industrial countries.

(b) Some of the Eastern European countries with their centrally planned economies do not fit easily into the developed/underdeveloped classification e.g. Bulgaria.

(c) Since the question specifically asks for named and located examples you should add some to the points made on page 100, e.g. India is a country where cultural factors have hindered change in rural areas.

The main point to make under (b) is that in order to measure objectively the degree of under-development of one country compared with others we use development indicators. These indicators may be concerned with economic, social and political development. Some of them are GNP, energy consumption per capita, consumption of steel per capita, illiteracy rates, calorific and protein intakes per capita, and life expectancy. You should show how each indicates development. You should also consider the extent to which they reveal the difference between the developed and underdeveloped nations, giving examples to support your statement.

Next, explain that the indicators are obtained by collecting statistics from all the countries of the world. They cannot be truly accurate or objective as not all are reliable. Usually, the less developed the country, the less reliable the statistics. You should point out the value of such indicators – they help us to recognize the degree of underdevelopment, the situation at a particular time in any country and the economic trends. Priorities for development can be worked out from the data.

Other ways of recognizing underdevelopment are:

(a) *The nature of a country's international trade.* Traditionally underdeveloped countries have exported raw materials and food to the advanced countries (tea from Sri Lanka, tin from Malaysia etc.). Manu-factured goods have to be imported and in densely populated regions food also has to be brought in (rice to Bangladesh).

(b) *Many underdeveloped countries have tried to achieve rapid development in recent years.* To achieve this they have borrowed capital from the rich nations and international agencies. So it is possible to recognize underdevelopment by examining the extent to which a country depends on overseas aid for financing major projects and programmes e.g. the Aswan High Dam project in Egypt.

Problems of underdevelopment have been tackled in a number of ways. In this final section of the answer it is very important that you should organize the information you intend to include in a systematic way. The following sub-headings provide one way of ordering (though they should not be used in your answer):

(a) *Investment of capital* Economic theories on trade suggest that countries are underdeveloped because they are too poor to invest capital in farming, industries, modern transport systems etc. which would lead to rapid development. Rostow's model of development for example, says that a country must attain a rise in the rate of investment of between 5 and 10 per cent per annum to succeed. Investment can be achieved in different ways: (i) by encouraging saving and investment within the country itself; (ii) through private investment from outside; (iii) investment by multinational companies.

The first is not likely to be very effective especially in the first stages of development. The under-developed countries are poor so there is little money available for saving and investment.

Private investment can greatly assist development. But (ii) also has disadvantages. For example, investment in mining in Sierra Leone in modern times created few jobs and the jobs that were made were low grade. Although the new mines brought earnings of foreign exchange and extra taxes to the govern-ment, most of the profits left the country and did not contribute to development. The same arguments apply to the modern bauxite industry of Jamaica.

Multinational companies can bring vast amounts of capital into a country in a short time and bring about rapid change. Again there are disadvantages – many of the new industries established do not need many workers, prices for the goods produced are frequently higher in a captive market than in the developed world. Powerful companies may also influence political independence e.g. ITT interfered in the economy of Chile to bring down President Allende. Some developing countries have offset the power of the multi-nationals by forcing them to give a major share in the industry to the government e.g. in the copper industry in Chile, the sulphur industry of Mexico.

(b) *Freeing trade* If international trade is free it allows the underdeveloped countries with their cheap labour costs to sell products in the rich countries e.g. the textile and electronic products of S. Korea and Taiwan. Rich countries however have imposed quotas and passed anti-dumping laws to stop such goods entering.

International agreements such as GATT (General Agreement on Tariffs and Trade), the Common-wealth Preference scheme, the EEC preference schemes for ACP states (46 African, Caribbean and Pacific countries) have been set up to free trade. In general they have made it easier to trade in raw materials needed by the rich countries but have not removed restrictions on trade in processed goods. So the agreements are of limited use.

(c) *Development plans* Most underdeveloped countries have drawn up development plans. The plans
  (i) identify the main problems and establish priorities and targets

(ii) provide a strategy for development, short-term and long-term targets can be used to support each other

(iii) the nation is given a sense of looking forward and gets a sense of progress and achievement when targets are met

(iv) well-designed plans make it easier to attract aid.

Two examples you could use in your answer are:

(i) India's 5-year plans. They have not been very effective because over-ambitious goals were set and the plans were not efficiently administered and led to bureaucratic corruption. The gap between the rich and poor grew wider and there was no single national sense of purpose to make them work.

(ii) Tanzania's national plan. This was outlined in the Arusha declaration. The President stated a belief in the equality of the citizens, the need for self-reliance, the importance of agricultural development and progress in health, education etc.

The plan has had limited results because Tanzania did not have the capital resources to carry out the tasks involved, and when it sought loans from the IMF it was expected to modify the plan. Also the educated and well-off in Tanzania were reluctant to make the sacrifices necessary to achieve greater equality.

(d) *International Aid* Aid is not given just to alleviate poverty. It provides impetus to modernization and development. Motives for aid vary e.g. the French and British feel responsible for former colonies; the Germans and Japanese want to develop trade; the Russians, Americans and Chinese are anxious to extend their political influence. Scandinavian countries usually see aid as a moral duty. Aid is co-ordinated by bodies such as the Development Assistance Committee (DAC) of the OECD. Some aid is for specific projects such as the multi-purpose river barrages on the Ganges. Aid may also be given to support a general development programme. There are two major disadvantages to this means of encouraging development:

(i) The rich donor nations often place conditions on the aid. For example Russian aid to the Egyptians for the Aswan High Dam project was accompanied by increased Russian military influence over Egypt.

(ii) Underdeveloped countries have to pay interest on the loans. Some countries fall deeper and deeper in debt and may become bankrupt e.g. Ghana under President Nkrumah.

(e) *Revolution* Some developing countries have attempted to achieve national development by changing fundamentally the economic and social system of the country. So development is based on revolution e.g. China, Cuba. Evidence suggests that this approach also creates problems – China has had to revise its policies and is now expanding economic links with the West. Cuba is dependent upon an artificial market and price for its sugar in the USSR and is now trying to get involved in the Caribbean tourist trade with rich western nations.

To conclude your answer, you should make a judgement of which of the approaches to tackling underdevelopment seem or seems to be the most effective in the light of the points you have included in your answer.

3 Comment upon the nature and location of manufacturing industries in any one country of the developing world (i.e. newly emergent countries)   (*in the style of London*)

Understanding the question This is a straightforward question but to gain good marks you need to think carefully about how you structure your answer. The question asks you to write about the nature and location of manufacturing industries but it also asks you to comment. You are expected to develop an analytical approach to the topic – to give an explanatory account.

Answer plan You should name the country you have chosen at the start of your answer. In this instance the example chosen is Nigeria, the biggest and most densely populated country in West Africa. The *location* of manufacturing industry is best shown by a map. The pattern shown in the map below (Fig. 23.5) is the result of two main influences:

(a) concentration at or near the coast e.g. Lagos, Port Harcourt.

(b) concentration in state capitals e.g. Kano, Kaduna, Ibadan.

In considering the nature of the manufacturing industries it is appropriate to point out the common features which Nigeria's industries share with those of other emerging nations:

(a) Large scale manufacturing is a recent addition to the economy – in Nigeria most development has occurred since 1956.

(b) The Government is anxious to encourage the growth of manufacturing industries. They see the close links between industrial development and a high standard of living in advanced countries.

(c) Some industries have been developed to save the cost of buying imports (this is called *import substitution*). Examples of such industries in Nigeria are: textiles, shoemaking, cement, steel rolling.

(d) Other industries have been developed to increase the value of exports by processing raw materials before they leave Nigeria. Examples: tin smelting, vegetable oil refining, oil refining, the making of plywood.

Compared with many other developing countries Nigeria has a number of important advantages:

(a) it has a very wide range of natural resources.

(b) it has important fuel resources – oil, coal, hydro-electric power.

(c) it has an internal market of 80 million for its manufactured goods.

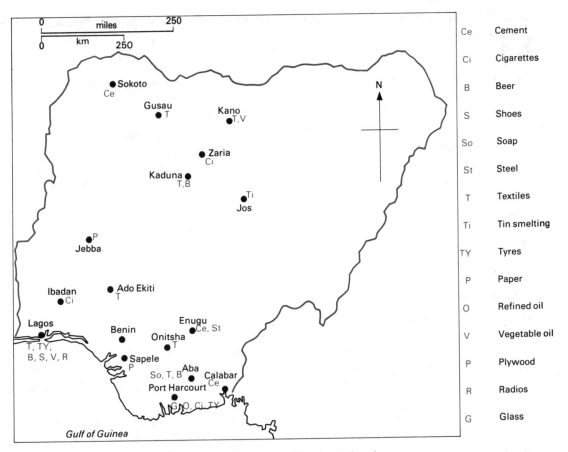

Fig. 23.5 Nigeria: manufacturing industries

(d) it is well located to develop exports to other West African countries.

In looking at the nature of the manufacturing sector in Nigeria it is also relevant to look at the *growth* of such industries. The development of manufacturing industries in the country falls into two distinct phases:

(a) *Pre 1950s* Manufacturing was introduced by Europeans who had the capital and skills in colonial days. The industries were concerned with processing cash crops e.g. cocoa for export. Small factories grew up to cater for local markets e.g. cigarette factories. Factories were built to make things which were expensive to import because they were bulky but were needed in large quantities e.g. beer, bricks.

(b) *Since the 1950s* There has been expansion of processing primary products for export: groundnuts and palm kernels for oil, tin smelting. There has also been import substitution – local materials used to make goods for Nigerian markets e.g. textiles, cement, shoes. Imports are processed for local markets.

Since further development depends upon the *expansion* of manufacturing it is important to consider the problems involved:

(a) Labour is still mainly unskilled. There is a shortage of people with managerial and technical skills, especially in the north of Nigeria.

(b) Capital investment from overseas mainly benefits the overseas countries. So Nigeria is still short of investment capital e.g. the shoe industry belongs to Bata.

(c) Since the average wage of many Nigerians is much lower than wages in the developed countries the size of markets within Nigeria for different products is limited.

(d) Emerging countries like Nigeria are more seriously affected by world economic depression and the fall in commodity prices than are advanced countries. A drop in oil prices for example would have very serious effects.

In your final section you should make a geographical judgement about the potential which Nigeria has to expand its industries e.g. iron and steel, oil refining, a chemical industry based on its natural gas. Such developments could give it the role of chief manufacturing country for the whole of West Africa.

23.7 FURTHER READING

Carr, M., *Pattern, Process and Change in Human Geography* (Macmillan, 1987)

Clark, J. I., *Population Geography and the Developing Countries* (Pergamon, 1971)

Hoyle, B. S. (Ed.), *Spatial Aspects of Development* (Wiley, 1978)

Jarrat, H. R., *Tropical Geography* (Macdonald and Evans, 1977)

Money, D. C., *Problems of Development* (Evans, 1978)

24 United Kingdom: regional problems

24.1 ASSUMED PREVIOUS KNOWLEDGE

A level studies of Britain require an understanding of the recent changes which have taken place in the spatial distributions of such things as population, communications and industry. They also require a detailed knowledge of regional strengths and weaknesses and the geographical significance of Britain's membership of the EC. Both before and during your A level course you should read articles in journals and newspapers which give information about geographical changes that are taking place. If you have not recently studied the geography of Britain you should read a recent GCSE book such as Martin, F. and Whittle, A., *The United Kingdom* (Hutchinson, 1986) or Marsden, W. E. and V. M., *Britain* (Oliver and Boyd, 1986).

24.2 ESSENTIAL INFORMATION

24.2.1 Why some regions have declined

(a) During the nineteenth century there was a period of rapid industrialization based on the availability of coal and the ease with which raw materials could be obtained, either locally or from overseas. As a result major manufacturing and urban centres developed in Central Scotland, the North-East, South Lancashire, West Yorkshire, the Midlands and South Wales. London also became an important industrial centre as well as the commercial and administrative capital.

(b) After World War I the attraction of the coalfields for industrial development steadily weakened. The more traditional exports of cotton textiles, coal, ships and heavy engineering products became less profitable as overseas countries developed their own industries. The pattern of world trade changed and the market for some products such as steam locomotives and rolling stock contracted severely.

(c) As people in advanced industrial societies have increased their personal wealth during the last three decades there has been a rapid expansion in the demand for consumer goods such as television sets, refrigerators, cars, washing machines, lawn mowers, typewriters and many other goods which are to be found in a modern home. The attraction of coalfield locations has weakened as consumer demand has changed. Proximity to markets or supplies of components is more significant and the emphasis has shifted to southern England, especially Greater London where the market and materials are available.

(d) World trade has also grown for such exports as aircraft engines, lorries, electrical products and electronic equipment. Commonwealth countries have become less significant as customers for British goods, but trade with Western Europe has increased.

(e) Improved transport facilities, particularly the door-to-door flexibility of road transport, the construction of a motorway network and the development of deep-sea container facilities have all strengthened the Midlands and the South-East and made them the most economic locations for new and expanding industries.

(f) The energy required by the growth industries can be obtained from the 400 kV grid of the generating companies, National Power and Powergen, the pipelines of the Gas Council and transmission pipelines from coastal refineries installed by the petroleum companies. The accessibility of energy supplies to practically all parts of Britain has given manufacturers the opportunity to locate their plants close to the major demand centres for their products, such as the London and South-East area where net personal incomes amount to 35 per cent of the national total, or in the Midlands which accounts for a further 16 per cent of the national total.

(g) The size of the labour market in London and the Midlands is also important. It means that the possibility of obtaining the right sort of skills was greater in these regions than elsewhere in the country. The skills required for the consumer industries, many of which use conveyor belt techniques, were not to be found in the traditional heavy industries of South Wales and the North. As industry declined on the coalfields labour from these areas migrated to the South-East and Midlands and has been trained to fulfil the needs of the new industries.

(h) Apart from employment in manufacturing industries, there has been a considerably enlarged demand for specialized services such as banking, insurance, government services and the retail trade which has increased employment opportunities in these occupations. The long-standing dominance by London of this sector of the economy facilitated expansion of services in this region rather than elsewhere.

The decline has not been confined to industrial areas; some rural areas have also become problem regions. The two largest rural regions which have experienced economic deterioration are the Highlands and Islands of Scotland and west and central Wales.

In these regions the economy is relatively unbalanced, there is little employment available and local incomes and opportunities are therefore limited. Remoteness has inhibited investment and the major primary activities, agriculture, fishing and forestry employ few people. As in the traditional industrial regions, the demand for the products of these rural regions has declined, not because consumer demand has changed (there is still a need for the primary products these regions can produce such as meat and timber), but because primary products can be obtained more cheaply from overseas. The lure of high wages and improved amenities in the industrial cities has encouraged migration from these areas.

24.2.2 Resultant problems

The regions which have declined during recent decades display a number of symptoms which classify them as problem regions:

High rates of unemployment which reach 20 per cent of the labour force in some towns. In October 1988 Merseyside had an unemployment rate of 16.2 per cent, Tyne and Wear had 13.6 per cent and Strathclyde 14.4 per cent (see Fig. 24.1). In Greater London the unemployment rate of 6.9 per cent was above the average for south-east England of 5.5 per cent. The worst hit regions have been those which grew up on the coalfields.

Net migration to other parts of Britain and overseas has resulted in a steady fall in the total population of these regions (see Fig. 24.2). Between 1971 and 1981 there was a 2.9 per cent

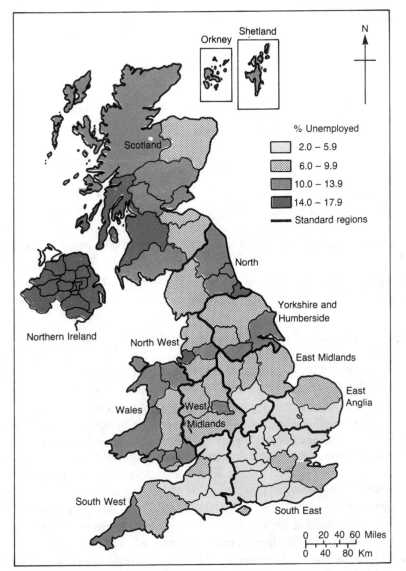

Fig. 24.1 Unemployment, October 1988

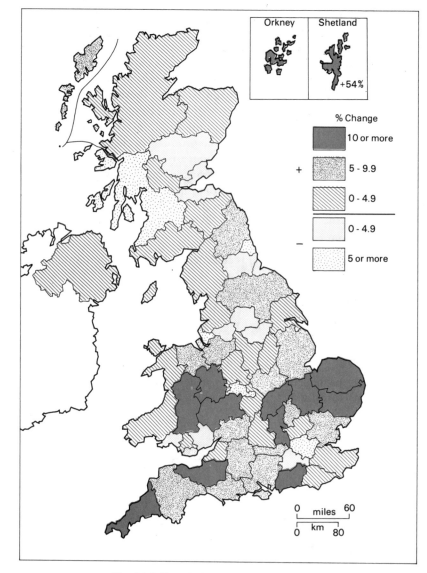

Fig. 24.2 Population changes 1971–1981

decrease in population in the North region, compared with a 0.5 per cent increase in the total population of England and Wales. It can be argued that there were decreases in the population of Greater London and the West Midland conurbation during the same period but these changes were caused by movement from the congested city regions to the neighbouring suburbs or rural hinterlands.

Higher unemployment and lower average wages have reduced the purchasing power in the problem regions with average weekly expenditure 20 per cent or more less than in the more prosperous regions of the south.

Economic stagnation and decline has affected the infrastructure of the less prosperous regions. Lack of investment and low levels of personal wealth have resulted in poor housing, fewer amenities, derelict mines and factories and a decline in the level of public services which contrast with the comparatively higher standards of the more prosperous areas.

Transport services are less viable because there have been few new developments and a decline in the economic structure. The improvement and extension of the existing transport network is discouraged except in higher population density areas such as London. Consequently transport costs are increased making the regions less attractive locations for new or expanding firms.

24.2.3 Government assistance

State assistance to the problem regions has been of considerable significance in the last thirty years. The nature and scale of the assistance changes from time to time and the areas shown on Fig. 24.3 are subject to modification as circumstances alter. The main forms of assistance are:

Assistance to traditional industries Considerable sums of public money have been used to assist the National Coal Board, the iron and steel industry, cotton textile firms and some sectors of

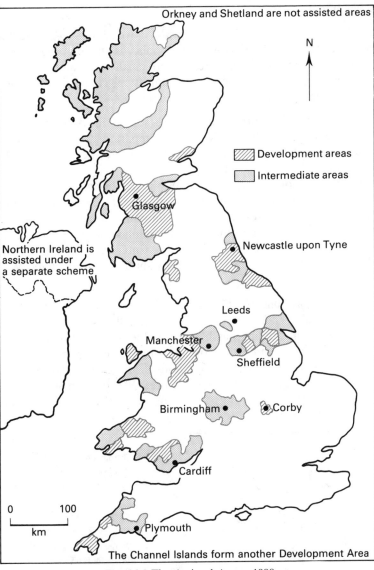

Fig. 24.3 The Assisted Areas, 1990

the motor-car industry. This money provides a form of subsidy to those industries enabling them to invest in new machinery and buildings and to keep uneconomic factories or pits operating.

Assisted areas The government has identified two categories of problem areas based on the scale of difficulty; they are the *Development Areas* and *Intermediate Areas*. The level of assistance is less for the intermediate areas. Money has been allocated to diversify and restructive employment. In addition a number of inducement have been offered to attract industries to these areas.

(a) Industrial estates with units ready for occupation have been built. These have rent concessions to encourage firms to move in.
(b) Grants and loans are available with tax concessions to new enterprises.
(c) Financial aid is available to train employees in skills required by new industries.

Selective assistance in the regions is given for certain investment projects undertaken by firms while regional Enterprise Grants are available to firms with fewer than 25 employees to support investment and innovation.

The English Industrial Estates Corporation provides industrial premises in England where private sector provision is inadequate.

New towns In order to accommodate residents of the assisted areas in new housing and provide a suitable urban infrastructure the government designated 14 new towns to these regions. They include Skelmersdale and Warrington in the North-West, Peterlee and Aycliffe in the North-East and East Kilbride and Glenrothes in central Scotland. These new towns have played a significant role in attracting manufacturing investment into regions of high unemployment.

24.2.4 Urban policy initiatives

Since 1981 there has been increasing action by the government to tackle the social and economic problems of inner city areas. Government policy is aimed at spending public money on inner city renewal by co-ordinating attempts to tackle environmental dereliction and encouraging private enterprise. The three main organizations concerned are the Departments of the Environment, Employment and Trade and Industry.

Enterprise Zones The first Enterprise Zones were set up in 1981 in city regions where unemployment was particularly high. The zones are comparatively small, ranging in size from 50 to 450 hectares and industrial and commercial activities have been stimulated by giving tax relief and financial aid.

There are 26 Enterprise Zones with no new ones planned because the scheme has probably outlived its usefulness. Enterprise Zones have been successful in revitalizing depressed regions but critics point out that businesses have shifted into the zones from elsewhere creating employment at the expense of the areas they have abandoned. Other areas may also have lost out. For example, in the West Midlands the Dudley Enterprise Zone has attracted several large retail schemes which threaten the future of city shops in Birmingham, Wolverhampton and Sandwell.

Urban Development Corporations (UDCs) Two UDCs were set up in 1981, in London Dockland and on Merseyside. The Corporations were modelled on the new town development corporations and government funding has been available to help provide the infrastructure (streets, roads, lighting, etc), and to encourage private firms who help to create new communities in these derelict areas by building houses, business premises, offices and retail centres. In 1987 five new UDCs were set up at Treforest Park (South Wales), Teesside, Tyne and Wear, Black Country and Cardiff Bay. Four more were added in 1988 at Leeds, Central Manchester, Sheffield and Bristol.

Fig. 24.4 Urban policy initiatives

The Urban Programme This programme gives grants to minor city projects, encouraging individual enterprise. Money has been used to renovate sites and buildings and there are special schemes to co-ordinate partnerships between neighbouring inner city regions such as those in Newcastle upon Tyne and Gateshead, Manchester and Salford. In 1985 these partnership schemes were strengthened by setting up city action teams. In Scotland the Glasgow Eastern Area Recovery Project (GEAR) is similar, while in Northern Ireland the 'Making Belfast Work' initiative was launched in 1988.

Target Areas In 1987, 57 Target Areas (Fig. 24.4) with special problems and needs were identified and government funds concentrated in those areas. Grants are given for specific projects to strengthen and revive the local economy, encourage local investment, improve the environment and rebuild confidence.

Garden Festivals A number of Garden Festivals have been funded with the aim of improving the environment and attracting tourists to inner city areas. The first was held at Liverpool in 1984 and was followed by Stoke-on-Trent, 1986, Glasgow, 1988 and Gateshead, 1990. The 1992 festival will be hosted by Ebbw Vale.

Welsh Valleys Programme In 1988 the Welsh Development Agency introduced a three year programme to improve the economic, social and environmental conditions in the valleys of South Wales.

24.2.5 Assistance from the European Community

Britain's membership of the European Community since 1973 has made a variety of EC measures available to the assisted areas. These measures are designed to ameliorate the problems of less prosperous parts of the Community and to create a market within which there is equality of competition, whatever the national or regional location. The aim is to see that similar regions with similar problems get similar assistance and no attempts are made to prop up enterprises which are out-of-date or inefficient. Membership of the EC imposes constraints upon the member governments since the Commission has the power to examine state aid and to rule whether or not it is in accord with EC policy.

Britain's assisted areas are eligible for a variety of financial aids administered from Brussels and the role and magnitude of this help is steadily increasing.

Loans from the European Investment bank (EIB) This bank will lend up to 40 per cent of the fixed capital costs of major projects at low interest rates. It will also lend money to modernize undertakings or to develop new economic activities. For example, the Bank is to help finance British Aerospace's share of the development of the new A-330 Airbus with £150 million of credit funding.

Grants and Loans from the European Coal and Steel Community (ECSC) Money is available for the coal and steel industries or for other industries which would create jobs for redundant coal miners or steel workers. When, in 1979 the British Steel Corporation closed its plant at Shotton in· North Wales, funds from the ECSC helped to improve the Corporation's redundancy scheme and support re-training schemes. In 1983 the National Coal Board borrowed £14 million to invest in the Nottinghamshire coalfield.

European Regional Development Fund (ERDF) The fund supports projects in less prosperous regions of the Community in the context of regional development programmes. About half of ERDF grants are for infrastructure projects such as motorways and industrial estates. The rest goes to industrial projects such as the modernization of a motor vehicle components factory in Antrim. Tourism and related projects also benefit. For example, ERDF money helped the development of the Merseyside Maritime Museum at the Albert Dock in Liverpool.

European Social Fund (ESF) The ESF was set up to assist vocational training, retraining and resettlement schemes in the EC, particularly for school leavers and the unemployed. The UK has received grants for handicapped immigrants, creating new jobs for young people and improving working conditions. Sometimes ESF grants are combined with funds from the ERDF. For example, Bradford received an injection of money to improve the environment and revitalize an inner city conservation area by bringing in new residents and constructing offices and hotels.

24.2.6 Case study—South Wales

In the nineteenth century, the economic vitality of South Wales stemmed from two basic industries – coal mining and iron and steel. Much of the coal was exported and steel furnaces provided the raw material for a number of specialized steel processing industries such as tin plating, galvanizing and the production of sheet steel.

Because of its narrow industrial base, South Wales was less diversified than its two main competitors, Central Scotland and the North-East. In 1970 there were 52 pits operating in the region employing 42 600 men. By 1989 only 10 pits were operating with a workforce of 8000. The number of blast furnaces dropped from 15 in 1970 to 7 in 1989, only those at Newport and Port Talbot remaining in use. In the same period employment in the iron and steel industry declined from 60 000 to 22 000.

As Fig. 24.1 shows, unemployment in the region in October 1988 was high with a corresponding out-migration to find jobs elsewhere (Fig. 24.2). Decline in mining and manufacturing has resulted in a fall in the demand for new housing and other buildings (Fig. 24.5.).

Government assistance to the region has been channelled in recent years through the Welsh Development Agency which is responsible for approving new schemes for funding. The largest amount of investment has taken place in Mid-Glamorgan, the county with the highest unemployment rate. Communications have been improved with the building of the M4 which reaches as far as Swansea and Llanelli. Improvements have also been made to the existing road network including the Head of the Valleys (e.g. through Merthyr Tydfil) road.

Good labour relations, high productivity and the financial incentives offered by Development Area grants have attracted a number of firms to move to South Wales or to develop there. Ford has opened an engine plant at Bridgend in Mid-Glamorgan, aided by a grant from the European Regional Development Fund. The project has brought 1800 new jobs to the area.

No fewer than eight Japanese manufacturers have moved to South Wales e.g. Panasonic and Sony in the Cardiff area and Aiwa at Blackwood to the north. In the Newport area, Plessey Marine is providing 500 jobs making defence electronics. Other firms with new factories in Gwent are Inmos (electronics), Ferranti (computer programming) and Mitel Telecom.

Cardiff, founded on iron production and the export of coal has been revitalized by new road schemes and the conversion of the steelworks site at East Moors into a factory complex.

24.2.7 Case study – The Highlands and Islands

The Scottish administrative areas of Highland, Western Isles, Orkney and Shetland (see Fig. 24.6) have a total population of only 277 000 compared with 5.1 millions for the whole of Scotland. The population declined steadily after 1861 and it is only in the last two decades that it has begun to increase. With the exceptions of Caithness and Nairn all districts recorded higher populations in 1981 than in 1971.

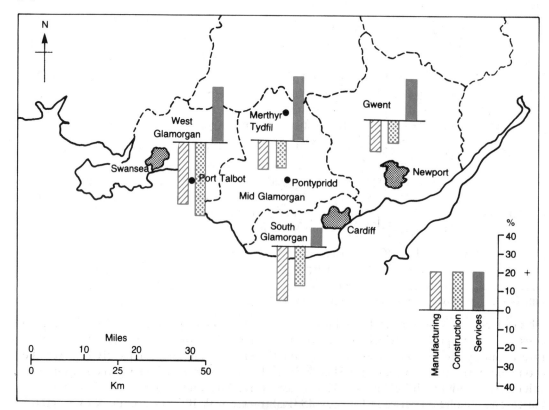

Fig. 24.5 South Wales: changes in the numbers employed 1977-1987

The main problems of the region are its remoteness, lack of resources, poor facilities for industrial development and limited industrial base. Although this rural region has been classified as a Development Area the exploitation of the North Sea oil and gas fields since 1974 has brought economic benefits to some parts of the region but not to others.

The Shetland and Orkney islands have benefited most and have low levels of unemployment, while Skye and the Western Isles have unemployment rates well above the Scottish average of 11.9 per cent, and, in the case of the Western Isles more than two times the British rate of 8.6 per cent (see Fig. 24.6).

The table below shows the number employed by companies whose output is wholly related to the North Sea oil industry.

Table 24.1 Numbers employed by North Sea oil industry

Administrative region	June 1980 (thousands)	Dec 1988 (thousands)
Highland	4.35	2.26
Islands	3.46	2.11
Fife	0.81	1.78
Grampian	32.32	46.42
Strathclyde	2.73	1.66
Tayside	1.81	1.17
Central & Lothian	0.86	0.59

Source: Scottish Economic Bulletin 1989

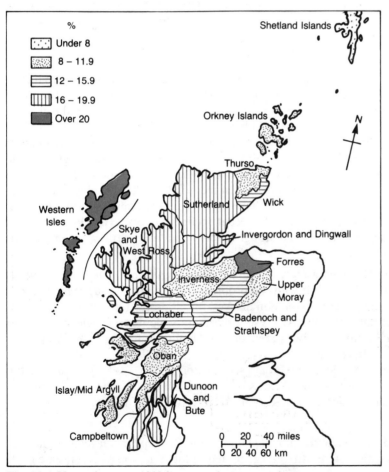

Fig. 24.6 Unemployment 1988 by unemployment exchange areas

Oil production has risen from 1.1 million tonnes in 1975 to 108.6m tonnes in 1988 but the rig construction programme which benefited the Clydebank yards has declined. Nevertheless investment by the oil companies has had a multiplier effect and has completely reversed the economic position of Orkney and Shetland which are in danger of having their traditional values swamped by the changes taking place there. Shetland experienced a 54 per cent increase in population between 1971 and 1981 compared with 10.7 per cent in Orkney and 6.3 per cent in the Western Isles. The increase in Shetland was related to the immigration of construction workers at the Sullom Voe terminal and once the work was finished there was a fall of 15.1 per cent.

Crofting, the traditional agricultural system of the islands and western coastlands has been assisted by the Highlands and Islands Development Board.

(a) There has been some consolidation of crofts to form viable farm units.

(b) Grants have been made for fishing and the development of fish-processing plants. This assistance is necessary since fishing is an important means of livelihood in most crofting communities.

(c) Money from the ERDF, the Highlands and Islands Development Board and the government has been used to improve the infrastructure, including roads, drainage and public transport. Small integrated programmes have been introduced involving work with the farmers and fishermen on such projects as a new pier on North Uist to improve facilities for shell fish and lobster boats.

Away from the crofting coastlands, *cattle and sheep rearing arable farming tourism* and *forestry* are the staple forms of livelihood. Assistance has been given to farmers and the tourist industry has been encouraged by the building of hotels and the development of Aviemore as an all-year-round tourist and sports centre.

Forestry has not provided the employment prospects which were once considered possible and the supply of timber for the paper and wood pulp industry of Fort William has proved costly, resulting in the closure of some plants. The Inverness area has proved attractive to firms which do not depend on local raw materials such as manufacturers of components used in telecommunications.

In recent years the problems of the Highlands and Islands have been overshadowed by large-scale investment in the oil industry which has not had widespread effects, and the rapid deterioration in employment and prosperity in the Strathclyde area where over 15 per cent of the workforce is unemployed.

24.3 GENERAL CONCEPTS

The government should adopt regional policies to make fuller use of the country's productive resources This concept grew up after World War II, partly because it was considered that public supervision of land use was in the best interests of the community, and partly from the growing belief that all members of the community should have the opportunity of employment. Jobs should be created in those parts of the country that suffered from economic stagnation and high levels of unemployment. There was a corollary to this idea, assistance to some regions could only be effective if limitations were imposed on growth in the more prosperous parts of the country. As a consequence restrictions were put on new factory and office developments in the Midlands and the South-East. Regional policies have had a significant influence on the spatial distribution of economic activity in Britain but the recent decline in overseas and home demand has resulted in setbacks to regional policy and brought about a relative deterioration in the once prosperous West Midlands.

The regional policy of the EC is increasingly affecting regional assistance. The Community is not concerned with national policies but with the problems of less prosperous regions within the Common Market with the goal of a 'common market' for all productive enterprises, whatever their national or regional locations. At present there are wide discrepancies in levels of prosperity within the Community. The average income in Hamburg, for example, is six times greater than that of southern Italy.

The European Regional Development Fund (see Section 24.2.5), is only available for projects in government scheduled assisted areas so the State can decide where the money is needed, although the amount allocated is based on EC assessments of material needs.

24.4 RELATED TOPICS

In many parts of the world national and regional planning have become increasingly significant ways of determining the location of industry and general development policy. Details are given in the unit on France about some aspects of that country's national plans.

Assistance to less prosperous regions and measures for national development are not confined to the more advanced industrial societies. Countries such as India and Brazil (see pages 183–190 and 191–198) also depend on central planning policy, implemented at a regional level, for the controlled development of their economies.

International organizations such as the World Bank and supra-national organizations such as the EC and ASEAN (Association of South East Asian Nations), all play important rôles in promoting economic development. The geographer studying countries belonging to one of these organizations must take into account the external assistance they provide.

24.5 QUESTION ANALYSIS

1 Use specific examples to illustrate how the economic geography of Great Britain is affected by government policy. *(in the style of Southern Universities' Joint Board)*

Understanding the question The term *economic geography* is not as formidable as it may appear at first sight. Economic geography is concerned with the spatial distribution of resources, activity and wealth. In the context of the question the emphasis is on the distribution of employment opportunities and the uneven pattern of prosperity in Britain. The principal means by which the government influences these distribution patterns is through the designation of areas to which various forms of assistance are given.

Answer plan Describe how government policy has attempted to restore prosperity to the non-prosperous regions of the country by a range of measures which have been geared to increasing employment and improving the infrastructure. Explain in detail the help given to the Development Areas (see Section 24.2.3) and inner cities (24.2.4) using examples from different parts of the country if possible, rather than from one particular region.

Give an account of how the prosperous regions i.e. the West Midlands and South-East England, have had controls to check new industrial building.

Describe how the government has also decentralized some departments of the Civil Service and for a time attempted to control office expansion in London by the use of Office Development Permits.

Finally, although much of your answer must be concerned with the industrial regions, write a section on the help given to rural areas such as the Scottish Highlands (see Section 24.2.7).

2 (a) Study Table 24.2 and present a geographical account of the changes in regional employment in Great Britain from 1979 to 1981. What factors may have been important in bringing about any spatial changes you can identify?

(b) Table 24.3 presents data on employment change in three Scottish regions in the 1970s. Evaluate the effect of North Sea oil upon these figures.

(c) With reference to Tables 24.2 and 24.3, consider the effects of scale upon the interpretation of geographical information.

Table 24.2 Employment changes in Great Britain 1979–1981

	% change in employment June 1979–Dec 1981			Average annual redundancies per 1000 employees 1979–81		Unemployment rate %	
	Total	Manufacturing	Services	Manufacturing	Services	Dec 1979	Dec 1981
South-East	−6.3	−14.5	−2.8	29.0	2.8	3.5	8.8
East Anglia	−8.4	−14.9	−5.3	31.6	1.8	4.2	9.8
South-West	−6.2	−11.3	−3.7	41.7	6.0	5.6	10.7
East Midlands	−7.2	−15.0	−0.5	41.3	3.0	4.6	10.6
West Midlands	−12.5	−22.1	−3.8	55.9	4.5	5.4	14.7
Yorkshire & Humberside	−9.8	−20.0	−2.8	60.2	4.0	5.6	12.9
North-West	−10.0	−17.3	−5.0	67.4	6.9	7.0	14.6
North	−11.9	−20.2	−5.5	68.0	3.3	8.5	15.8
Wales	−12.4	−23.4	−4.9	86.5	4.1	7.8	15.6
Scotland	−9.5	−21.6	−2.6	70.0	4.2	7.9	14.4
Great Britain	−8.6	−17.7	−3.3	49.0	3.8	5.5	12.0

Table 24.3 Employment and employment changes in Shetland, Orkney & Grampian (including Aberdeen) 1970s

	% of employees 1971			Change in employment in		
	Shetland	Orkney	Grampian	Shetland 1971–81	Orkney 1971–77	Grampian 1971–77
Agriculture, forestry and fisheries	4.8	19.2	7.3	−70	−90	−1738
Other primary	0.8	0.1	0.3	n/a	n/a	n/a
Food, drink & tobacco	15.9	7.0	10.1	−279	−48	−807
Textiles	6.9	1.0	2.9	−182	n/a	−339
Other manufacturing	3.3	2.0	13.2	+66	+77	+2709
Construction	9.8	10.7	8.8	+417*	+143	+3367
Services	58.5	60.0	57.4	+3457	+1045	+22109
Total	100	100	100	+3409	+1127	+25301
Total employees	4985	4789	150981	8394	5916	182198

*excluding Sullom Voe oil terminal construction workers.

(Northern Ireland Schools Examination Council, 1986)

Understanding the question The two tables complement each other. Table 24.2 shows employment changes during the two year period 1979–81 by standard regions for manufacturing and service industries. The table also shows the regional unemployment rate changes for the same period. Table 24.3 shows changes in employment by major occupational groupings in two Scottish regions – Orkney and Grampian for the period 1971–7 and in Shetland in 1971–81.

Answer plan Tackle each part of the question in turn, noting carefully the table to which it refers. *Part* (*a*) asks for a 'geographical account'. This means that you should describe changes between regions and also changes between manufacturing and service industries. Thirdly, you should describe changes in unemployment rates between regions. For example, whereas the rate in Wales doubled between 1979 and 1981, in the West Midlands it increased by 172 per cent.

In *part* (*b*) there is a marked difference between the decline in employment in agriculture, forestry and fisheries and some areas of manufacturing such as textiles, and the striking increase in employment in construction (such things as oil rigs, installations and housing) and the service industries. These increases can be attributed to the development of the North Sea oilfields during this period, particularly affecting the neighbouring islands and the Aberdeen region. List some of the services for which demand will have increased such as transport, communications and banking.

Part (*c*) The comparison in this section should be between the figures in Table 24.2 for Scotland and those in Table 24.3 which are for specific regions in Scotland. The impression given by Table 24.2 is that in Scotland the unemployment rate increased uniformly by 122 per cent between 1979 and 1981, slightly above the national average of 118 per cent. Table 24.3 (which is for a slightly different period of time) indicates that, whereas the figures for Scotland in Table 24.2 showed rising unemployment, those for the specified regions within Scotland in Table 24.3 show a fall in unemployment. Scale is therefore highly significant when interpreting geographical information.

24.6   Further reading

Balchin, P. N., *Regional Policy in Britain. The North–South Divide* (Paul Chapman, 1990)

Fothergill, S. and Gudgeon, S., *Unequal Growth: Urban and Regional Employment Change in the UK* (Heinemann, 1982)

Johnston, R. J. and Doornkamp, J. C. (ed), *The Changing Geography of the United Kingdom* (Methuen, 1982)

Kirby, D. and Robinson, H., *Geography of Britain* (UTP, 1981)

Manners, G., Keeble, D., Rogers, B. and Warren, K., *Regional Development in Britain* (Wiley, 1980)

25   France: population and regional development

25.1   Assumed previous knowledge

GCSE coverage of France consists of a survey of the major regions such as the Paris Basin and the North-East, together with systematic studies of such aspects as energy resources, viticulture and industrial regions. Many recent developments in France are changing the geography of the country. Unless you have studied France for GCSE recently you should spend some time reading up-to-date textbooks, published or revised in recent years.

25.2   Essential information

25.2.1   Population distribution and change

(a) Total population In 1987 France had a population estimated at 55.5 millions, an increase of over 2.6 millions in ten years. By European standards the population is growing relatively fast, with the result that 20 per cent of the population is under fifteen years old. Another 38 per cent is in the 15 to 39 category which includes the years when child-bearing is at its peak. The number of elderly people has also increased as a result of a reduction in the death rate.

(b) Distribution and density of population As Fig. 25.1 shows, although there are considerable variations in the population density the overall density is remarkably low. In 1987 there were 122 persons per square kilometre compared with 246 in West Germany, 325 in Belgium, 359 in the Netherlands, 190 in Italy and 233 in the United Kingdom.

The greatest concentration of population occurs in the Île de France where the Paris metropolitan area dominates the rest of France as well as the surrounding region. There are also high densities in the lower Seine valley and the Pas-de-Calais. The other major zone of relatively high densities is in the Rhône valley and along the Côte d'Azur. On Fig. 25.1 the distribution

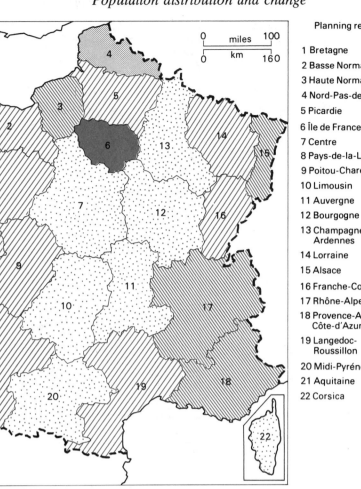

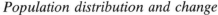

Under 60	
60–99	
100–199	
200–399	
Over 800	

Density km²

Planning regions

1 Bretagne
2 Basse Normandie
3 Haute Normandie
4 Nord-Pas-de-Calais
5 Picardie
6 Île de France
7 Centre
8 Pays-de-la-Loire
9 Poitou-Charente
10 Limousin
11 Auvergne
12 Bourgogne
13 Champagne-Ardennes
14 Lorraine
15 Alsace
16 Franche-Comté
17 Rhône-Alpes
18 Provence-Alpes-Côte-d'Azur
19 Langedoc-Roussillon
20 Midi-Pyrénées
21 Aquitaine
22 Corsica

Fig. 25.1 Population density 1982 by planning regions

within the Rhône-Alpes and Provence-Alpes-Côte d'Azur regions is more concentrated than it appears when more refined statistics are examined. The central and southern Alps have very low densities whereas the Mediterranean coastlands from Marseille to the Italian frontier and the region around Lyon and Grenoble are areas of relatively high population densities.

The regions with the lowest densities include the upland regions of the Massif Central, the Alps and the Pyrénées, together with the Les Landes region of Aquitaine. These regions of low density are essentially rural areas not influenced by neighbouring urban agglomerations. However, even in rural areas the more productive farmland supports higher population densities – the main examples being fruit and vegetable growing along coastal regions of Brittany and the Rhône valley, the viticulture of Bas-Languedoc and the arable farming of Alsace.

The dominant feature of the population pattern is the distribution of urban centres, particularly those which have absorbed smaller rural communities to become major urban agglomerations. (The word *agglomération* is used in the French Census to describe clusters of urban communes – districts with some self-government and a mayor.) The fifteen largest urban agglomerations are shown in Fig. 25.2. The map indicates the size and significance of the Paris urban region, nearly eight times as large as the Lyon region which is the next largest agglomeration. Four of the agglomerations are coastal ports which have rapidly increased their share of French industrial development and trade in recent years. They are Nantes, Bordeaux, Toulon and Marseille, the last-named agglomeration being the largest port complex in France.

Urban agglomerations have increased in size at the expense of rural areas and there has been a significant shift towards the polarization of the French economy within urban centres.

There is a number of reasons for the migration from rural areas.

(i) The consolidation of agricultural holdings and increased farm mechanization has led to a fall in the demand for agricultural labour.

(ii) The growth of industries and tertiary services in urban areas has attracted people from rural areas where work is more difficult to obtain.

(iii) The movement of young people to the towns to obtain work leaves rural areas with low birth rates. This results in an excess of deaths over births in many areas such as the Massif Central.

(iv) Positive efforts have been made by the State to develop urban communities by investment

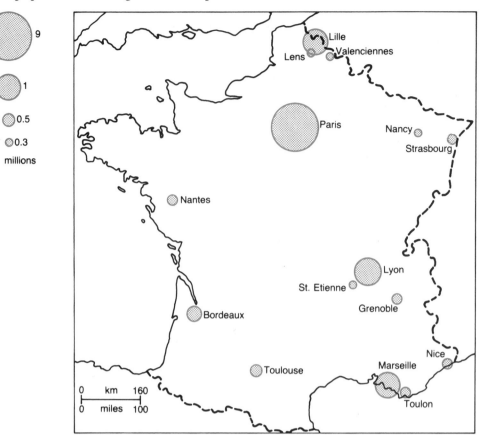

Fig. 25.2 Relative size of the fifteen largest urban agglomerations 1982

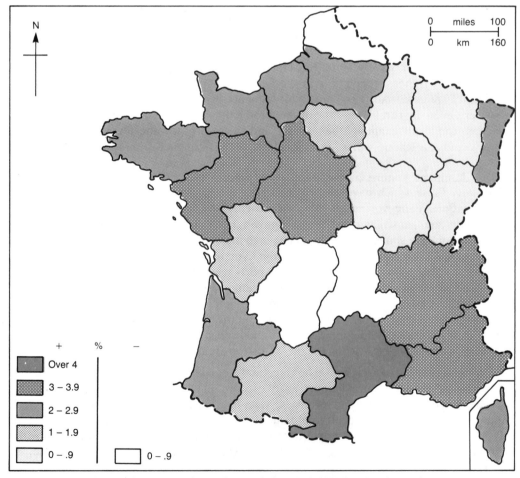

Fig. 25.3 Percentage change in population 1982–1987 by planning regions

to equip the towns to meet the material and cultural needs of urban society.
(v) The depletion of rural population results in the erosion of services, particularly those such as clubs and recreational amenities which enrich the quality of life.

(c) Movement of population Fig 25.3 shows the percentage change in population which took place in the five years between 1982 and 1987. The trend shown on the map has been caused by inter-regional migration in addition to natural increase. The Paris agglomeration is no longer the magnet for migrants which it used to be. Decentralization policy to relieve some of the congestion in Paris has resulted in population movement to some of the surrounding départements, particularly in the Centre region.

More significant, however, has been the migration to the planning region of Rhône-Alpes and the neighbouring region of Provence-Alpes-Côte d'Azur and to Languedoc-Roussillon. The increase in these regions is particularly significant since it consists to a large extent of migration from other regions rather than natural increase which is lower here than in parts of northern France. At the time of the 1982 census, 10.3 per cent of the people in the Provence-Alpes-Côte d'Azur region had lived elsewhere in 1975. The climate of the Midi with its warm winters and long hours of sunshine has attracted specialist industries and tertiary services as well as retired people. Furthermore the area has also been attractive to overseas immigrants such as Algerians. At the same time the older industrial regions of the Nord and Lorraine have become zones of exodus with a decline in the overall population in Lorraine and virtually no change in Nord-Pas-de-Calais. In the Massif Central the populations of Limousin and Auvergne showed a slight decline.

What Fig. 25.3 does not show is the continual decline of many of the rural areas, with the loss of young people, and the rapid increase in the population of the urban areas where work is available. In 1982, 74 per cent of the French population lived in settlements with populations of over 2000, whereas in 1931 the urban population only just equalled that of the rural communes. In recent years the population of many French cities, except Paris, has increased rapidly. Marseille, Bordeaux, Toulouse, Nantes and Toulon have expanded at rates above the national average. Even higher growth has been experienced by small cities such as Aix-en-Provence (+14.4 per cent) and Perpignan (+17.2 per cent). Most of these increases are the result of net migration and not natural increase.

25.2.2 Agriculture

The agricultural labour force Over 40 per cent of French agricultural holdings are farmed by owner-occupiers who depend on family labour to work the holdings. Little more than 16 per cent of the labour force consists of paid workers. The family unit is the mainstay of French agriculture and the high proportion of female labour is supplied by grandmothers, wives and daughters who work part-time. Fig. 25.4 shows the three main categories of the agricultural labour force and the very large reductions in the numbers working the land between 1962 and 1982. Nearly half the farms are held by tenant farmers under a variety of schemes.

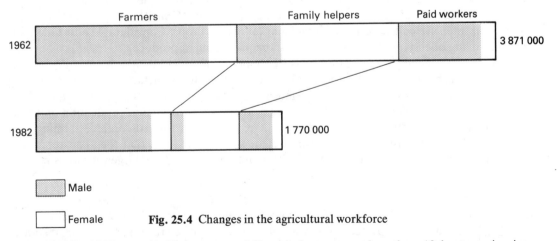

Fig. 25.4 Changes in the agricultural workforce

Farm size In 1987, nearly 35 per cent of French farms were less than 10 hectares in size. Many of the smaller farms were little more than large gardens providing specialized products such as vines, flowers, fruit and market garden crops. The highest proportion of French farms are between 10 and 50 hectares in size. As Fig. 25.5 shows, about 48 per cent of the holdings are in this category, with farms of over 50 hectares forming nearly 17 per cent of the total. The relatively small size of French farms has encouraged co-operative schemes for bulk purchasing and marketing.

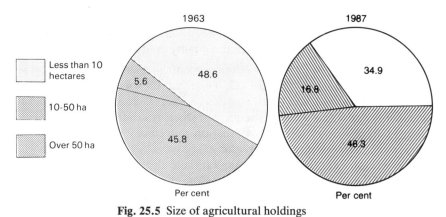

Fig. 25.5 Size of agricultural holdings

Agricultural problems There have been strenuous efforts by the government to rationalize the structure of farming. The modernization programme has been particularly concerned with the following aspects of farming.

(a) *Redistribution and consolidation of property* Many farms are fragmented with small fields restricting the use of machinery. The consolidation programme funded by the government comprises two organizations to promote consolidation. The first, formed in 1960, consists of the *Sociétés d'Aménagement Foncier et d'Établissement Rural (SAFER).* These societies buy land offered for sale and use it to enlarge farms which are too small to operate efficiently. The second, formed in 1963 is known as *Fonds d'Action Social pour l'Aménagement des Structures Agraires (FASASA).* This fund offers incentives to farmers to retire at 60, subsidizes farmers moving to areas where operators are scarce and retrains operators who wish to leave non-viable farms. Fig 25.4 shows that there has been a remarkable drop in the size of the agricultural labour force since 1962. Fig 25.5 shows changes in the sizes of holdings between 1963 and 1987 which indicate that there has been progress towards larger and medium sized holdings.

(b) *Increasing the level of technology* The State has invested considerable sums in rural electrification and the improvement of water supplies. Teams of agronomists and engineers promote agricultural advancement, but the individual co-operation and initiative of the farmers is essential. In remote areas and on smallholdings where little capital accumulation is possible, investment in fertilizers and new machinery is extremely limited.

(c) *Meeting the needs of the market* Membership of the Common Market has heightened the need for French agriculture to increase productivity to remain competitive with the other members of the Community. Marketing boards for commodities such as grain and wine have been set up and a network of regional markets serves the main consuming areas. French farms are still geared to polyculture (the growing of a variety of crops) together with livestock rearing, and it has proved difficult to encourage farmers to change to new crops. Specialization is only to be found in some regions such as the limon-covered plains of the north-east (grain) and Languedoc (viticulture). In the lower Rhône valley irrigation schemes have vastly increased the area under intensive cultivation of fruit and market garden crops. Considerable effort has gone into improving quality, maximizing yields and the packaging of produce.

(d) *Resistance to change* Sporadic outbursts by French farmers against the importation of, for example, British lamb or Irish beef are evidence of the reluctance of many farming communities to accept rationalization and adopt larger farm units. Historically these communities of small farmers have been the backbone of French food production and they have retained considerable political influence. The movement towards organic farming may help to prolong the existence of many inefficient small farms provided they can retain the quality of output which has been sacrificed for quality by many of the larger agro-businesses.

25.2.3 Energy resources

France is unable to meet her energy needs from domestic resources and just under half the total energy needed must be imported, mostly as oil. Domestic power resources are long distances from many of the consuming areas. Western France is particularly weak in energy resources. Both domestic and imported energy is located around the periphery of the country and distribution is therefore costly.

In Fig. 25.6 the domestic resources from which energy was produced in 1987 are shown, expressed in millions of tonnes of oil equivalent. The distribution pattern is as follows.

Coal The Monnet Plan has modernized the coal industry but competition from oil and imported coal has drastically reduced output. In 1964 France produced 55 million tonnes of

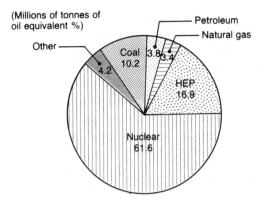

Fig. 25.6 Production of energy from domestic sources

coal, in 1987 only 17.5 million tonnes were mined. Furthermore, the most important area of production is no longer the Nord-Pas-de-Calais field. In 1987 only 1.4 million tonnes were produced there compared with 9.9 million tonnes in Lorraine. The third area of production is the Centre-Midi where the output of 5.2 million tonnes comes from a number of small fields.

Petroleum The major source of petroleum is at Parentis in Les Landes which is linked by pipeline to refineries at Bordeaux. There are also small deposits in the Brie region of the Paris Basin and near Pechelbron in Alsace but the total domestic output is only 3.8 per cent of the petroleum refined in France. The imported oil is processed at large refineries located at deep-water sites such as the Seine estuary and the Étang de Berre to the west of Marseille, or at inland sites supplied by pipeline from the coast. The South European pipeline from the Étang de Berre to Karlsruhe in West Germany has resulted in refineries being built at Feyzin near Lyon, and at Reichstett and Herrlisheim near Strasbourg. A pipeline from the Seine estuary to the south-east of Paris is linked to a refinery at Grandpuits. Similarly a pipeline from the Loire estuary to Rennes has resulted in a refinery being built near that city.

Natural gas There was a major find of natural gas in 1951 at Lacq in the south-west of Aquitaine ten years after a small gas field had been found at St Marcet to the east. These fields are now joined by pipelines to Toulouse, Dijon, Paris, Rennes and many other centres en route. Natural gas supplies are very important as a source of power for industry, particularly the chemical industry, and 84 per cent of the gas used is imported, much coming from the Netherlands.

Hydro-electric power The Alps, the Pyrénées and the Massif Central provide extensive water resources which are harnessed to make hydro-electricity. Multi-purpose schemes have been developed on the Rhône and the Durance. Great engineering skill, suitable sites and water resources have been utilized extensively to provide hydro-electricity in large quantities. Eighteen per cent of the electrical energy is derived from water power with half of the amount coming from the Rhône valley and the neighbouring Alpine region, one-fifth from the Massif Central and the remainder from the Pyrénées and Rhine valley.

Nuclear power The deficiencies in indigenous energy resources have stimulated the building of over 50 nuclear power stations. These are located in areas where sufficient power to meet the demand is not available from local resources. For example, there are sites at Chinon at the confluence of the Loire and Vienne, St. Laurent-des-Eaux between Orléans and Blois, and Brennilis in central Brittany. Heavy investment in the nuclear energy programme in recent years has increased the amount of electricity generated by nuclear power stations dramatically. In 1987 nearly 70 per cent of electricity came from nuclear power.

The siting of nuclear reactors near border zones and the Channel coast has caused some concern in Germany, Luxembourg and the United Kingdom. Local resistance to atomic power sites has become more evident since the Chernobyl disaster in 1986.

Other power sources Fig. 25.6 shows that more energy is produced from 'other' sources than from either domestic natural gas or petroleum. These sources include the tidal barrier on the River Rance at St. Malo, small solar power stations in the Pyrenees and thermal power in the Massif Central.

25.2.4 Industrial development

Mineral resources The most important mineral resource is *iron ore* which, apart from a small quantity in Normandie near Caen, comes from the Moselle area of Lorraine. In 1981 the Lorraine iron-ore field produced 11 million tonnes of ore with an iron content of about 31 per cent. A further 15 million tonnes is imported to meet the needs of the industry. Lorraine also possesses vast deposits of *rock salt* which, in association with local coal and petro-chemicals, is the basis for the extensive chemical industry at Dombasle, Sarralbe and Dieuze. There are also deposits of *bauxite* in Provence at Brignoles with smaller deposits at Villeveyrac and Bedarieux in Languedoc. Nearly 1.7 million tonnes was mined in 1983, reduced to alumina using nearby lignite deposits, and then smelted in the Alpine valleys using HEP, or at Noguères, near Lacq using cheap electricity generated from natural gas.

Industrial regions France possesses very few extensive industrial regions on the scale to be found in Britain or West Germany. The largest industrial agglomerations are: (a) the Paris region; (b) the Nord-Pas-de-Calais coalfield with extensions around Lille, Tourcoing and Roubaix; (c) Lorraine with discontinuous industrial centres between Longwy in the north and Nancy in the south, extending eastwards to the German frontier; (d) Lyon and the neighbouring scattered industrial centres in the Loire coal basin.

In addition to these industrial regions there are port industries based on the import of raw materials such as crude oil and iron ore, facilities for re-export and transport inland, and local demand. The ports which have developed as industrial centres are Marseille, Bordeaux, Nantes, Le Havre and Dunkerque. The inland ports of Rouen and Strasbourg have similar characteristics.

Finally there are a number of minor industrial centres, some with a wide and others with a narrow range of industries. Their locations are dispersed and they draw on local skills and obtain raw materials and energy resources from elsewhere. They include Toulouse, Grenoble, Rennes, Clermont-Ferrand and Dijon.

25.2.5 Relocation of industry

Industrial development in France took place in the nineteenth and early twentieth centuries on or near the coalfields and iron ore resources. These developments, together with good routes, particularly those converging on Paris, encouraged industrial growth in the north and east of the country. As a result three-quarters of the industrial capacity was east of a line from the département of Normandie to Haute-Alpes. Except for the seaports there was an absence of large industrial plants in the centre, the south and west and a corresponding deficiency of labour, capital and business enterprise. Since the end of World War II there have been considerable changes in the distribution of industry. These changes have been the result of two major forces. The first was the *decline of the basic industries* such as wool and cotton textiles, iron and steel, heavy engineering and shipbuilding. At the same time there has been an *expansion in demand* for aerospace products, electronics and consumer goods which are not dependent on nearby coal or raw material supplies.

The second influence on industrial location has been the State which has set up *planning regions* (see Fig. 25.1) and introduced a succession of national plans. State assistance to industry in certain areas and State measures to improve the quality of life in the poorer regions have helped to bring dramatic changes to industrial location. Between 1954 and 1975 the numbers employed in manufacturing increased by 99.8 per cent in the Pays de la Loire region, 91.1 per cent in Normandie (Baisse), and 65.6 per cent in Centre. In the Nord-Pas-de-Calais there was a decline of 19 per cent and there was virtually no change in Île-de-France.

Some industries which were concentrated in a few localities (for example, car manufacture) have been encouraged to disperse, in some cases to areas of declining industries such as the Nord coalfield, to prevent large-scale unemployment.

25.2.6 Regional development – The Massif Central

The planning regions of Limousin and Auvergne which cover much of the Massif Central form the most extensive rural problem area in France. The main difficulties of the area can be summarized as follows.
(a) Physical conditions and a limited natural resource base which restrict the agricultural and industrial potential of the region.
(b) Rural depopulation, leaving the older people behind.
(c) Lack of comprehensive road and rail networks.
(d) Small farm units, uneconomic to work – over 90 per cent of the farms are less than 50 hectares in size.
(e) Lack of urban development except in cities like Clermont-Ferrand and Limoges which are peripheral to the region.
Improvements to the area are taking the following forms.
(a) Consolidation and enlargement of farm holdings and the encouragement of older farmers to retire (see Section 25.2.2).
(b) Setting up of agricultural co-operatives.
(c) Introduction of improved livestock breeds.

(d) Afforestation of upland regions.

(e) Encouragement of tourist industries, particularly winter sports.

(f) Improvements in communications.

In other problem areas similar measures are being introduced, where appropriate.

25.3 GENERAL CONCEPTS

France is a country of urban communities This is a relatively recent concept. France used to be considered as essentially an agricultural nation with a relatively static population which was concentrated in the northern half of the country. But changes which have been taking place since 1945, in particular, the rapid increase in the total population, have rendered this picture inaccurate. These changes have brought about new patterns in French regional and national life.

The French Government is a major instrument of planning and change This concept is also new. France was the first country in western Europe to introduce national economic planning. This planned economy has resulted in extensive changes to the spatial distribution of people and employment. An urban hierarchy has emerged and this, together with the existence of distinctive regions, forms the basis for planned development at national level.

French society has proved to be more dynamic in the last three decades than previously. With so much rapid change, it is important to make sure that, when studying the geography of France, you have up-to-date information.

25.4 RELATED TOPICS

Some of the problems associated with France today, such as rural depopulation, industrial change, urban agglomeration and the need for central planning, are to be found in many other industrialized countries, including Britain and France's neighbours in Western Europe. The causes, for example, of rural depopulation in France are similar to those experienced in Norway and Scotland, as well as elsewhere. The remedies, allowing for differences of emphasis, are also similar. When studying the problems of other advanced economies it is well worth relating what is happening in France to the area being studied.

Industrial decline in the traditional industries, including the coal industry, has not been confined to France. It is a phenomenon which has its counterpart in Britain, the United States, Belgium and the majority of countries which developed as industrial nations in the nineteenth century.

25.5 QUESTION ANALYSIS

1 Why has there been a significant migration from rural areas in France in recent years?

(in the style of Cambridge)

Understanding the question Although the question does not identify the time span precisely, the expression *in recent years* can be assumed to cover approximately a twenty year time span. Statistics only become available once or twice each decade so a 'significant' migration would not become apparent in less than about ten to twenty years. The question does not refer to the attraction or 'pull' of urban areas, but do not limit your answer to the negative attraction of the rural areas. The positive gains from living in towns and cities are strong inducements for migrants from the countryside. When giving your reasons for the rural exodus be careful to distinguish between immediate causes such as greater job opportunities in the towns, and the long-term effects such as the gradual reduction in the birth rate resulting from the decline in the number of young people in the countryside.

Answer plan The question requires an orderly presentation of the reasons for rural migration and care should be taken to maintain a balance giving each reason sufficient attention.

Before answering the question make a list of the reasons and then classify them into those connected with farming, transport, job opportunities and social opportunities. Explain the problem of farming in many rural areas (see Section 25.2.2 and Section 25.2.6), using specific examples whenever possible. Describe the limited transport facilities, job opportunities and social provision in rural areas using examples instead of generalisations. Contrast these facilities with those available in urban areas. Again, you must be as specific as possible, drawing on either case studies you have read or places and regions you have visited.

The final section of the answer should be concerned with the long-term consequences of rural migration, such as the decline in village communities to the point where shops and other services become uneconomic. You should point out that these long-term consequences have a 'snowball' effect by persuading more people to leave the rural areas until the point is reached when the region is depopulated.

2 Fig. 25.7 shows regional variations in the scale of government aid to industry in France. Explain the pattern shown.

(Oxford, June 1986)

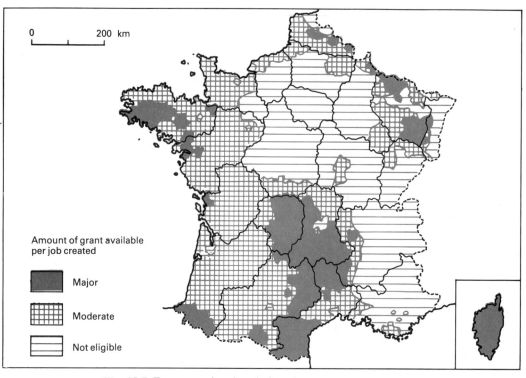

Fig. 25.7 France: regional variations in government aid to industry

Understanding the question Look carefully at the explanation of the key shown on the map. It gives two levels for grants awarded by the government for each job created in different parts of the country. The largest grants are given to areas shaded most heavily, while more moderate grants go to areas shaded less heavily. The third grouping, which is least heavily shaded but which must not be ignored, shows those areas that are not eligible for a grant. You should note that although the boundaries of the planning regions are shown, the choropleth is based on smaller and more precise areas in many cases.

Answer plan The first paragraph of your answer should explain briefly the basis used in France for regional aid. As in Britain, industrial expansion is the keystone of regional development. Grants are given to firms for each job created within a fixed period. The underlying policy is one of decentralization away from Paris and the Paris Basin – the core region. Growth is encouraged in the peripheral regions, the intention being to implant self-sustaining industrial growth in the provinces which will, in the long run, be beneficial to the Paris conurbation as well as the rest of France.

A further paragraph should examine the areas receiving the highest amount of grant. These fall into two broad categories: peripheral rural regions such as Corsica, south Brittany and the Massif Central, and old industrial regions where heavy industries have closed down, such as the north-east and Lorraine. A survey of the problems of the rural regions should be given (see 25.2.6), followed by sentences describing the decline of the basic industries (see 25.2.5).

Another paragraph should examine the areas not eligible for grant, i.e. the Paris Basin, Alsace and the growth areas of Rhône-Alpes and Provence-Alpes-Côte d'Azur. Do not forget to mention the smallest areas which include the industrial cities of Bordeaux, Toulon, Clermont Ferrand, Rennes and St Étienne.

The final paragraph should describe the areas obtaining moderate grants. These are mainly rural regions where there are few industries, such as Aquitaine and the Cotentin Peninsula.

25.6 FURTHER READING

Clout, H. D., *The Geography of Post-War France* (Pergamon, 1973)

Clout, H. D., *The Massif Central* (OUP, 1974) (This is one of a series 'Problem Regions of Europe', which includes several other useful titles on areas of France.)

Flower, J. E. (Ed.), *France Today* (Methuen, 1987)

House, J. W., *France: an Applied Geography* (Methuen, 1978)

Thompson, I. B., *Modern France: A Social and Economic Geography* (Butterworth, 1970)

Winchester, H. and Ilbery, B., *Agricultural Change: France and the EEC* (John Murray, 1988)

26 West Africa: population issues

For GCSE, you will already have studied the climatic and vegetation zones of West Africa, and the nature of the desert land, savanna and rain forest regions. You will have some knowledge of related agricultural economies – nomadism, shifting cultivation, cattle-rearing in the Tropics, plantation agriculture and the production of tropical food crops. If you studied Africa or West Africa as areas for detailed regional study, you should be well-informed about specific countries such as Nigeria or Ghana. You will also have studied other relevant topics such as the exploitation of mineral resources in the Third World and the growth of world population.

26.2 ESSENTIAL INFORMATION

26.2.1 Definitions and formulae

Optimum, under and overpopulation See Section 26.3.

Demographic transition See Section 26.3.

Urbanization There is no simple definition of urbanization. Broadly speaking it is the proportion of the population living in towns and the increase in the population of town dwellers compared with that of rural dwellers. As a result of this process there is an increase in the number and size of towns and cities. The rate of urbanization is accelerating. Between 1950 and 1970 the proportion of the world's population living in towns and cities increased from 28.2 per cent to 38.6 per cent. It is now the dominant factor causing changes in the distribution of population. Urbanization is particularly rapid in developing countries, including those in West Africa.

Birth rate This is the number of live births per thousand of population during a particular year. Birth rates vary from about 12 to 50 throughout the world. The birth-rate is over 40 in all the countries of West Africa. We need to know the birth rate together with the death and net reproduction rates because together they indicate whether a population is likely to grow or decline.

Death rate This is the number of deaths per thousand of population during a particular year.

Migration Spatial movements that involve a change of place of residence and the crossing of a political boundary. Residential moves which do not result in crossing such a boundary are usually referred to as mobility rather than migration. *Out-* and *In-migration* refer to internal

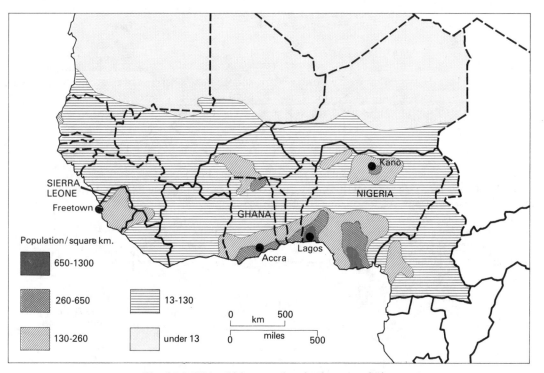

Fig. 26.1 West Africa: rural and urban population

migration from or to a given area within a country. *Internal migration* refers to change of residence within a country.

Net reproduction rate This is the average number of female children born to every woman in the population. This measure recognizes that many children do not live long enough to become parents, many women die when they are still young enough to have children and some who are alive are too old to have children. NRR is a measure of whether a society is reproducing itself. If NRR = 1 the population is stable; more than 1 indicates a growing population; less than 1 means that the population is in decline.

Population density This measure relates the size of population to the area of land on which it lives.

$$\text{Pop. density} = \frac{\text{number of people}}{\text{unit of area}}$$

It is expressed as the number of people per square kilometre or square mile, e.g. the present density in Britain is about $229/km^2$. This figure is the *average* number of people per square kilometre of the country. It does not recognize the difference in density between city areas and the Highlands of Scotland. Fig. 26.1 shows that in West Africa the density of population increases southwards from the Sahara to the Gulf of Guinea.

Primate city The largest city in a country or region. It is also the centre of political affairs, trade and economic, social and cultural activities. According to the *rank-size rule* the primate city should be twice the size of the second largest city. However the ratio between the two cities is often considerably higher.

26.2.2 Main features of the population of West Africa

The distribution pattern shown on the map (Fig. 26.1) reflects the influence of climatic factors upon settlement. Human life is concentrated south of a line which approximates with the 200mm (8 in) isohyet. This line runs from the lower Senegal valley to the north-east of Lake Chad. To the north of this line conditions are too dry for cultivation so permanent settlements are found only at oases (see Fig. 26.2). Thirty-three per cent of the population of Africa lives south of this line.

The colonization of West Africa by European nations intensified the pattern of increasing concentration near the coast of the Gulf of Guinea. The rain forest climate of the area encouraged the growth of commercial crops such as cocoa and rubber. Ports were established to handle trade with Europe.

The population of West Africa is growing rapidly but since the density of population was low compared with other parts of the world the problems caused by the rapid growth are not as severe as in Asian countries. The rates of growth in West Africa do not differ as much *between* countries as they do between districts *within* the same country.

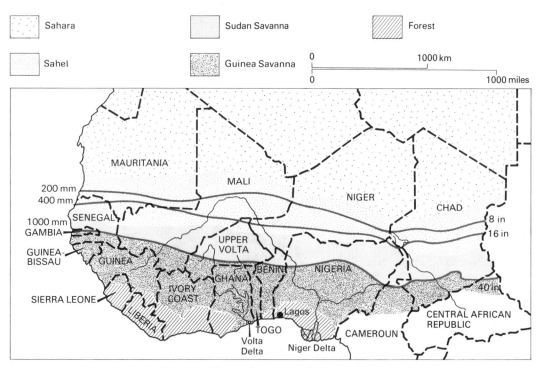

Fig. 26.2 Vegetation zones of West Africa

The basic picture today is of concentration of intensive economic activities and of people in scattered 'islands' which are separated by vast areas in which little change has occurred.

Before the establishment of colonies in the nineteenth century there were few towns in West Africa except in the lands of the Hausa in Northern Nigeria (e.g. Kano) and in Yorubaland (e.g. Ibadan and Oyo). When colonies were established existing towns became administrative centres. Those which were developed as river or coastal ports or rail termini grew fastest. The largest became colonial capitals. Since independence national capitals have grown rapidly as political and administrative centres. They now attract immigrants from other parts of the country.

Urban growth is now the most widespread process. The number of town dwellers doubles every ten to twelve years. Urban populations are characterized by a high proportion of young people (15–45 years old) who have migrated from the rural areas.

26.2.3 Rural-urban migration

In 1975 only 25 per cent of the people of Africa lived in urban regions. By 2000 AD it is estimated that 42 per cent of the population will be urban. This change involves the massive re-distribution of people to certain key receiving regions. The primate cities and their surrounding regions are the major magnets.

Reasons for this movement include:

1 *The influence of colonial contacts* In colonial times the spatial structures of the economies of different territories were focused upon a small number of port cities. These cities became the centres of newly established transportation systems and gradually became the largest internal markets and the centres of manufacture. Rural-urban migration focused upon these nodes. So urbanization was not common to all Third World urban centres but specifically to those which became centres of the western form of 'modernization'.

2 *The effects of independence* Since independence the domination of these cities has increased. Multi-national companies have been attracted by cheap labour e.g. females for the textile industry so factories have been located in or near the largest urban agglomerations. The 'developed' urban regions have therefore become even more attractive to migrants.

3 *Perceptions of high wage opportunities* These have attracted rural people who also believe they will have a better life style in the big city. The migrants have also acted as sources of information about job opportunities and wages for families and friends they have left at home. If they feed back positive information to their home areas it stimulates further migration.

4 *Transfer of money* Money sent back to their families also stimulates migration. The money reinforces optimistic perceptions of the opportunities offered by the city. If the money is used to educate the young they have wider horizons and are more likely to move away.

26.3 GENERAL CONCEPTS

Optimum, under- and overpopulation These concepts relate to the law of diminishing returns. It is argued that as long as the techniques used remain the same, the application of additional capital and labour will lead to a proportional increase in output. This applies both to agricultural land and to industries. There is a point of maximum return, which is the optimum (best) situation.

Optimum population is therefore the point at which a country has achieved a density of population which, with the given resources and skills, produces the greatest economic welfare (the maximum income per head). If the population is greater than this the country may be said to be overpopulated. If it is less, it is underpopulated. In the case of both underpopulated and over-populated regions the standard of living is lower than it would be if the optimum prevailed. In West Africa it could be argued that the Sahel region is overpopulated. Liberia could be said to be underpopulated because mining development is held back by a limited supply of labour.

Demographic transition This is the change from a low total population experiencing high birth rates and high death rates to a high total population experiencing low birth and low death rates. There are four main stages recognized in this process.

1 *Initial or pre-industrial stage* This is a period of high birth and high death rates. Since both are high there is little natural increase in population size.

2 *Early transitional stage* This is a period of high population growth because the death rate is declining much more rapidly than the birth rate.

3 *Late transitional stage* The population continues to grow overall but at a slower rate. The birth rate is now declining more rapidly than the death rate.

4 *Industrial or post-transitional stage* This is a period in which both birth and death rates are low so population growth is small, or there may even be a decline in numbers.

Most of the countries of the Third World, including those in West Africa, are in the second stage of the demographic transition. Modern farming techniques produce more food and modern medicine has been introduced, so death rates have declined. Birth rates are still high, partly because large families are socially prestigious.

26.4 DIFFERENT PERSPECTIVES

The Malthusian perspective When the population of Britain was increasing rapidly, Thomas Malthus published an essay on population in which he argued that the rate of increase in food supply could not keep up with the rate of growth in population. He forecast that unless population growth was checked there would inevitably be starvation, disease and war. But the agricultural revolution and the spread of farming to the grasslands of the southern hemisphere increased food supply dramatically. The fate outlined by Malthus seemed to have been put off indefinitely.

In the Third World today the supply of food has increased e.g. the Green Revolution but the increase has not kept up with the rapid growth of population. So in areas like West Africa the relationship of food supply to population is a crucial problem.

The demographic transition theory outlined above provides a comparative perspective. The theory is based on observations and descriptions of what has happened in Europe and North America in particular. There is no guarantee that what has happened in other parts of the world in the past will be repeated in the Third World today. For example mortality declined in Europe and North America because housing and health conditions improved. In West Africa it is the result of the application of modern medicines. At present there is little evidence of a decrease in fertility which is a key part of the transition from stage 2 to stage 3.

26.4.1 Models of migration

The *Todara model* (1969) seeks to explain and predict the volume of rural-urban migration in terms of the difference in income between the formal wage earning urban and rural sectors. The effects of this difference are modified by the expectation of migrants about getting a job in the urban sector. It is criticized because it only considers economic motivation and ignores the fact that many new immigrants survive by earning money in the informal sectors of the economy.

The *Amin model* (1974) of labour migration sees internal migration as a form of involuntary transfer from the rural to the coastal regions of Africa as a result of the unequal investment of international capital into export crop producing areas. It is criticized as being crude and over-simplistic.

Mabogunje's model (1970) described rural-urban migration as a circular, interdependent and increasingly complex system. This model recognizes how personality and other individual characteristics influence decisions as well as the physical and socio-economic environment.

Clarke and Kosinski (1982) developed an *alternative opportunity* model. They suggest that in order to understand the causes of the flow of people this flow has to be put in the context of flows of capital, information, innovations profits and information supplied by returning migrants (Fig. 26.3.) The over-riding goal of the migrant is to maximize his opportunities – first in his home area, but if this is not possible, elsewhere.

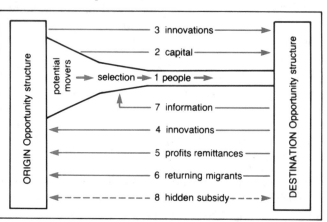

Fig. 26.3 Spatial flows associated with population mobility

26.4.2 Results of migration

The process of migration is self-reinforcing and the population redistribution that results can cause regional economic change. Migration affects the cities to which the migrants move and the areas from which they move. The economically most productive people and the demographically most fertile move to the cities. This creates massive housing and employment problems as well as the associated social problems. Rural areas may be left with the elderly and sick and have severe labour shortages for harvesting. In some areas farmers have changed to a cash crop economy because migrants can act as their selling agents in the city. So migration can significantly change the character of rural economies.

26.5 RELATED TOPICS

You should look at the following units:

14 *Growth and distribution of population* will help you place West Africa in a wider context.
23 *Less-developed countries* outlines the economic and social problems faced by newly emergent countries.
16 *Rural settlement patterns* and 17 *Central place theory* provide theoretical background on the processes involved in the evolution of settlements.
15 *Movement of population* deals with the effects of migration on the distribution and density of population.

26.6 QUESTION ANALYSIS

1 How valid is the statement that 'the varying density of population in West Africa south of the Sahara is primarily a reflection of the climatic conditions'?   (*in the style of the Joint Matriculation Board*)

Understanding the question The question is about *density* of population, not distribution, although the two are obviously inter-related. To show that you are aware of this fact it would be a good idea to define density in the early part of your answer.

To answer the question well you need to have good knowledge of the climatic conditions of the region, the pattern of population density, the key factors which have determined this pattern and West African examples and facts to back up the arguments. You are not being asked to show how climatic factors determine density but to judge how important climate is in comparison with other factors. *Primarily* means chiefly so the statement really implies that climate is the most important though not the only factor.

Comparison of Fig. 26.1 and Fig. 26.2 will bring out the relationship between population density and climate very clearly.

By referring to the diagrams you will be able to make the following points briefly:

The Sahara is largely uninhabited. Permanent settlements are located at oases and on highlands where water is available. Nomadic herdsmen make use of sporadic vegetation growth after rain storms as grazing for their animals. Density of population is very low.

The Sahel is a semi-arid desert margin. Its rainfall is 200–400mm per annum and the vegetation is acacia trees and tall grass. It is able to support more animals and people than the Sahara and there is a greater density of population. In recent years the density has increased and the region now seems overpopulated. After some years of above-average rainfall farmers were encouraged to settle and new wells were sunk. But rainfall is unreliable and after poor years some areas have been overgrazed and reduced to bare sand. Many people are now kept alive by international aid.

The savanna lands are well settled by farmers who grow millet, sorghum and groundnuts and keep cattle. The density of population varies in the region because in addition to the rural settlements there are large towns which are the administrative centres.

The Guinea savanna and rain forest areas near the coast provide the greatest opportunities for agriculture. A wide range of tropical foods is grown e.g. cocoa, bananas, rice, and farming is supplemented by fishing. So this region has been capable of supporting a denser population than other parts of West Africa.

Having established the basic climatic/farming pattern you can now consider how important other influences have been. It would be worth making the point that the picture described above is too simple and that the pattern of population density is the result of the interaction of a number of factors of which climate is only one. Other important ones are:

(a) *Historical* The effects of the colonial period in particular. The pattern of density of population in West Africa in colonial times reflects the fact that the greatest locational factor was access to the colonial powers which had acquired the region. Port locations and places on or near new railway lines became areas with the greatest densities of population. This in turn emphasized the differences in density which had first been created by climatic factors. The pattern was further intensified by the development of plantation agriculture and the exploitation of minerals near the coast.

This was not a uniform factor throughout West Africa however. Different colonial powers had different development policies. Some colonies were left underdeveloped e.g. Chad and Niger had no railways so in those regions the pre-colonial population density patterns remained unchanged.

(b) *Cultural* There were few towns and cities in West Africa before colonial times so the simple relationship between climate and population density was basic because the economies were rural. This simple relationship broke down in Yorubaland and the Hausa region of northern Nigeria. Both peoples had a tradition of living in towns and cities and so created pockets of very dense population and significant difference in density between areas which lay within the same climatic belts.

(c) *Economic* Economic development has led to the redistribution of people according to the new opportunities which exist for work and wages. In the first place development has increased the densities of the coastal region as mining, manufacturing and trade have grown. Modern development projects have also created new pockets of population and increased densities in some other areas e.g. the Volta river project and the development of the Niger basin.

The cities created in colonial times have become centres of greatest population densities as people are attracted from the rural areas by job opportunities, the highest wage rates in the country and opportunities to take part in national politics and government. Migration from the rural areas is not creating the greatest variations in density within individual countries. Towns and cities have become 'islands' of development and pockets of dense settlement. The process of urbanization continues to intensify this feature.

Those who live in the cities tend to be younger and of child bearing age. Medical and other facilities tend to be better in urban areas. So the live birth rate is greater and this in turn further increases the density of population compared with rural areas.

The final section of the essay should be devoted to weighing up the relative importance of the factors you have discussed. The key question to ask yourself is 'Are the climatic factors more important than any other'? Points for consideration are:

(a) Since West Africa is still relatively underdeveloped the density of population still basically reflects the agricultural possibilities of the land which are fundamentally the result of climatic factors.

(b) You can argue that climatic factors *seem* basic because political and economic factors in modern times have concentrated development and settlement near the coasts thus reinforcing the pattern originally due to agricultural economies. The apparent relationship between climate and density of population today is more coincident than causal.

(c) To what extent are the present densities the result of West Africa not having moved into the third stage of the demographic transition model? If you feel this is vital then you can argue that stage in development is more important than climate.

(d) Today urbanization seems to be a fundamental determinant of population density in the Third World. Towns and cities act as magnets for economic reasons irrespective of the climatic – farming environment.

Finally you should commit yourself on the evidence you have provided as to how valid the statement is.

2 (a) Outline the reasons which have led to large-scale international migrations of population.
(b) With reference to case studies, comment on the ways in which international movements have affected the social and economic geography of areas to which movement has taken place.   (*Cambridge, 1987*)

Understanding the question This type of question is fraught with danger for some candidates who tend to write all they know on a broad subject such as part (a). It is vitally important therefore that you: confine yourself precisely to answering what is asked, e.g. in (a) the *reasons* for *international* migrations; discipline yourself over time–this question requires no more time than any other; *pattern* or structure your answer–this shows the examiner that you can handle a wealth of material in a systematic way.

You should also draw case studies from a variety of areas to show a good spread of geographical knowledge.

Answer plan (a) It would be useful to have a brief definition of migration to set the framework for the rest of your answer.

Set the structure for your answer by stating that the causes of migration can be grouped under three main headings: natural hazards, social and political factors, economic factors.

Natural hazards include climatic changes such as the desertification of the Sahel, locust plagues in East Africa and recurring floods in countries like Bangladesh. These can all encourage movement away from the stricken areas to other countries.

Social and political factors include wars–the flight of Afghan people into Pakistan; refugee movements of people who are being persecuted on religious or political grounds, e.g. Jewish emigration from the USSR; and racial prejudice, e.g. the Turkish minority which left Bulgaria.

All the above examples are 'push' factors but there are also social and political 'pull' factors, e.g. the attraction of Israel as a Jewish homeland; the movement in the past of refugees from Nazism and Communism to the western democracies for greater personal freedom.

Economic factors should also be dealt with in terms of reasons for leaving the home territory and reasons for the magnetism of other lands. Appropriate migration models should be used to help explain particular examples.

(b) You are asked to use case studies so one example is not enough. Three different examples you could use are:

1 *Migration to Israel* Points to include–the setting up of a Jewish state brought an influx of people with common goals; skilled energetic immigrants with the backing of rich nations such as the USA were able to transform the economy of the land introducing modern farming and industry; the establishment of Israel has made its indigenous Arab population second class citizens and people have been permanently displaced. The open door immigration policy for Jews has created social problems with the 'westernized' Jews becoming a

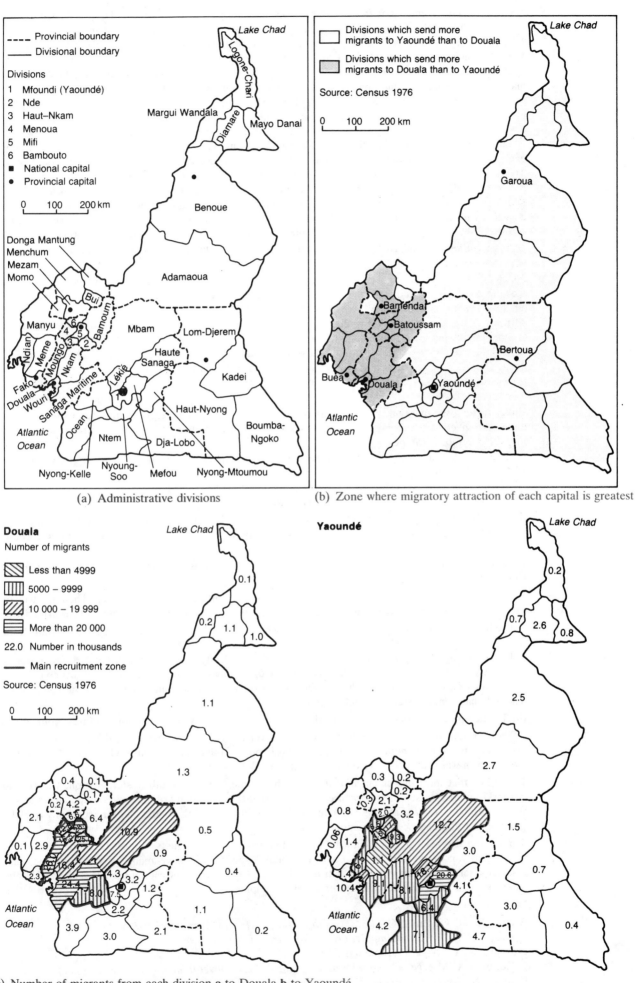

(a) Administrative divisions

(b) Zone where migratory attraction of each capital is greatest

(c) Number of migrants from each division **a** to Douala **b** to Yaoundé

Fig. 26.4 Cameroon

minority; large-scale emigration from Eastern Europe has encouraged the settlement of disputed territory and has bitterly divided the nation politically. Heavy defence costs have caused severe economic problems.

2 *Migration from Mexico into the USA* The opportunity model of migration has particular relevance here. Migration has provided south-western USA with a large low cost labour force for unskilled and semi-skilled work. The social costs include: the proved ineffectiveness of government agencies to stop illegal movement, massive employment and education problems in parts of the south-west borderland; the expansion of the drug trade; the increase in the proportion of the population who speak Spanish so that English now has to be legally protected in California; the building-up of Third World type problems in the cities to which migrants flow.

3 *Migration from Pakistan to Britain* The Pakistani immigration began as a result of the need for low cost labour in the textile industry so that it could retain world markets. Hardworking immigrants brought low expectations of adequate working conditions and settled in concentrated areas of northern industrial towns. This is one of the migrations that has made Britain multiracial and has contributed significantly to the growth of the Muslim religion here. Racial prejudice and racial tensions have resulted from reactions of indigenous people to the establishment of a distinctive culture in their midst. At times of economic difficulty the immigrant community tends to suffer most from unemployment and social disadvantage. It has forced British people to reconsider the values by which they live by confronting them with alternative views on education, etc.

In *conclusion* you should draw together the main points you would wish to identify from the case studies you selected.

3 Cameroon is a developing country in West Africa. It has a population of 9.26 million and a population growth rate of over three per cent. The GNP per capita is 880 US dollars (UK: £490). Life expectancy at birth is 53 years and there is an infant mortality rate of 98 per thousand (UK: 11). It is unusual in West Africa in that it does not have a primate city but has two major magnets for in-migration: the economic capital Douala and the political capital Yaoundé.

(a) The figures on the maps in Fig. 26.4 show the location of the two cities and features of migration to them. Outline the pattern of migration illustrated by these figures.

(b) Discuss possible explanations for the pattern you identify.

(in the style of the Joint Matriculation Board)

Understanding the question The first part of the question can be answered entirely from the data you have been given. The second part requires good knowledge of alternative explanations of rural-urban migration in modern Africa.

It is important that you spend a reasonable amount of time examining **all** the maps carefully. Analyse what they show in terms of:

(a) features common to both cities.

(b) distinctive features that apply to one city only.

Answer plan **(a)** Spend about 40 per cent of your time on this part of the question. So when you have examined the figures list the chief points that have emerged and select the most important ones. These would include:

1 Both cities receive migrants from all over the Cameroons. This is a reflection of their status as the two chief cities.

2 Both cities receive the bulk of their migrants from the south-west of the country. Their joint main catchment area forms a roughly quadrilateral area with the two rivals on the western and eastern edges.

3 The borders of the main recruitment zone for each are marked by a fall to less than 5000 migrants from a division.

4 The percentage of migrants from divisions received by each city depends upon distance. In-migration is inversely proportional to the distance from the city of the division from which the migrants come.

The main distinction is that whereas Douala's attraction is essentially confined to the south-west, Yaoundé has a more extensive catchment area.

(b) In attempting to explain the pattern there are three issues to which you should address yourself:

1 The fact that in-migration is occurring from all over the country. In exploring this point you should draw upon the various models of migration—'push/pull', Todara and Amin models and Mabogunje's model which was especially developed in relation to West Africa.

2 The fact that most movement into the cities has come from the south-western region. Clarke and Kosinski's model is useful here because it suggests that the availability of information and awareness of the opportunities offered are critical. You should also draw upon other parts of your geography work by pointing out that distance is a fractional cost.

3 You have also been given a significant piece of information about the two cities—their complementary rôles. You can argue that as the political capital Yaoundé will have many government posts and associated service activities which will attract migrants. As the economic capital Douala is likely to be the chief focus for modern industrial and financial development.

26.7 FURTHER READING

Allen, C. and Williams, G., (Eds.), *Sub-Saharan Africa* (Macmillan, 1982)
Grove, A. T., *The Changing Geography of Africa* (Oxford, 1989)
Morgan, W. T., *Nigeria* (Longman, 1983)
O'Connor, A. M., *The Geography of Tropical African Development* (Pergamon, 1971)
Pounds, N., *Success in Economic Geography* (John Murray, 1981)

27 India: development issues

27.1 ASSUMED PREVIOUS KNOWLEDGE

Even if you did not study India as one of the special regions chosen for regional study for GCSE, you will almost certainly have built up a certain amount of background knowledge from other parts of your course; e.g. the monsoon climate; the cultivation of rice, cotton and tea; the modern scientific and technological developments such as the 'Green Revolution', the construction of multi-purpose development projects (see Lines and Bolwell, *Revise Geography* (Letts, 1989, p. 103)) and river barrages. You may also have studied population problems in the sub-continent.

27.2 ESSENTIAL INFORMATION

27.2.1 India's agricultural economy – general points

(a) It is not possible to cover in any detail the physical background against which agricultural development has taken place. It is therefore important that you study this aspect of the geography of India by reading, for example, pages 44 to 60 in Johnson's book, *The Physical Environment of Agriculture* (see Section 27.7).

(b) India is one of the chief agricultural countries of the world and its agricultural produce is once of its main sources of wealth. Agriculture is likely to be the mainstay of the Indian economy for the forseeable future.
(c) Seventy per cent of all Indian workers depend on farming for their livelihood and farming provides 46 per cent of the national income.
(d) Subsistence is still an important factor in the farming economy. Small farms retain about two-thirds of their crops for food for the family, large farms sell about two-thirds of their produce.

(e) The crops which are grown are therefore determined by two main factors. They are (i) the need to provide food for the family, (ii) market factors – the saleability and profitability of different crops.

(f) Forty-five per cent of the country is cultivated; 15 per cent is double-cropped. So the possibility of extending the cultivable area is very limited. Much of the remaining 55 per cent of the land area is made up of mountains, deserts and land which has been built over. Efforts therefore have to be concentrated on intensification of agriculture.

(g) Population pressure combined with unscientific farming methods has meant that much land is being overworked. Nearly 25 per cent of the cultivated land suffers from soil erosion, 66 per cent of the arable land needs soil conservation measures (see Section 27.3).

27.2.2 Traditional practice and farming methods

Traditional practices and expertise built up over the centuries in different parts of the sub-continent reflect physical environmental factors, especially climatic and hydrological (water supply) rhythms; and the growing conditions required by the crops. Traditional methods include cattle used as draught animals; an emphasis on grain production; the 50/50 division of the crop between landowner and farmer; the farmer's division of his own share – some sold to pay the interest on debts, some kept for seed; hand-sowing; transplanting, weeding and harvesting with a primitive plough; little or no fertilizer. These traditional practices are now being challenged by pressures for change.

Irrigation is the means whereby essential water is provided in farming areas which have marked dry seasons and/or extreme variability. Although temperatures throughout the year in most parts of India permit crops to grow, rainfall is strongly cyclical and monsoon rain unreliable in amount and occurrence. So irrigation has always been a vital factor in the agricultural economy of India. Fig. 27.1 shows the areas which are most prone to drought.

Fig. 27.2 shows the distribution of the traditional methods of irrigation. Some of them date back for thousands of years – the canal system of the north-west was part of the Indus civilization of 4000 years ago. The tanks of the south and south-east are also part of India's cultural inheritance. The type of irrigation is dependent on certain physical factors.

(a) The availability of surface water in the form of river systems has allowed the development of canal irrigation e.g. in the Punjab.

Table 27.1 India: Regional variations in use of irrigation and irrigation techniques

Region	Canals % Indian canal irrigated area	Tanks % Indian tank irrigated area	Wells % Indian well irrigated area	Others % other irrigated areas of India	Total % of Indian irrigated area
North West Haryana, Himachal Pradesh, Jammu & Kashmir, Punjab Rajasthan, Delhi	27	5	28	5	22
North Centre Bihar, Uttar Pradesh	26	12	39	42	31
North East Assam, Manipur, Meghalaya, Nagaland, Tripura, West Bengal	11	8	–	21	7
West Centre Gujarat, Madhya Pradesh, Maharashtra	10	10	20	6	13
South and South East Andhra Pradesh, Kerala, Karnataka, Orissa, Tamil Nadu	26	65	13	26	27
Totals	100	100	100	100	100

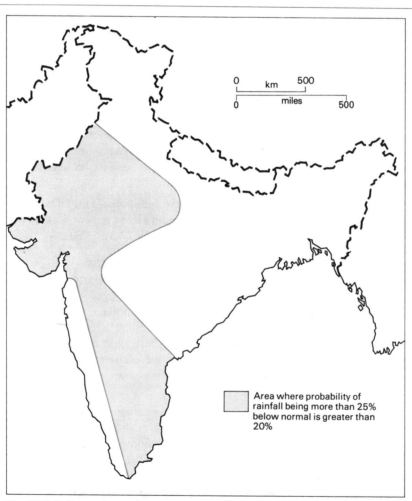

Fig. 27.1 India: main area prone to drought

(b) Relief and the nature of the terrain may also determine the techniques used, e.g. in Kashmir nearly all the irrigation is by *kuls* – leads led off from mountain streams along terraced areas with wooden aqueducts taking the water across ravines.

(c) Floods also provide a source of irrigation with flood channels diverted to lead the water into the fields.

(d) The hard rock terrain of south India has encouraged the construction of tanks dammed by earth or granite blocks in shallow valleys.

(e) The depth of the water table is also important. Ancient wells exist in areas where the water table is high and easily reached but a low water table makes this technique useless.

The use to which the water is to be put also influences the type of irrigation method used, e.g. in the south the water is stored in tanks to provide water for use immediately after the rainy season to allow rice and sugar crops to grow. The tanks are not intended for long-term use during the dry season.

In recent years, new methods of irrigation have been developed (see Section 27.2.5).

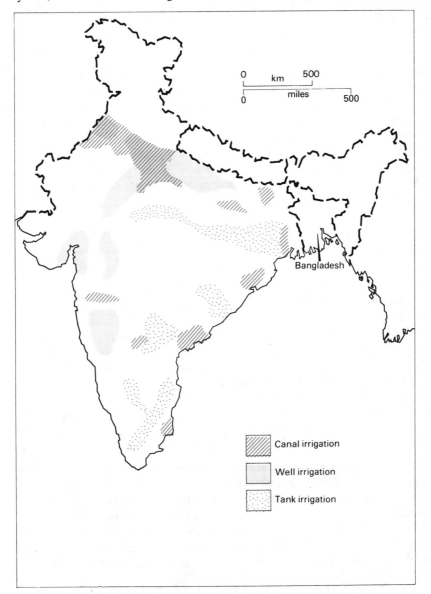

Fig. 27.2 Distribution of main types of traditional irrigation

27.2.3 Agricultural regions

There are four main regions:

The Himalayan zone (rainfall 1000–2500 mm per annum) The main crops are wheat, maize and rice. Seed potatoes and fruit are also grown.

The dry zone (less than 700 mm rainfall per annum) The chief crops are millets, oil seeds, wheat, maize, groundnuts, cotton and gram (chick peas).

The sub-humid zone (rainfall 700–1250 mm per annum) Sugar cane, tobacco and rice are important crops in addition to those grown in the dry zone.

The wet zone (rainfall more than 1250 mm per annum) The main crops are rice, tea, jute, sugar cane, spices, oil seeds, gram, millets and wheat.

27.2.4 Modernization of agriculture

The aim of modernization is to achieve greater efficiency in agriculture. Modernization is therefore concerned with both the physical conditions of the environment such as water supply and soil fertility, and with socio-economic conditions e.g. the size of landholdings, availability of credit to buy new equipment, education levels which give the people skills necessary to put new ideas into practice.

In order to increase efficiency in Indian agriculture four major problems had to be faced.

The need to increase the availability of a guaranteed water supply In recent years major developments have occurred in the extension of irrigation. Major barrage schemes such as the Bhakra–Nangal scheme provide water to irrigate 2.6 million hectares. The sinking of tube wells and the increasing use of small electric pumps by farmers to tap deeper wells than could otherwise be reached have also made important contributions to the expansion of the irrigated lands. Fig. 27.2 shows the regional variations in the use of irrigation; the farmers choosing the type most suitable for their particular area.

The need to increase yields The increasing provision of irrigation has been accompanied by the introduction of high yielding variety (HYV) crops. The HYV programme began in 1966/67 and by 1973/74 more than half of the wheatlands had been sown with new varieties. In that period wheat production in the Punjab (the 'granary of India') rose from 2.5 million tonnes to 5.4 million tonnes as a result of increases in yield.

In regions such as the upper Ganges where water depth can be carefully controlled high yielding rice strains have been very successful. By 1973/74 25 per cent of the rice land was sown with HYV seeds and in some areas yields increased by 93 per cent in the first six years. HYV strains of millets and maize have also been developed.

These 'benefits' however also created new problems. Because of the high yields production rose and process fell. Many small farmers became bankrupt though some have now formed cooperatives to help them cope with changing market conditions better. Another example of problem-solving innovations creating fresh problems is shown in Section 27.2.6.

Essential land reforms Rural areas are overpopulated and there is not enough land. The operational landholding worked by a family is often too small to provide the food they need. In rice lands a family holding may be less than 1 hectare and even in the less productive and less intensely cultivated dry lands of West Gujarat the holdings are no more than 6 hectares. This situation is worsened by conditions of tenancy. Many farmers are sharecroppers and landlords expect their rents but are only prepared to make minimal investment in the land. This problem applies to about 20 per cent of the total farmland.

Attempts have been made to achieve land reforms but well-intentioned laws have proved difficult to enforce. Attempts have been made to transfer land ownership to the farmers who work it, to limit rents to 20 per cent of the output etc. But in a large country in which tenants have traditionally regarded landowners as their social superiors real changes are difficult to achieve.

Farmers' need of capital for modernization The modernization of agriculture is dependent upon the availability of capital for investment. Farmers need money to sink tube wells, buy electric pumps to raise water, buy the HYV seeds and the fertilizers and pesticides which go with them. The importance of investment is shown in the model overleaf (Fig. 27.3).

Despite the establishment of rural co-operatives and efforts by central government the problem of capital has not been solved. As prices for fertilizers, pesticides etc. have increased poor farmers have applied less to the land and this in turn has decreased HYV yields.

27.2.5 A model of agricultural improvement (Fig. 27.3)

The model shows the relationship between agricultural improvement and the general economic development of a Third World country. Apply the model to the factual information on agricultural development in this section. Work out how the changes in agriculture relate to (a) the industrial development of the country (b) the urbanization of the population.

27.2.6 Development projects

As part of its economic strategy India has pursued major development projects as a means of increasing national wealth and standards of living. Fig. 27.4 shows the Narmada Valley in which such a project is now proceeding. The Narmada river has a drainage basin of 98 000 sq. km and this basin at present supports 20 million people.

The development of the plan consists of the construction of 30 major dams, 135 intermediate size dams and 3000 minor ones along 1300 kms of the river. The total effect will be to flood 6000 sq. kms—an area about the size of Devon. More than a million people are to

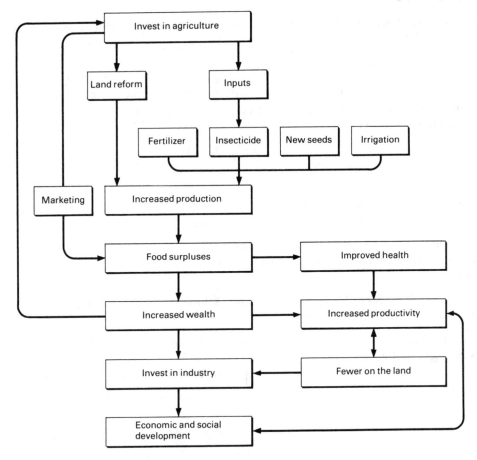

Fig. 27.3 A model of agricultural improvement

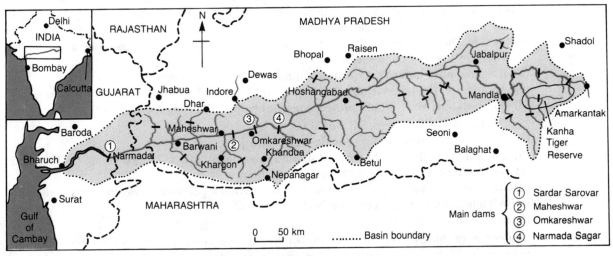

Fig. 27.4 Narmada Valley project

be re-settled. The project is intended to provide three things: drinking water, water for irrigation, and hydro-electricity for the state of Gujarat. It is being supported by international aid from the World Bank.

Until recently comprehensive multi-purpose development projects were seen as the key to rapid social and economic development in the Third World. There was little or no opposition to proposals which made significant contributions to regional development. Today feelings are more mixed. Reported attitudes to this project include:

'The most significant development milestone of the decade' (an Indian political commentator)

'One of the world's greatest man-made disasters' (Indian editor of an Environmental magazine)

'We are not fighting to stop a dam, we are fighting for social justice' (a sociologist)

'Environmentalists are anti-progress, anti-Gujarat and anti-national' (a dam builder)

Cost benefit analysis When a project of this type becomes controversial vested interests overstate their cases. A less emotional view can be gained by analysing the benefits which can reasonably be expected as a result of the development and the social and economic costs that will have to be paid.

Narmada Valley project

Benefits	Costs
1 Bring good quality drinking water to Gujarat to replace a diminishing supply from silted-up tanks and rivers	**1** 2000 sq. kms of farmland will be lost through flooding
2 Vast increase in water available for irrigation which will increase agricultural production and farmers' incomes	**2** 3500 sq. kms of forest will be lost by flooding
3 Additional power available for industrial development in Gujarat	**3** 1.5 million rural population will be displaced
4 Provides a permanent improvement in water supply, dams are planned to last 200 years	**4** Traditional ways of life of three peoples destroyed
5 Re-housed villagers will have more modern homes	**5** Flood irrigation schemes of this type cause waterlogging and soil salinity problems
6 Timber felling provides ready money for development and jobs	**6** Wildlife will be displaced. Rescue operations as the water rises will not be able to save all the animals living in the area
	7 Man/animal conflicts will become serious as traumatized wild animals seek food in fields and villages

Alternative solutions   In place of large-scale prestigious development projects which have drastic effects upon existing environments and communities, some scientists suggest that less capital intensive, smaller scale improvements would be more effective. For the Narmada Valley it has been suggested that significant and lasting improvements could be achieved by:
(a) The wise use of groundwater.
(b) Building more water tanks and de-silting existing tanks, ponds and lakes.
(c) Adoption of dry-land farming techniques so that there is less demand for irrigated water.
(d) Afforestation which would protect soils and help soak up monsoon rains.
   The *benefits* of this approach would be:
(a) The de-silting work would provide many local jobs.
(b) Local communities would not be displaced, traditional ways of life would not be disrupted.
(c) Afforestation would reduce the rate of siltation.
(d) New forests could be properly managed to provide a permanent income for the region.
(e) Wildlife habitats remain undisturbed.
(f) India would incur less debt.

Associated problems   Decision-making in developing countries in relation to major development plans is often complicated by issues not strictly related to the direct benefits and costs incurred. These include:
(a) Political pressures – local and national politicians may gain support as a result of a prestigious scheme.
(b) The influence of important absentee landowners.
(c) The size of the profits that major companies can get from large capital intensive development projects.
(d) Pressures from organizations such as the World Bank which has considerable expertise in supporting large scale projects but no comparable infrastructure for handling small loans.
(e) The lack of vast sums of capital available within the country for investment in development projects.
(f) The buying-off of local opposition by commercial and political agents.
(g) The use of the media to influence local attitudes to the project.

27.3 GENERAL CONCEPTS

The Green Revolution was an attempt to intensify agriculture. Many authorities believe that the best hope of improving food yield in the Third World is through the intensification of farming in existing cultivated lands rather than by attempting to extend the agricultural areas. The revolution did not merely involve the introduction of HYV seeds. It was a complex process whereby low-yielding traditional agricultural practices were transformed by the introduction of new technology and national economics. The revolution also involved irrigation developments, flood control schemes, drainage and erosion control, mechanization and the introduction of fertilizers and pesticides. In addition efforts were made to provide guaranteed markets, and efficient transportation links between the suppliers and consumers had to be established. A final significant element was the creation of credit facilities for farmers.

Population pressure is the condition of disequilibrium between the size and rates of growth of population on the one hand, and the availability and rate of development of resources on the other. It is not synonymous with population growth. If technological advance and social and political development occurs, this may result in increased resources becoming available, thus, although there is an increase in population, there is a decrease in pressure.

Internal disequilibrium The degree of this that exists between people, economies and resources is directly related to the amount of population pressure. It is partly the result of increased heterogeneity which has occurred as a result of the introduction of modern sectors into the economy and the social and political structures of the country. In India, for example, modern development has resulted in the creation of a small, highly privileged sector of the population localized geographically in the major cities. They are the government officials, politicians, professional people etc. Their ways of life and standards of living are related to international links more than to traditional culture and their policies and activities are dependent on foreign aid, investment and overseas markets. The existence of this sector in the primate cities magnifies the extent to which those cities attract migrants through their wealth and quality of life. India's major cities account for two-thirds of her urban growth.

27.4 DIFFERENT PERSPECTIVES

The economic perspective The agricultural development of India may be placed in the context of underdevelopment (see Unit 23, pages 148–55).

The demographic perspective The Malthusian view of the relationship of population to resources is relevant to the present situation of rural population in India. In western Europe the technological advances of the Industrial Revolution made the projected situation described by Malthus seem unlikely. Some hope that the present agricultural and industrial development of India will have a similar effect to the Industrial Revolution in Europe.

The technological perspective This is symbolized by the Green Revolution (see above). This perspective sees India's economic and social salvation in the implementation of rational and scientifically-based development programmes.

27.5 RELATED TOPICS

The units on underdevelopment and the distribution of population should be read in association with this section. You should also study the climates of India, the agricultural conditions necessary for the production of the major crops, some of the major development projects and the place of agricultural development in the economic geography of India today.

27.6 QUESTION ANALYSIS

1 With reference to specific areas, examine the problems of the use and management of water.
(*Associated Examining Board, June 1987*)

Understanding the question A straightforward question but one in which you have to draw your own limits to make it manageable in the time allowed. So establish the framework within which you intend to write–the use of water for industrial, agricultural and domestic purposes. Think about how many areas you wish to include–reference to too many could lead to a descriptive and superficial answer. A suitable response, but not the only one, would be to refer to one area in the developed world and one in a developing country. This would bring out the different priorities these regions have.

Answer plan Use a brief introduction to specify what you intend to write about, the areas to which you will refer and your reason for choosing them.
There now follow the main points to include using two contrasting examples.

England and Wales
(a) The vastly increased demand for water and leisure uses which has occurred as standards of living have risen has stretched water supply systems to their limits.
(b) This has been worsened by autumn and winter droughts and hot summers recently.
(c) Least water is available in the regions with the greatest demand and the greatest increases in demand–London and the south-east and East Anglia.
(d) The lack of a national grid system for water means that each region has to find its own solution.

(e) Lack of capital investment means that a high proportion of water is lost in distribution through leakages (up to 30 per cent in some regions).

(f) Rising costs which would result from expanding water supply sources are resented by an ageing population that resists higher taxes.

(g) Plans to build new reservoirs in appropriate locations generate opposition from environmental groups and local residents who may be displaced.

You may wish to comment upon the expected effect of greater efficiency as a result of the recent privatization programme.

India: the Narmada Valley

(a) The monsoon rainfall régime makes it difficult to manage run-off while heat and drought cause demand problems at other times in the year.

(b) Deforestation has greatly reduced the water holding capacity of the natural environment.

(c) Existing supplies are diminishing through neglect (siltation, etc).

(d) Large scale development which would bring significant additional supplies, demand large capital investment and increase the national debt.

(e) Increasing population and new industries have increased the need for more drinking water and water for irrigation.

(f) Increased irrigation can lead to diminishing returns as soil salinity increases.

(g) Development can have major ecological effects on local communities, wildlife and vegetation.

Your conclusion should:

(a) Bring out similarities–the influence of climate, the increase in demand, the need for major capital investment etc.

(b) Point out that the developing region's needs for water are basic ones–drinking supplies, food production, etc, while in Britain the high level of consumption is the result of general affluence.

2 Discuss, using examples, the significance of land tenure, farming methods and agricultural productivity in the context of India.
                                                                        (*in the style of Oxford*)

Understanding the question This is a fairly general question but you are given good guidance on how to structure your answer by the listing of the three features of land tenure, farming methods and agricultural productivity. You should use examples from your regional work. The biggest danger in answering the question is that you write a descriptive account. In asking you to discuss the significance the examiner requires an analytical approach.

Answer plan As an introduction you could point out that in a country which is still largely agricultural and is attempting to achieve rapid economic development, the need to increase agricultural productivity is critical. It is even more critical in a country with a large and rapidly growing population with migration to the cities further increasing demands for more food to be produced. There is little hope for increasing food production by bringing new lands under cultivation. The hope is to use existing cultivated lands more intensively and efficiently.

Discuss the problems that have to be faced in attempting to achieve greater productivity and efficiency. Two of these are land tenure (see Section 27.2.4) and farming methods. Give a brief account of traditional methods (see Section 27.2.2) and contrast this with the developments associated with the Green Revolution (see Section 27.3).

Intensification of agriculture and increased productivity also depend upon ensuring that the water supply in drought-prone areas is reliable and sufficient to grow the preferred crops–food or industrial raw materials. Use the information on irrigation in this unit to elaborate on this point giving relevant examples.

Make the point that all three, land tenure, farming methods and agricultural productivity, are therefore interlinked and that agriculture is not an isolated sector of the Indian economy. For the foreseeable future the economic development of India hinges substantially on agricultural progress. Use the model above (Fig. 27.3) to illustrate this. Emphasize that the entire economic programme depends on agricultural development which will help create the wealth required for investment both in industry and in new agricultural projects.

27.7 FURTHER READING

Bradnoch, R. W., *Agricultural Change in South Asia* (Murray, 1984)

Farmer, B. H., *An Introduction to South Asia* (Methuen, 1983)

Johnson, B. L. C., *India The Physical Environment of Agriculture* (Heinemann, 1979)

Mountjoy, A. B., (Ed.), *Developing the Underdeveloped Countries* (Macmillan, 1985)

Spate, O. H. K., Learmouth, A. T. A. and Farmer, B. H., *India, Pakistan and Ceylon* (University paperbacks, 1972)

28 Brazil: regional strategies

28.1 ASSUMED PREVIOUS KNOWLEDGE

For GCSE you may have studied Brazil as an area for special regional study, dealing with the main features of the physical and human geography of the main geographical regions and chief cities. You may have encountered Brazil when looking at the major climatic and vegetation belts of the world. This will probably have focused your attention on Amazonia. You may have studied Brazilian examples of important aspects of human geography e.g. the integrated steel plants (Volta Redonda), the shanty towns, the new capital, Brasilia.

28.2 ESSENTIAL INFORMATION

28.2.1 Basic economic facts

(a) Brazil is the fifth largest country in the world by area, the sixth by population. It takes up nearly half of the continent of South America but much of its territory remains empty or thinly populated.

(b) Despite its emptiness Brazil is now the major manufacturing country in South America. Its major industrial region is the Sao Paulo–Rio de Janeiro region (south-east Brazil).

(c) Brazil does not have the developed energy resources necessary to match its size and ambitions. Coal production is the equivalent of that produced by two large efficient pits in England; its oil production is only 4 per cent of Venezuela's; it has only 4 per cent of the hydro-electricity available per head of population that Norway has.

(d) From the late 1960s until recently Brazil's economic growth rate was one of the most rapid in the world. Its population growth has also been rapid so the gross domestic product (GDP) is still very low compared with the western industrial nations.

(e) The 'economic miracle' of the 1970s is now over and Brazil is one of the heavily indebted nations of the developing world. Urbanization has continued and with the economic down-turn the largest cities face severe social and economic pressures.

(f) The newest phase of development, the large scale exploitation of Amazonia, has aroused considerable international concern. Agencies and major countries to whom Brazil is in debt are exerting considerable pressures on Brazil to review and alter its policies of exploitation.

28.2.2 Economic cycles

The economic development of Brazil has been characterized by a series of cycles. In each cycle a particular resource has been intensively developed in one part of the country. The intensive development has led to the exhaustion of the resource and the region in which it was located has been abandoned as a centre of economic interest. The main cycles have been (in chronological order):

(a) **Brazil wood** This first attracted the Portuguese to the north coast of Brazil. The wood yielded a dye which was sent back to Europe. This was followed by the exploitation of the Amazon Forest, especially for its wild rubber trade. The *rubber cycle* brought great wealth to those who controlled the trade until alternative sources of high quality plantation rubber in South East Asia became the world's suppliers, e.g. Malaya and Indonesia.

(b) **Sugar cane** This cycle developed in the north-east in the Recife–Santos area. For 150 years up to 1700 this was Europe's chief source of sugar. It was unable to meet the competition of the plantations set up in the West Indies and the sugar economy of north-east Brazil declined.

(c) **Gold and precious metals** Deposits were found in the eighteenth century in Minas Gerais state and exploited until they were largely exhausted. Only in recent years has Minas Gerais become an important mining area again with iron ore and bauxite mining.

(d) **Coffee** The coffee boom of the nineteenth century began in the Paraiba valley and then spread to Sao Paulo. This boom came to an end in the 1960s when much of the stock was diseased, there was no suitable empty land upon which to extend coffee growing and emergent countries such as Kenya competed successfully with Brazil for western markets. Land used for coffee growing at the peak of the cycle has now reverted to poor quality farmland.

(e) Industry Much of the capital generated by the coffee industry was invested in the growth of modern manufacturing industries in the south and south-east.

(f) Forests The newest of the cycles which is still in operation. Amazonia and its tropical rain forest is seen as a massive economic resource from which vast profits may be made quickly. Deforestation, the development of mining, the introduction of heavily subsidized cattle ranching and building of roads and airstrips has brought fundamental changes to Amazonia and its people.

28.2.3 Regional differences

Economically the country is dominated by the coastal areas. Ninety per cent of the people of Brazil live within 550 miles of the sea. The 'core' region is the Sao Paulo–Rio de Janeiro–Belo Horizonte triangle in which the new industrialization and other aspects of modernization have been concentrated. This region contrasts dramatically with other regions which are largely undeveloped or having to cope with high population growth rates unmatched by economic development.

Economic regions may now be distinguished which are different from the traditional geographical regions (Fig. 28.1). Each economic region is characterized by a particular stage and type of economic development. (See chapter 4 of Henshall and Momsen listed at the end of the section.)

Eastern heartland The economic core of the country including Sao Paulo state, Minas Gerais state, and Rio de Janeiro. The region contains 10 per cent of Brazil's area, 40 per cent of its population, and its average wage is 150 per cent that of the country as a whole. Sixty-six per cent of Brazil's industries are located here and it provides the market for 75 per cent of Brazil's manufactured goods.

The north-east When plantation agriculture was important and prosperous this was the chief economic region of Brazil. It is now in long-term decline and is getting steadily poorer than other regions in the country.

Fig. 28.1 Economic regions of Brazil

The north is still a resource frontier not fully integrated into the Brazilian economy. It is being transformed rapidly and fundamentally but in ways which are attacked by many as ill-conceived and unacceptable at national and international levels.

Eastern periphery The southern sub-tropical half of this region is an area of established family farms and small industries. The northern tropical half has been isolated and underdeveloped. It is now beginning to develop as part of Rio de Janeiro's sphere of influence.

The rimland This frontier region is a transition zone between the heartland, the north-east and the north. The traditional cattle rearing economy is being improved and, as new roads are built across the rimland, economic development is commencing.

28.2.4 Modernization of agriculture

Agricultural productivity is increasing as a result of technological change. The government is subsidizing this development. As farm labour has decreased (as a result of rural-urban migration) mechanization has increased. Farms have also changed from a grazing to an arable economy in many areas and this has stimulated mechanization further. A national plan for the manufacture of tractors, the production and distribution of fertilizers through agricultural service centres, and Federal and State programmes of technical assistance have encouraged greater production. Financial bodies have been established to arrange loans for farmers. New marketing arrangements and new roads mean that the additional produce can be marketed more efficiently.

Nevertheless, some regions remain agriculturally backwards, especially the north-east. The expansion of commercial agriculture in many areas is severely limited by remoteness from suitable markets.

28.2.5 Industry

Industry began to develop in Brazil in the middle of the last century, on the proceeds of the coffee boom. The first industries, concerned with the processing of widely available raw materials, were the:

Traditional consumer industries Originally they began where the raw materials were produced, e.g. flour mills were built near the wheat fields. However, they have gradually become market-located – in the cities and in the Sao Paulo – Rio de Janeiro region especially. For example, the cotton industry began in the 1860s in Bahia where raw cotton was grown. By the 1880s two-thirds of the production was based in Rio de Janeiro and the cotton was grown in the south-east on former coffee plantation land. Now the industry is even more market-located – 71 per cent of textile production is in the heartland.

Traditional heavy industry Like cotton, iron and steel were originally produced close to the source of the raw materials. Until 1945, iron ore, charcoal and limestone were obtained chiefly from the iron quadrilateral of Minas Gerais. In 1946 the building of Volta Redonda saw the move towards a market location when the new works were located midway between the mineral fields and Sao Paulo. By the end of the 1960s 25 per cent of the iron and steel production came from integrated plants which were essentially market-located – mainly around Sao Paulo and Rio de Janeiro. The new locations were influenced by three main factors: (a) the continued availability of Minas Gerais ore (b) the possibility of transporting coal by sea (c) the markets for steel in the Sao Paulo–Rio heartland.

Industrial 'take-off' after 1964 In 1964 a military government took over Brazil. It was determined to bring economic and political stability to the country and to achieve economic growth. Inflation was checked, overseas investors encouraged to invest in Brazil and great incentives were given to manufacturing industries. By the 1970s Brazil had one of the highest economic growth rates in the world. At first, existing industries were expanded. Soon, however, economic growth involved the introduction of new industries – in particular, cars and petrochemicals – of which Brazil is the leading Latin American producer. The government encouraged foreign firms to establish plants in Brazil and the import of finished vehicles etc. was discouraged by high import duties. The plants were established in the heartland, close to the major internal markets. As well as new industries, new roads, hydro-electric power stations and a telecommunications network were developed. There is now a dualism in the manufacturing economy of Brazil – modern efficient plants in the heartland, traditional small industries, closely related to the availability of raw materials, elsewhere.

The rise in oil prices, the world recession of the early eighties and falls in commodity prices as demand decreased had serious economic consequences for Brazil. High levels of inflation and indebtedness and increasingly difficult social problems produced changes in political attitudes and policies. In 1985 democratic elections were held and the country moved from a military dictatorship to elected civilian rule. The present government hopes to achieve more permanent progress and to move away from unplanned 'stop-go' policies. Prestigious but expensive projects such as the new road to Peru for the export of Amazonian timber have now been revised or abandoned.

28.2.6 Regional development programmes

Attempts were also made to decrease regional disparities. One such attempt was the building of a new capital city.

Brasilia The capital of Brazil·used to be Rio de Janeiro but it ceased to be so in 1960. The Brazilian government decided to build a new capital for their 'new' nation, and Brasilia was built on a site 1600 km inland from Rio. This decision signified the Brazilian determination to develop the interior – the focus of economic development was switched from the coastal areas to the centre of the country. The establishment of Brasilia has provided a major magnet on the edges of the undeveloped areas, deflecting population flow from the Rio de Janeiro– Sao Paulo axis and encouraging economic growth in new areas. New land transport links between Brasilia and the heartland of the south-east assist in this.

Attempts have been made to develop other parts of Brazil. For example, new highways were begun in the less developed regions such as Amazonia. Regional development administrations, financed by national taxes, have been established. There are incentives for private firms to set up works in the less developed areas. SUDENE in the north-east and SUDAM in the north are two of these organizations. Their programmes include road building, the installation of power stations, settlement schemes for new farmlands, building schools, developing ports and encouraging new industries. Although progress has been made, however, these regional programmes have not lived up to expectations.

28.2.7 Development problems (see also Unit 29)

The less well developed regions have not shared fully in the economic boom and the development gaps between the regions have not been narrowed. The differences between the rich and poor which are most obvious are the differences between cities such as Rio de Janeiro and the least developed areas. The main economic problems are:

(a) The north-east has a large population but little employment.

(b) The need to mobilize the resources of the Amazon basin.

(c) The west is not fully settled and should take people from the overpopulated north-east.

(d) Continued expansion of the south and south-east will depend on the creation of bigger markets within Brazil. This means that other parts of the country have to acquire more purchasing power.

(e) Urbanization – the movement of people to the cities, and rapid population growth within the cities has resulted in the development of shanty towns in which poverty, crime and other social problems are concentrated (see Section 15.6).

Environmental degradation e.g. the large scale destruction of the Amazon forest by tree felling, the expansion of cattle ranching, the development of mineral resources, is fundamentally altering the geographical character of the region. Environmentalists claim that the destruction of the forest would deplete the earth's supply of oxygen, huge areas would suffer from soil deterioration, wild life would be destroyed and traditional Indian cultures would disappear. International concern has now been aroused. This concern is not confined to issues of conservation or the fate of the Amazon peoples but is also related to the question of the global atmospheric and climatic effects of the destruction of the largest tropical forest in the world (see Section 29.6).

28.3 GENERAL CONCEPTS

(a) Look again at the concepts and models introduced in Unit 21 *Location of industry* (pp 135– 41).

(b) Look too at the concepts and models introduced in Unit 23 *Less-developed countries* (pp 148–55).

(c) Look at the concepts and models contained in Unit 15 *Movement of population* (pp 101– 106).

(d) Perceptual frontiers Brazil still has a relatively small population for its total area and in some respects an exploitive 'frontiersman' attitude still prevails. No region within Brazil has had a concentration of natural resources and locational advantages sufficient to make it the permanent supreme focus of the country. Brazil has a history of different discoveries in different locations offering apparent possibilities of great wealth. But these opportunities have never quite lived up to expectations and over time this has led to a speculative approach to economic development once summed up as the search for 'El Dorado'.

(e) Core/periphery phenomenon This is the situation in which major contrasts in economic progress have developed between a core region in the country and the underdeveloped interior which is almost totally outside the activities of a modern industrialized country.

(f) The empty heart This is best illustrated by the distribution of population (Fig. 28.2). It is

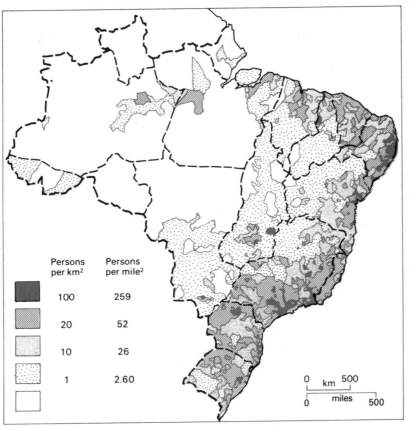

Fig. 28.2 Brazil: population density

important however to distinguish between areas with very little potential for development e.g. Andean areas of South Chile and areas such as Amazonia with major untapped resources. The concept is also complicated by the fact that north-east Brazil is an area of difficulty. In terms of its potential for development it is already overpopulated, although it appears on the map to be thinly-populated.

(g) The concept of economic cycles has already been dealt with in Section 28.2.2.

28.4 DIFFERENT PERSPECTIVES

For GCSE you may have studied Brazil in a regional fashion i.e. by examining the geographical features of each region such as the north-east, Amazonia etc. At A level this approach is less appropriate; it is necessary to focus upon the processes which are at work in changing the economic and social structure of the country. These processes must be examined against the background of the resources available as a result of the country's physical make-up and the factors which have influenced its geographical evolution to the present time.

Views about the present economic activities and future potential of Brazil vary enormously. Some geographers see the progress made in recent years as an 'economic miracle' similar to that of Japan. Other geographers see a different picture – the empty interior, the decaying older agricultural communities of the north-east, the shanty towns, the widespread poverty. This second viewpoint sees the economic progress occurring as a result of ruthless exploitation by the rulers and upper classes, with a maldistribution of income, high inflation and heavy borrowing from foreign commercial banks. So Brazil is an enigma.

28.5 RELATED TOPICS

The chief topics linking with this chapter are the locations of industry; underdevelopment and the movement of population. You should also try to apply ideas contained in the section on transport and transport networks to Brazil.

28.6 QUESTION ANALYSIS

1 To what extent is the location of manufacturing industry in Brazil related to the distribution of industrial raw materials?   (*in the style of Cambridge*)
Understanding the question This is not a very complicated question as long as you make certain that you answer the question asked. You are not being asked to describe and account for the location of industry but to judge the extent to which the location of manufacturing is dependent upon the distribution of raw materials. Since the question is about *manufacturing* industry answers about mining and forestry are not relevant.

Answer plan Begin with a factual introduction which provides a suitable background for the remainder of your answer. Two major points are:

(a) Brazil is an important industrial country and manufactured exports are now more important than exports of coffee. In terms of industrial development São Paulo and Guanabara states are roughly similar to a country like Spain.

(b) Major categories of manufacturing industries are textiles, food, engineering and electrical goods, clothing and footwear.

At the start of the main part of your answer it is important that you establish that the question can best be answered by classifying manufacturing industries in Brazil into three main categories:

(a) traditional consumer industries, e.g. textiles, food, cement.

(b) traditional heavy industries, e.g. iron and steel.

(c) new industries, e.g. cars, petrochemicals.

If you deal with each of these in turn (using the information given earlier in this unit), you should be able to make a judgement of the importance of the distribution of raw materials as a factor of location. In the case of (a) and (b), consider the change in location from source of raw materials to market. New industries, on the other hand, have been established in the heartland.

Having discussed these three categories, you should then go on to consider other significant factors. You should mention the dualism in Brazil's manufacturing economy. You should also point out that the concentration of industries in Sao Paulo and Rio de Janeiro is not just because of the availability of markets, but also because of advantages of industrial linkage, external economies, availability of skilled labour and availability of capital and business expertise.

You could state that the precise location of some industries is the result of direct government action, e.g. a law passed in 1958 set up the Merchant Marine Fund, thus leading to the establishment of modern shipbuilding in Guanabara Bay. Also, government determination to develop the remoter areas has led to subsidies for new industries outside the heartland, e.g. food and textile mills in Recife. Investors have also been given grants, tax exemptions and other concessions for financing developments in Amazonia.

On the basis of the evidence you have presented you should conclude by making some judgement of how important raw materials are in determining industrial location in Brazil at present.

2 Study the map below showing the urban development strategy for Brazil, based on the Second National Development Plan, 1975–79.

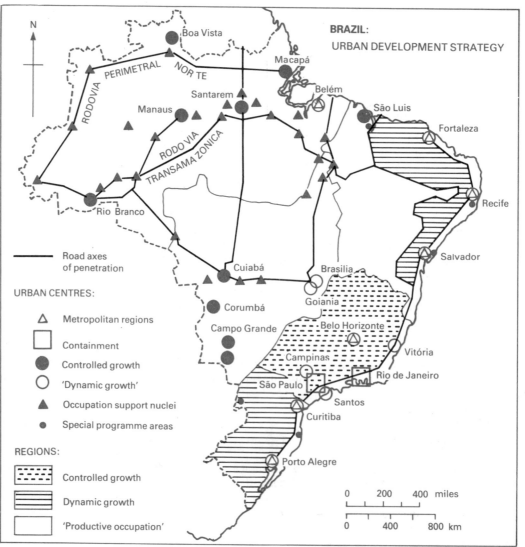

Fig. 28.3

(a) Discuss the extent to which the type and location of urban centres is related to the three-fold regional division show.

(b) Examine the extent to which map evidence reflects the attempt to redistribute urban development away from the densely populated eastern coastlands, and assess how successful this attempt has been in practice.

(WJEC, June 1984)

Understanding the question This is not a very complicated question, but to earn good marks you must address yourself to the precise questions asked, and you need a good general knowledge of the geography of Brazil.

The question does not ask you to describe the urban development programme, but to analyse the relationship between the regions defined on the map and the changing urban system.

Answer plan You should answer in two distinct parts, as defined by the question. Part A is a little more descriptive than part B, which is entirely analytical. It is reasonable to assume, therefore, that more marks will be allocated to the second part.

Part (a) A sound way of answering this section would be to examine each of the urban categories listed in the key. This would enable you to establish the following:

(i) The *metropolitan centres* reflect earlier economic development in Brazil, with an emphasis on coastal location, exploitive agricultural cycles, and mining. Most of those centres for which dynamic growth is planned are located in regions of dynamic growth, and will be key centres for regional development which, if successful, will reduce the overwhelming supremacy of the core region of south-east Brazil.

(ii) The two greatest metropolitan areas, Rio de Janeiro and São Paulo, are designated for *containment*. The further economic development of these cities and their region is being restricted in order to shift economic development from the core to the periphery.

(iii) Cities, other than metropolitan regions, for which controlled growth is planned are located in the interior, and particularly in Amazonia. This is also true of the occupation support nuclei. This reflects a national determination to open up the interior to exploit natural resources and to achieve a more dispersed pattern of urban development. The location of the cities in relation to the new road network does not conform to the Taaffe model of network development (see page 144), but the map is incomplete in that it does not show the river system, which also provides routes in the interior.

(iv) The Brasilia area is indicated as an area of dynamic growth. Brasilia has grown rapidly since its establishment in 1960, but still does not rival Rio or São Paulo in economic and social significance. So the attempt to shift the centre of gravity away from the coastal axis continues.

Part (b) The growth of Brasilia symbolizes a determination to develop the interior and shift the national economic focus away from the coast. The map reflects the attempts to revitalize the depressed north-eastern region, which has suffered from earlier economic cycles (see 28.2.2). Planned dynamic growth in the south also reflects a determination to spread economic activities away from the south-eastern heartland region.

The establishment of Brasilia has provided a major 'magnet' on the edge of the underdeveloped regions, and this has increased in influence with the development of the new road network. It has therefore redirected interest and economic activity towards the empty lands. Critics maintain, however, that this development is an extension of the exploitive frontiersman attitude which has prevailed in the past in Brazil and has caused some of the current problems.

This criticism has been substantiated by the ways in which the development of the Amazon is now proceeding and by the efforts international agencies and other countries have made to encourage Brazil to manage the process of development more sensitively and efficiently. At present the development of the Amazon appears to be yet another exploitation cycle rather than the planned, permanent economic and social development of the largest region of the country.

The 'economic miracle' intensified the magnetic attraction of the cities of the core region. This has added to the social and economic problems which Brazil has to solve. The end of the economic boom and inflation followed by a stricter monetary policy have caused social unrest and further heightened the contrasts between rich and poor.

The 'economic miracle' of the 1970s was based mainly upon the introduction of new industries into Brazil by foreign firms. These firms required prime locations and therefore concentrated in the south-eastern core region or on its edges – near the major internal urban markets. As a result, the focus of rapid industrial growth was the traditional core region, which counteracted efforts to redistribute people and work to the interior.

Attempts to solve the problems of depressed regions by agencies e.g. SUDAM, SUDENE have only been partially successful. In 1987 Brazil, with one of the largest foreign debts in the world, suspended payments of interest on its $68 billion debts.

3 'São Paulo should have succeeded Rio de Janeiro as capital of Brazil.' On the basis of your geographical knowledge of Brazil how far do you agree with this statement?

(in the style of the Welsh Joint Education Committee)

Understanding the question In asking you to say how far you agree the examiner expects you to look at both sides of the argument before you come to a decision. You are also expected to express your ideas as to what makes a capital city.

Answer plan Briefly state the background in your introduction. Rio de Janeiro ceased to be the capital in

1960. Its only rival in terms of size, economic importance and international reputation was São Paulo City. The Brazilian government however had decided to build a new capital city 1600 km inland from Rio called Brasilia.

The main body of your answer might fall into three parts:

(a) *What do we expect of a capital city?* It should be a large and economically powerful city so that it is not completely overshadowed by rival cities e.g. London. It is usually the focal point of the country and a city that belongs to the whole nation e.g. Warsaw, Moscow. It symbolizes the character and values of the nation e.g. Paris. It is the centre of government and national administration.

(b) *How far could São Paulo satisfy these conditions?* It is the largest city in Brazil, larger even than Rio de Janeiro. It is the chief industrial and commercial centre of Brazil. It symbolizes the industrial development of the country. As one of the two major nodes of Brazilian transportation networks it could quickly have become the centre of administration and government. It is 640 km from the coast and therefore a little more central to the country than Rio de Janeiro was. It is well known internationally and has world-wide business, trading and cultural links. In many ways it was the natural successor to Rio de Janeiro which it had outgrown as the largest city and chief economic centre.

(c) *What were the arguments against São Paulo?* Brazil is a 'new' national and has needs which are different from those of well-established developed nations. The USA and Australia as new nations had set up new federal capitals in the past. A new city could be planned as a city of the future – a fitting symbol for a nation of the future and in contrast to the older cities where the quality of urban life was deteriorating.

You could then go on to explain the significance of Brasilia in relation to the development of Brazil (see Section 28.2.6).

Finally, in making your judgement you have to weigh the strong economic and cultural reasons for São Paulo to succeed Rio de Janeiro against national political goals which favoured a new capital. Remember too that São Paulo and Rio de Janeiro are virtually twin cities. So a move to make São Paulo capital would have achieved nothing new.

28.7 FURTHER READING

Brumley, R. D. F. and R., *South American Development* (Cambridge, 1982)
Dickenson, J. P., *Brazil* (Longman, 1982)
Henshall, J. D. and Momsen, R. P., *A Geography of Brazilian Development* (Bell, 1974)
Morris, A., *South America* 2nd edition (Hodder and Stoughton, 1981)
Odell, P. R. and Preston, D. A., *Economies and Societies in Latin America* (Wiley, 1978)
Vaughan-Williams, P., *Brazil A Concise Thematic Geography* (Unwin Hyman, 1988)

29 Exploitation and conservation

29.1 ASSUMED PREVIOUS KNOWLEDGE

Work at GCSE level will have included study of tropical rainforests, issues relating to conservation and case studies of economic development in the Third World. Regional studies may also have provided you with more detailed information about particular areas.

29.2 ESSENTIAL INFORMATION

29.2.1. Definitions

Ecosystems See Unit 13.

Conservation The protection of natural resources and the natural environment for the future. This includes the effective management of resources such as soils, minerals, landscapes and forests to prevent their over-exploitation and destruction. Conservationists are increasingly concerned with the preservation and protection of whole habitats as well as of endangered individual rare species. They practise careful environmental management to achieve and maintain an ecological balance and do not concentrate on passive protection.

Exploitation The unwise or careless utilization of hitherto unused or under-used natural resources for commercial purposes. Exploitation involves freedom for the operation of market forces with profit a prime motive and only limited concern for the effects of either the scale or rate of economic change involved. In developing lands exploitation has been essentially a short-term commitment to a region to make the most profitable use of minerals, land and other valuable resources, with little regard for the need to establish sustainable economic activities which over the long term could bring about comprehensive development without too great an ecological or social cost.

Development, developing countries see Unit 23.

Tropical rainforest A type of forest dominated by very tall trees that grow near the Equator. The plant cover is rich, varied and quick growing in response to the climatic conditions of all the year round rainfall (above 1500mm) and consistently high temperature (25-30°C). These forests contain commercially valuable hardwood such as mahogany.

TFAP The Tropical Forest Action Plan of the United Nations Food and Agriculture Organization (FAO). It is jointly sponsored by the United Nations and World Bank. It is designed to establish global tropical forest conservation and development programmes. It aims to obtain finance for these programmes from national governments, private industries and international organizations.

29.2.2 Exploitation and conservation–an example in the Developed world

The rapid industrialization of Britain in the 19th century created similar impatience to create economic wealth and similar conflicts of interests as those encountered in the Developing world today. At that time there was also a lack of awareness of and concern for the environmental effects of large-scale exploitation of previously unindustrialized areas. As a result, in recent times work has had to be done to repair the environmental damage caused more than a hundred years ago. A prime example of the reclamation of land laid waste by industrial exploitation was the Lower Swansea valley project. The area was the site of major smelting and refining factories in the 19th century. The works were abandoned and decayed. They were surrounded by waste heaps containing traces of arsenic and other poisonous materials so there was little or no vegetation cover. The project was based on scientific research at Swansea University and the conservation programme designed to reclaim this industrial wasteland involved the removal of major eyesores, the grassing of waste tips, afforestation and the education of the children of the area in conservation issues and concern for the environment.

Present day environmental concern and pressure for effective conservation policies have not however removed the problems caused by industrial exploitation in this part of Wales. A major industrial raw material need in Britain today is for stone and aggregate for the building industry. The extension of the M4 motorway westwards in South Wales created an intense local demand for aggregate and construction companies were eager to extend quarrying operations. In some cases permission for quarrying was granted before planning controls came

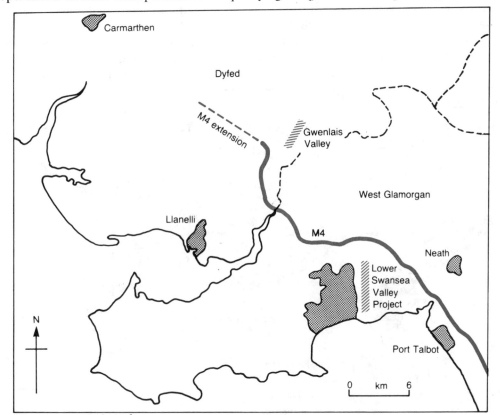

Fig. 29.1 Development projects in the South Wales region

into existence. In the 1940s, Intermediate Development Orders (IDOs) were issued which can be activated today without any fresh enquiries. The Gwenlais Valley in Dyfed is threatened by an IDO. The valley ridge is an area of Special Scientific Interest (SSI) and in the valley there are five working farms. A major construction company has the legal right to destroy the entire valley through quarrying.

This is a major issue in other parts of Britain too. It was estimated that 280 million tons of aggregate would be needed each year by the year 2005. This target was passed in 1988. The most suitable materials and quarries are located in regions of beautiful scenery and even in National Parks. As pressure for building materials increase many counties have to make decisions about conflicting interests. You would find it useful to make a cost benefit analysis similar to the one in Section 27.2.6 to demonstrate conflict and pressure on the environment resulting from this type of exploitation in an area known to you.

29.2.3 An example from the Developing world – general features

Concern over exploitation and conservation issues currently focus upon what is happening to the tropical rainforests. Most tropical rainforests are found in developing countries. These countries are deeply in debt to international banks and the governments of wealthy developed countries. They are therefore desperately anxious to exploit new sources of income in order to tackle major internal economic and social problems.

The destruction of the rainforest is the result of a multi-faceted process of exploitation. Valuable timber is a source of immediate profit but the land which is cleared also offers possibilities for profitable cattle ranching until the soil is exhausted. The heavy rainfall and nature of the terrain makes possible massive hydro-electricity projects in some locations. The forest areas also contain valuable mineral deposits. The entire programme of exploitation is underpinned by the creation of a transportation network of new roads along which timber, minerals and meat may be exported, and airstrips which make even the most remote areas accessible. Construction and transportation in turn offer the possibility of huge profits.

The tropical rainforests are being destroyed at an alarming rate (Fig. 29.2). Half the world's tropical forests have now been removed, mainly in the last 40 years. Every 60 seconds 40 hectares is said to be destroyed and it has been forecast that at the present rate of removal all the rainforests will have disappeared within 30 years. Accelerated cutting rates were reported in 1988 in the Philippines, Sabah, Sarawak, Thailand, Madagascar and the Ivory Coast. So the problem is not confined to the Amazon.

The exploitation of the rainforests has become the subject of much criticism by countries of the developed world which have become more conservation conscious recently. However the advanced countries have already destroyed most of their natural environments and they play the major rôle in creating pollution through their industrial activities. They also use up the world's natural resources most rapidly. Despite this, the developed countries have the finance and the expertise in science and technology to work with the developing lands to conserve and manage the remaining forests. Some of this work is now being fostered by international

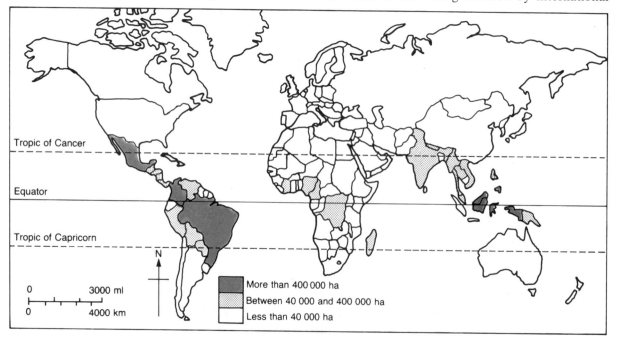

Fig. 29.2 Annual destruction of the rainforests

agencies and by the implementation of agreements to international conventions designed to set good standards of utilization and conservation. Working in parallel with official organizations are voluntary associations such as the World Wide Fund for Nature (WWF) and Friends of the Earth (FOE) which share the concern for the maintenance of our natural environmental heritage.

29.2.4 The destruction of the Amazon forest

The rate at which this forest is being removed is indicated by the fact that in three months in 1989 a rainforest the size of England, Scotland and Wales combined was cleared. Much of the clearing is by burning which scientists have calculated contributes 7 per cent of the entire world emission of carbon dioxide which contributes to the greenhouse effect (see Unit 30).

There are currently four main aspects of the economic exploitation and industrialization of the Amazon:

1 The clearance of forest to make space for commercial cattle ranching This has until recently been supported by the government of Brazil which subsidized clearance and did not tax farm incomes. This activity has attracted farmers from other regions who see better prospects, e.g. the farmers of the south who produced soya beans as a cash crop but have been replaced by mechanized techniques. It also attracted land speculators who saw huge profits in the heavily subsidized process of clearing and using the land.

2 The use of timber for commercial purposes Logging camps, new saw mills, forest tracks along which felled timber is dragged to mills and collection points and the creation of.new roads along which the timber is exported have all contributed to the destruction of the forest. The timber has a ready market in the furniture and building industries of the developed countries.

3 The development of hydro-electric power resources The prime mover in this has been Brazil's northern region electricity company, Electronorte. The Amazon region is seen as Brazil's chief future source of cheap power which will contribute significantly to the development and prosperity of the country. Virgin forest has been flooded and new dams built. Tucurui, built in 1984, is now Brazil's largest man-made lake (the size of Dorset) and has the world's third largest dam.

4 Mineral exploitation Fig. 29.3 shows the distribution of important mineral resources within the Amazon region. Mining has involved forest clearance and the displacement of indigenous peoples. Industrial processes have also damaged the environment. In gold mining, for example, mercury is used in the extraction process and mercury poisoning has affected rivers and streams.

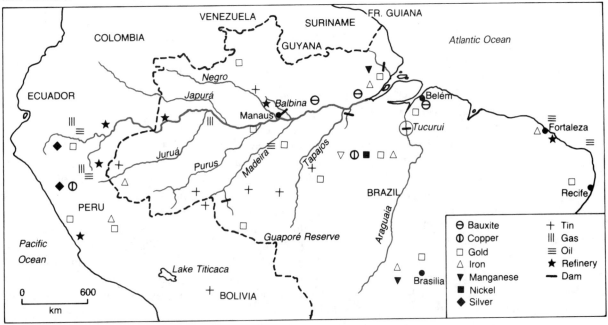

Fig. 29.3 Distribution of mineral resources in the Amazon region

29.2.5 The effects of government policy

Until recently the central government strongly favoured the rapid industrialization of the Amazon. This was seen as the way to tackle Brazil's international debt problems and to ease social and economic pressures internally by offering new opportunities to the unemployed and poor. Consequently policies were adopted which favoured developers and speculators. These included: the subsidising of ranchers who cleared forest; exempting farm income from tax; approving major schemes such as the B364 road designed to take timber out through Peru, and the hydro-electric projects; paying little attention to the interests of the indigenous peoples who lived in small numbers throughout the forest; and paying little attention to the need for conservation programmes.

Recent changes in government has led to a re-evaluation of the Amazon development. Subsidies have been withdrawn from ranchers and tax concessions ended. Speculators see investment in the Amazon as less profitable while strict anti-inflation measures have made it more expensive for investors and speculators to borrow money. The government is also attempting to improve the effectiveness of the surveillance services which are supposed to control forest burning and to ensure that native people are protected from miners and ranchers when there is a conflict of interests.

The effectiveness of the new policies is limited by: the vast amount of money needed to implement them; the lack of coordination between government agencies (when the central government withdrew subsidies INCRA, the colonization agency continued to pay them); and the size and remoteness of the region which makes it difficult to supervise and control.

29.2.6 Possible solutions

Solutions which have been put forward by the Brazilian government, international agencies and other interested groups include:

(a) The zoning of the Amazon into economic activity regions so that mining, ranching, etc, will only be permitted in parts.

(b) Establishing a National Park or a number of Parks which would conserve the unspoiled forest that remains (this has been done in Cameroon).

(c) Setting aside reservations for the native peoples.

(d) Establishing more carefully planned and managed economic activities, e.g., *selective logging policies* which would mean that only saleable trees were felled and were replaced by planting to give the industry permanence; and *new farming techniques* designed to replace ranching with crop production so that the need for new farmland is diminished.

(e) The establishment of tourism which would bring new income to the region and would also sensitise visitors to the need to conserve what remains of the forest (this is being developed in the forests of Costa Rica).

29.3 GENERAL CONCEPTS

Ecosystems See Unit 13.

Less-developed countries, development See Unit 23.

29.4 DIFFERENT PERSPECTIVES

(a) See this section in Unit 23: Less-developed countries.

(b) **The free market perspective**. This sees the Amazon as one of the few remaining regions of vast and untapped economic wealth which could be created by the application of modern industrial and financial processes to the region. It argues that Brazil should use the development of the Amazon to create work, raise standards of living and meet its international debts even if this means the disappearance of existing habitats and displacement of people.

(c) **The conservation perspective** Conservationists see the tropical rainforests and the Amazon in particular as important areas of the natural environment that have to be protected for the future. They emphasize the need for effective management and control of rapid economic exploitation. They favour development plans which minimize the effects upon the ecosystem and people and work towards international cooperation on conservation programmes.

(d) **The sociological perspective** Many sociologists are concerned with the protection of the traditional ways of life of indigenous groups in the Amazon and ensuring their welfare as miners, ranchers, etc, move into their lands.

(e) **The nationalist perspective** Prominent Brazilians resent the interference of other nations in issues relating to the development of the Amazon. They see the Amazon as their own business and reject well-meaning proposals by developed countries to inject finance needed to undertake some of the conservation programmes, e.g. Norway suggested a *debt swap* which meant that Norway would forget what it was owed if Brazil spent the equivalent amount on conservation. It is claimed that conservation pressure on Brazil is the result of the fact that developed nations covet the Amazon and wish to interfere in Brazil's affairs.

29.5 RELATED TOPICS

The closest links are with Unit 23, Less-developed countries and Unit 28, Brazil: regional strategies.

29.6 QUESTION ANALYSIS

1 Explain and illustrate why the question of exploitation of the world's tropical forests has become a matter of global concern. (*Northern Ireland Schools Examinations Council, June 1989*)

Understanding the question This is an essay style question and you will be judged by the quality of the argument and the extent of knowledge you show in providing illustrative material as evidence to back up the points you make. The danger is to over-write in a descriptive way–the key word in the question is to *explain* and you must adopt an analytical approach. Since the question is not about a specific forest area it is advisable to illustrate your answer from more than one region of the world.

Answer plan You need to plan your answer so that a systematic argument is provided with some indication of which you consider to be the most significant reasons for concern. These should include:

(a) The possible contribution to global warming that results from the destruction and burning of large forest tracts. This is of global concern because it could bring about major weather changes and a significant rise in sea level and so affect the whole world.

(b) The effects on indigenous people whose economic systems were sustainable with very little damaging effect within the forest ecosystem.

(c) Thoughtless waste produced by rapid clearance which could be to the detriment of everyone–the loss of potentially valuable plants for food production and medical research. Friends of the Earth, for example, claim that over 2000 tropical forest plants are being investigated for their anti-cancer properties.

(d) The rate at which the forests are being destroyed and the irretrievable situation this presents–over 50 per cent have disappeared in the last 40 years.

(e) Exploitation often takes the form of activities that are not sustainable over a long term and so bring little permanent benefit to the developing lands, e.g. only 6.2 per cent of current timber production in the tropical forests is sustainable, i.e. trees being planted to ensure future supplies. The major sources of hardwood could disappear for ever.

(f) The destruction of the forest in turn leads to the exhaustion of the soils after initial fertility has been lost which means that vast areas will contribute little to meeting the food needs of the developing world.

(g) The disappearance of rare species–the tiger in south-east Asia, the gorilla in the Congo as well as many species of plants and flowers.

(h) Some developing countries like Brazil suggest that developed countries are concerned because they do not wish to face increased competition in world markets for meat and crops which may be grown on the cleared lands.

(i) Debt-ridden weak developing countries may be pressured by powerful international companies and banks to agree to forest exploitation which is not in their national interest and which may eventually lead to an increase in social and economic tension and even political instability.

2 The map below (Fig. 29.4) shows those countries who have signed one or more of the following four international conventions:

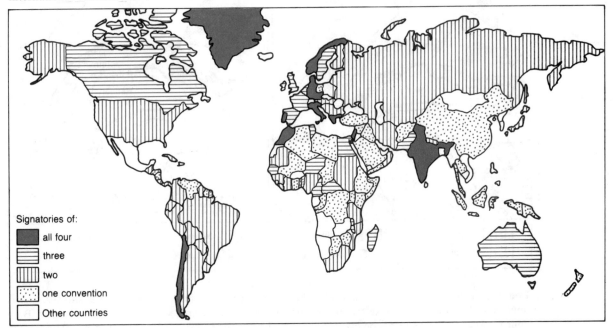

Signatories of:
- all four
- three
- two
- one convention
- Other countries

Fig. 29.4 (from Michael Kidron and Ronald Segal, *The New State of the World Atlas*, Pan, 1984)

(i) The World Heritage Convention.
(ii) Convention on International Trade and Endangered Species of World Fauna and Flora.
(iii) The Bonn Convention on the Conservation of Migratory Species of Wild Animals.
(iv) The Convention on Wetlands of International Importance.
(a) Suggest one possible reason for setting up each of the Conventions listed (i)–(iv).
(b) What are the reasons which might cause the states in north–west Europe, Greenland and Canada to be signatories to a majority of these conventions?
(c) Outline three reasons why such conventions are not signed by some governments.

(Associated Examining Board, June 1987)

Understanding the question This is a straightforward question which does however draw upon your general geographical background and awareness of the operation of international agencies and national governments in the field of conservation. Although you are not required specifically to study the map you should do so. Quickly note how few countries have signed all, the countries that have signed none and mentally divide the others into more or less enthusiastic.

Answer plan Work systematically through the precise questions asked.

(a) (i) To ensure the protection of the world's most important treasures of architecture, archaeology, art etc.
(ii) To discourage trade in these species so that they do not become extinct.
(iii) To foster cooperation amongst nations through whose territories wild animals migrate.
(iv) To preserve the delicate ecological balance of these regions now threatened by economic development.

These are not the only answers, you may think of better ones but they need be no longer than this. It is likely that this section is allocated few marks so do not spend too long on it.

(b) This has three distinct regions named but you do not have to give different answers for each. Your answer could include:
1 They have a common attitude to conservation problems.
2 They have particular interests in most of the organizations.
3 They recognize that species, habitats, and sites which are the subject of particular conventions are threatened within their own territories.

(c) This is where you might show you have studied the map by referring to countries indicated as non-signatories. Suggestions could include:
1 Some non-signatories are the least developed nations–they have more pressing social and economic problems than questions of conservation.
2 Some of the criticized activities may be a significant source of income, e.g. the illegal trade in ivory and hide from the endangered Asian elephant in which parts of Burma are engaged.
3 A belief that such conventions are useless and that no worthwhile action may result.

3 Study the sketch maps of an estuary in 1930 and 1985.
(a) What have been the changes in land use in the area between 1930 and 1985?
(b) With the aid of a sketch map or diagram, explain the siting of the oil refinery.
(c) (i) What conflicts occur between refinery operation and:
    1 Marine activity.
    2 Environmental conservation?
(ii) In what ways might the refinery aid conservation in the area?

(London, June 1989)

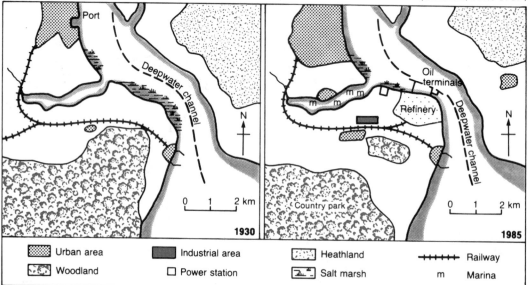

Symbol	Legend	Symbol	Legend	Symbol	Legend	Symbol	Legend
	Urban area		Industrial area		Heathland	++++++	Railway
	Woodland	☐	Power station		Salt marsh	m	Marina

Fig. 29.5

Understanding the question This is a straightforward question which sets you four main tasks. It is important that you do not make too many assumptions about the area shown on the sketch maps. Base your reasoning and arguments on evidence provided by the maps. This evidence has to be related to the theoretical knowledge you have gained about the siting of modern refineries and to your awareness of conservation issues which may arise from industrial development of a coastal area.

Answer plan (a) Do not pick out specific details in a random fashion but put the evidence you find into a systematic geographical framework. This could include:

1 Reduction in the areas of both heathland and woodland with the major woodland area designated a country park by 1985.

2 Expansion of the major urban area but the decline of its function as a port.

3 Industrialization of estuarine areas away from the major port–the oil refinery and deepwater terminals, the power station, a new industrial area.

4 The appearance of leisure facilities and industries–the country park and the marina.

5 The establishment of residential areas near major developments–the industrial estate and the marina.

6 Significant reduction of the salt marshes, some of which have been used for the refinery site.

7 Decrease in the land devoted to railway lines.

(b) The sketch map is essential. Annotate your sketch to emphasize key factors of location. In your written section do not repeat what the sketch shows but elaborate on what you have provided visually.

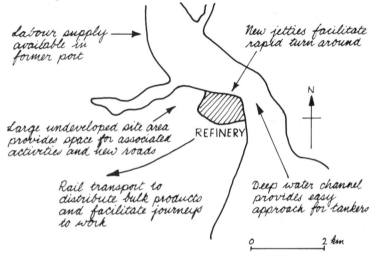

The siting of the oil refinery

Modern refineries require:

1 Large areas of coastal lowland or estuarine flats which can be easily developed for the refinery itself and associated industrial plant, e.g. the power station.

2 A deep water routeway for large modern tankers and a site for terminals which speed up turn-around time so reducing costs.

3 A transportation network to enable distribution of refined and finished products and to facilitate the journey to work from surrounding urban areas.

Check that your map illustrates these factors in relation to this specific site.

c (i) 1 You need to think about what is meant by 'marine activity' in this subsection–it means more than the movement of tankers. You should consider the possible conflict between leisure sailing from the nearby marina, the possible presence of a traditional fishing industry; remaining port activities, if any.

2 The building of the oil refinery took up a large salt marsh area which effectively destroyed an estuarine habitat; the industrial development may also have seriously affected the scenic quality of the coastal area; extension of urban and industrial areas may also have created pressures on existing open spaces both for further residential development and also in terms of more intense use as the local population has grown. Air pollution, waste emissions into the estuary and the use of water for cooling and other industrial purposes may also cause conflict.

c (ii) 1 The oil company will be very sensitive to local opinion and possible attacks on environmental grounds and may have agreed a carefully designed package that enhances conservation goals in the area.

2 A large company may sponsor conservation projects.

3 The disappearance of some of the open space may have made local people more aware of the need to value and protect the remainder.

4 Migrants who moved to the area with the new industries may bring new energy to the work of local conservation groups.

5 Increased revenue from the industrial development may provide local authorities with income to undertake conservation programmes.

29.7 FURTHER READING

Goudie, A., *The Human Impact: Man's Role in Environmental Change* (Blackwell, 1981)

Hart, C., *Worldwide Issues in Geography* (Collins, 1985)

Knowles, R., Johnson, C. and Colchester, M., *Rainforests. Land Use Options for Amazonia* (Oxford, 1988)

Money, D. C., *Tropical Rainforests–Environmental Systems Series* (Evans, 1980)

Tivy, J. and O'Hare, G., *Human Impact on the Ecosystem* (Oliver and Boyd, 1981)

30 Pollution

30.1 ASSUMED PREVIOUS KNOWLEDGE

Many GCSE Geography syllabuses include sections on environmental issues, particularly the causes and effects of air and water pollution (see Lines C. and Bolwell L., *Revise Geography*, Letts 1989). Take every opportunity you get to read topical articles about pollution of the environment in local, national daily and Sunday newspapers. Remember that some environmental pressure groups may not give an objective assessment of the situation, ignoring other points of view and some government statements may also give an unbalanced assessment. Take particular care when presented with statistics and check that their source is reliable and unbiased. Build up a file of newspaper and magazine cuttings relating to pollution and its consequences. Date the cuttings and classify them according to the type of pollution, for example air-borne sulphur and nitrogen oxides.

30.2 ESSENTIAL KNOWLEDGE

30.2.1 Definitions

Acid rain A somewhat misleading term, 'acid deposition' is more accurate as this form of pollution may fall as dust as well as precipitation. Normal precipitation has a pH of 5.6 whereas acid rain can have an acidity as high as pH 2.4. The acidity is caused by sulphur dioxide (SO_2) and nitrogen dioxide (NO_2). The process by which these gases form part of precipitation is discussed in detail later in this chapter.

Eutrophication The nutrient enrichment of a body of water which frequently results in a range of other changes. Among these are the increased production of algae, the deterioration of water quality and the reduction of fish numbers. The enriching nutrients are usually phosphates and nitrates which reduce the oxygen content of water and lead to the death of aquatic plants and other living organisms. This decaying matter falls to the bottom, increasing the silt layers and slowly filling up the lake. Eutrophication ages a body of water to the point where it cannot support life.

Global warming The increase in the global temperature which results from the build-up of 'greenhouse' gases, for example methane and carbon dioxide in the atmosphere. Like the panes of a greenhouse, these gases let in solar heat and then trap it when it is reflected back from the earth's surface. This process has become popularly known as 'the greenhouse effect'. Carbon emissions from fossil fuels–coal, oil and natural gas–have increased the amount of carbon dioxide in the atmosphere, trapping more heat and causing global warming. Plants absorb carbon dioxide but as the rainforests are destroyed by burning, stored carbon dioxide is returned to the air and there is less vegetation to absorb CO_2.

Pollution by carbon dioxide from fossil fuels has increased global warming with a measured temperature increase of half a degree centigrade in the past 100 years. The consequences of global warming include melting ice-caps, flooding of low-lying land areas and the growth of deserts.

Ozone A form of oxygen (O_3). At low levels near the earth's surface ozone is produced by the action of very strong sunlight on air particles in the presence of nitrogen oxides and volatile organic compounds, including emissions from car exhausts and power stations. Ozone is a pollutant and a key component of photo-chemical smog. It can affect health with symptoms which include running noses, coughs and asthma. Research in the United States indicates that ozone may interfere with the body's immune system. Although ozone is most likely to occur in industrial areas, particularly in large cities, readings at monitoring stations during the July 1990 hot spell in England show that some rural areas were more heavily polluted than central London as a result of winds blowing the ozone pollution away from the cities.

The ozone layer is a zone within the atmosphere between 20 and 40 kilometres above the earth's surface where ozone is at its greatest concentration. The ozone layer prevents most of the potentially damaging ultraviolet radiation from the sun from reaching the earth's surface, so protecting life forms and helping to maintain the earth's heat balance. (The difference between the amount of the sun's heat trapped in the atmosphere and that which is radiated back into space.)

Aerosols and refrigeration plants emit chlorofluorocarbons (CFCs) which, with halons and other industrial gases, set off chemical reactions in the atmosphere, destroying the ozone layer faster than it can be replaced. In 1985 British scientists discovered that there was a 'hole' in the ozone layer over Antarctica, supposedly due to pollutants such as CFCs. Any increase in ultraviolet radiation is known to cause skin cancer in humans, hence the urgency at the present time to reduce the amount of CFC and other pollutant gases into the atmosphere.

Pollutants Key pollutants include carbon dioxide, sulphur dioxide, oxides of nitrogen and chlorofluorocarbons. Apart from these gases toxic metals such as lead, copper and mercury are also responsible for pollution as are radioactive materials, oil, nutrients, hydrocarbons, heat and noise.

Pollution The release of substances and energy as waste products of human activities which result in changes, usually harmful, within the natural environment. Pollution is caused by people and can harm living organisms as well as reducing the amenity value of the environment.

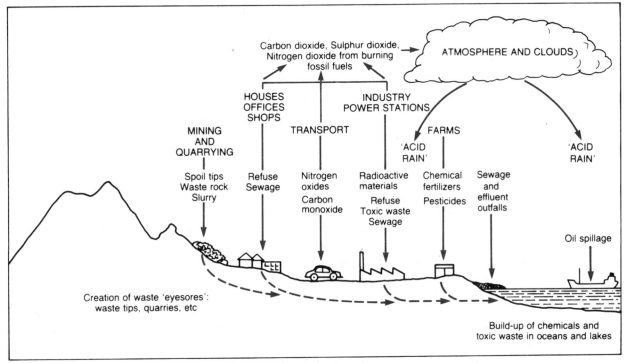

Fig. 30.1 Main sources and forms of pollution

30.2.2 Air pollution

Gas emissions Pollution of the air is the result of the emission of gases such as carbon dioxide, sulphur dioxide and nitrogen dioxide. These gases are found in high concentrations, particularly in cities, as the result of the burning of fossil fuels and their derivatives. In addition there are particulates in the air. These take the form of dusts such as fine particles of clay or limestone. Also present are effluents from industry in the form of volatile compounds such as fluorocarbons. Metals such as lead also pollute the atmosphere.

The table below shows the types of air pollutants and their source as measured in the United States in 1986. Air pollution in the US is estimated to cause up to 50 000 deaths a year and cost £24 billion in health care and lost working days.

| Pollutant | SOURCE *per cent by pollutant* | | | |
	Transportation	**Fuel combustion**	**Industrial**	**Miscellaneous**
Carbon monoxide	70	11.8	7.4	10.8
Sulphur oxides	4.2	81.1	14.6	0.1
Nitrogen oxides	44.0	51.8	3.1	1.1
Volatile organic compounds	33.3	11.8	40.8	14.1
Particulates	20.6	26.5	36.8	16.1
Lead	40.7	5.8	22.1	31.4

Table 30.1 Air pollutant emissions by pollutant and source, 1986
(*Source: Statistical Abstract of the United States, 1989*)

About 75 per cent of the lead in the air is derived from lead in petrol while 85 per cent of the carbon monoxide is also emitted by road vehicles.

The rôle of industry and power generation as major air polluting agencies is evident from the table below which shows that in the European Community the industrialized countries have the highest amount of air pollution.

Air pollution 1987	'000 tonnes
Germany (Fed Rep)	2 969
UK	2 439
France	1 652
Italy	1 570
Spain	937*
Netherlands	560
Portugal	303
Belgium	371
Denmark	266
Greece	217*
Irish Republic	68
Luxembourg	22

*1980

Table 30.2 Air pollution in EC countries (*Source: OECD*)

In the United Kingdom power stations are responsible for 71 per cent of the sulphur dioxide, 32 per cent of the nitrogen dioxide and 33 per cent of the carbon dioxide emissions. By comparison road transport is responsible for one per cent of the SO_2, 45 per cent of the NO_2 and 18 per cent of the CO_2.

Acid rain Sulphur dioxide and oxides of nitrogen are released into the atmosphere in large quantities as a result of burning fossil fuels, particularly coal and oil. About 50 per cent of the gases fall in the immediate area of the discharge as dry fallout–microscopic particles which do not cause acid rain. The remaining gases combine with water in the atmosphere. Negatively charged sulphate ions cause water in the atmosphere to become enriched with positively charged hydrogen ions which then falls to the ground as acid precipitation. The pollutants drain into the earth and release poisonous metals such as aluminium, cadmium and mercury from their compounds in the soil.

Trees use stocks of nutrients such as calcium and magnesium as a defence mechanism against acidity but these nutrients become depleted in acid rain conditions and the trees are prey to attack from ozone pollution, fungus, insects and disease. Some species of trees are more sensitive than others to air pollution. In West Germany, 7.7 per cent of the country's

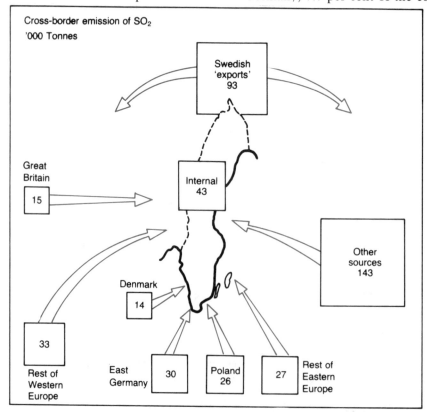

Fig. 30.2 Sulphur dioxide pollution in Sweden

trees have been affected by acid rain. Of these 75 per cent are fir, 41 per cent spruce, 26 per cent beech and 15 per cent oaks.

Acids and metals in the soil move in solution to the nearest lakes and streams where they concentrate. The combination of low pH values in the water and the increasing concentration of metals (notably aluminium) affects most aquatic species and insufficient food results in the death of fish and other living organisms. The process is complex, much depending on local soils, geology and the size and shape of the water body.

Buildings are also affected by acid rain. The Acropolis in Athens, the Taj Mahal in India and the Statue of Liberty in New York harbour are all being eaten away by industrial smog–sulphur dioxide in the air mixed with acid dust and water droplets.

Old fashioned and inefficient sites and power stations in Eastern Europe are responsible for high levels of air pollution in neighbouring countries. The belief in the early 1980s that Swedish acid rain was mainly caused by Britain has been proved wrong as Fig. 30.2 shows. Nevertheless, Britain is responsible for a considerable amount of cross-frontier pollution and has not joined the other industrial countries of Western Europe and Scandinavia in agreeing to at least a 30 per cent desulphurization programme for its power stations, although it has made its own counter-proposals.

30.2.3 Water pollution

Pollution of rivers and lakes by acid rain was described in the previous section because the underlying cause was atmospheric pollution by gas emissions. Rivers and lakes are also particularly vulnerable to two other types of pollution: (a) plant nutrients (b) toxic waste.

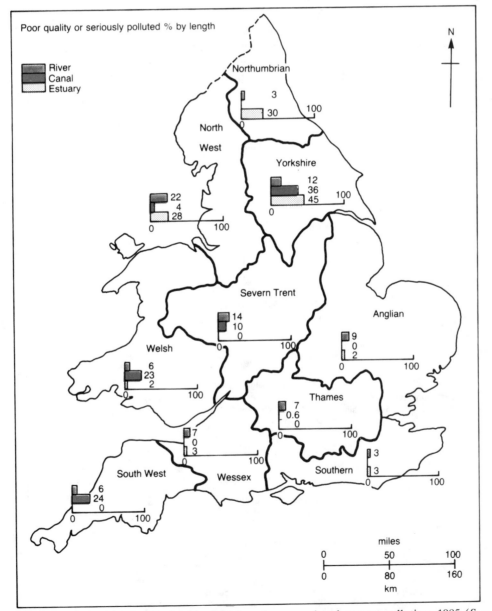

Fig. 30.3 Water authorities for England and Wales: river, canal and estuary pollution, 1985 (*Source: River Quality in England and Wales, 1985, a report of the 1985 Survey, HMSO*)

Plant nutrients The enrichment of rivers and lakes by plant nutrients such as nitrates and phosphates is partly caused by run-off from agricultural land where these chemicals are used as fertilizers. It is also caused by effluent from sewage works. The annual human release to sewage is about 630 gm of phosphorus and 5 kg of nitrogen per person per year. Added to these are phosphates from other sources such as detergents. Sewage effluent is a rich fertilizer, stimulating the growth of algae and other photosynthetic organisms and leading to the overloading of plant nutrients in rivers and other water bodies which in turn leads to the over-production of algae. This forms a green scum on the water surface and decays to add organic matter on the lake bed. Some blue-green algae produce toxic substances which are poisonous to animals and people. In coastal waters the discharge of untreated sewage into the sea close inshore is a further source of pollution. Beaches are affected and Britain will have to spend large sums of money to improve many coastal areas to the standard set by the European Community.

Toxic wastes Effluent from industry and sewage can be deadly. Some metal treatment processes produce an effluent containing cyanide and although pollution laws in many countries are strict, accidents do happen killing fish and aquatic plants over a wide area. Some toxic materials can become concentrated in organisms and passed on to form further concentrations in the higher members of the food chain. DDT became concentrated in coho salmon introduced into the Great Lakes. Mercury has also become concentrated in fish in some lakes, often because compounds of mercury used on seeds as fungicides have been leached out of farmland into rivers and lakes. Lead is yet another toxic pollutant liberated by car exhausts as well as industrial processes.

Oil spillage is usually associated with accidents to tankers at sea but waste oil from industry and shipping is also a pollutant in many rivers and lakes. Fig. 30.3 shows the extent of pollution in the rivers, canals and estuaries controlled by the 10 water authorities in England and Wales. The highest percentages of pollution are in the Midlands and North where industry has inherited the legacy of the industrial revolution.

Fig. 30.4 The effects of pollutants on a Broadland ecosytem

Pollution of an ecosystem The Norfolk Broads have their own distinctive fresh water ecosystem developed in the lakes (Broads) which were made by excavations for peat in the Middle Ages. The area is a mosaic of reedswamp, grazing marsh and open waters linked by six rivers and surrounded by arable land. In this century the area has suffered from extensive pollution with only four of the 52 Broads still supporting their former wealth of plants. The basic cause of the deterioration of the Broads, as Fig. 30.4 shows, is nutrient enrichment of the rivers and lakes by sewage effluent and fertilizers. The situation has been exacerbated by the growth of tourism with 250 000 visitors each summer. Many of the tourists' interests underline the need to check pollution and large sums must be spent to restore the Broadland ecosystem. One example of what can be done is Cockshoot Broad which has been dredged to remove silt and cut off from the River Bure to keep out further pollution. Aquatic plants have flourished and the water is clear and clean. Elsewhere scientists are working with water fleas which eat the algae and help clean the Broads.

30.2.4 Other sources of pollution

Noise Noise is recognized as a major environmental pollutant and prolonged exposure to noise levels in excess of 80 decibels is considered a hazard to mental, physical and social well-being. The decibel scale which measures the intensity of a sound is logarithmic so that the noise of a heavy truck rated as 90 decibels is 10 times that of the noise inside a small car rated at 80 decibels. In Britain two-thirds of the complaints about domestic noise concern the loud playing of music and the barking of dogs. The persistent noise of motor vehicles on main roads and motorways is the cause of most noise pollution and the main target of legislation and pressure groups such as the Noise Abatement Society.

Radioactivity The accident at the Chernobyl nuclear power station on 26 April 1986 has raised serious doubts about the safety of nuclear reactors. Radioactive fallout from Chernobyl affected pastureland in Britain and many other parts of Europe. This resulted in a concentration of radioactivity in milk, lambs and reindeer which made them unfit for human consumption. People living near the disaster and other areas, such as Sellafield in Cumbria where radioactive material has been leaked, may suffer from the long-term effects of irradiation.

Sweden has subsequently committed itself to a complete shutdown of its nuclear energy programme by 2010. The Netherlands and Belgium have stopped building nuclear power stations but, by contrast, France continues to expand its nuclear power programme (see 25.2.3). The dumping of nuclear waste is a form of pollution that is also causing concern and the attempt to identify four new sites in Britain in 1986 resulted in vigorous local protests.

Waste materials The creation of dumps containing unwanted man-made materials such as paper, discarded household goods and industrial rubbish pollutes the landscape both physically and visually. Much discarded rubbish is made from finite resources such as iron and oil. In some cases the cost of recycling these materials may be high but public awareness of the need to reduce the waste of non-renewable resources is encouraging recycling, particularly of paper and aluminium. The table below shows what is being done in the United States.

Gross waste generated *(millions of US tons)*		**Per cent recovered**
Paper and board	64.7	22.6
Ferrous metals	11.0	3.6
Aluminium	2.4	25.0
Glass	12.9	8.5
Plastics	10.3	1.0

Table 30.3 Waste in the USA, 1986
(*Source: Abstract of the United States, 1989*)

30.3 GENERAL CONCEPTS

Pollution is a form of environmental degradation which can result from natural phenomena such as a volcanic eruption, but is made, in the vast majority of cases, by people acting thoughtlessly, deliberately or in ignorance. Pollution usually has a harmful effect on living organisms and some forms of pollution such as radiation can cause illness and death.

Ecosystems are complex organizations with inter-dependent components and pollution of one part of an ecosystem can set up a chain reaction which disrupts the stability of the whole system.

Pollution is an international problem since some forms of pollution such as acid rain may be generated in one country but transferred in the atmosphere to other countries some distance away.

Pollution is also a global problem because the preservation of the ozone layer which is threatened by the build-up of harmful gases is world-wide and one of the long-term effects is to break down protection from ultra-violet rays.

Many pollution problems are political issues. Governments must decide between conflicting interests, for example, the farmers' need to increase output by using artificial fertilizers and the polluting effects these fertilizers may have on drainage basins and ecosystems. Governments must also decide on the allocation of public money to meet the high costs required to check or prevent pollution. For example, the cost of building sewage outfalls some distance out to sea to prevent beach pollution or the cleaning of emissions from power station chimneys to limit the escape of SO_2 and NO_2.

International diplomacy is also an important aspect of global pollution issues. There is some conflict between the rich nations and the Third World as to the phasing out of CFCs and other industrial gases. Developing countries want to expand industry and manufacture refrigerators on a large scale even though CFCs and other industrial gases are involved. Money from the more developed countries is available to help find suitable alternatives but how the money will be shared out, and persuading developing countries to co-operate, is a contentious issue. The rôle of the multinationals in increasing pollution in the Third World was highlighted in 1984 when a deadly gas escaped from a tank owned by the Union Carbide Corporation at Bhopal in India, killing over 2500 people.

30.4 Different perspectives

Growth of interest in the environment and lack of extensive research in the past has resulted in the publication of a variety of conflicting statements and reports. For example, until the late 1980s Britain was believed to be the main source of acid rain pollution in Sweden. Recent research by Swedish scientists (see Fig. 30.2) has shown that the bulk of Sweden's sulphur pollution, which causes acid rain, comes from Eastern Europe. Our knowledge of the holes in the ozone layer is still in its infancy but this should not result in complacency about global warming since recent research suggests that the problem is more acute than was first thought. When studying pollution and its causes try to obtain up-to-date material since our knowledge of the subject is increasing rapidly.

Human beings and other life forms have lived with pollutants throughout their evolution and have adapted to their presence at levels which occur naturally. For example, organisms are continuously subject to low levels of radiation from natural sources but the dosage is very small. Only in recent times as a result of urbanization, industrial development and scientific discoveries have the levels of pollution become very high and in some cases dangerous to life. Any further addition to pollution levels increases the potential of biological damage.

30.5 Related topics

Unit 13, Plant communities: ecosystems and Unit 29, Exploitation and conservation should be read in conjunction with this unit.

30.6 Question analysis

1 Examine the possible causes and consequences of river pollution.
       (*Welsh Joint Education Committee A/S Level Specimen Paper for the June 1991 examination*)
Understanding the question The first page of the question paper reminds students that they should make the fullest possible use of examples in support of their answers and that sketch-maps and diagrams should be included where relevant.

The question is really in two parts, the first part dealing with river pollution causes and the second with the consequences. Probably marks would be similar for each part so it is essential to spend as much time describing the consequences as is taken outlining the possible causes.

Answer plan First define pollution and then list the main pollutants of rivers. These include acid rain (see 30.2.2.); chemicals from farmland and industrial sites; sewage discharge; seepage from rubbish dumps and farmyard manure and rubbish dumped directly into the river.

Acid rain will fall on a river catchment area and will be carried to lakes or the sea, but the concentration in rivers is far less significant than that of nitrates and toxic chemicals from farms, industry and sewage. There is heavy pollution of the River Avon below Coventry from sewage discharges which amounts to 19.5 million litres per day of only partly treated sewage from the city. The consequences have included the death of large numbers of fish and a warning not to eat eels taken from the river. Few swans are to be seen at Stratford-upon-Avon because of river pollution and canoeists on the river have been warned not to practice the 'roll'. Toxic chemicals are poured into many rivers including the Rhine in West Germany and the Netherlands. Salt from France's fertilizer industry in Alsace corrodes water pipes in Holland and damages irrigation. 'Green' political candidates are winning seats in France and West

Germany, partly because of pollution of the river systems of those countries by toxic waste. Accidents have had disastrous results. For example, in 1987 British Sugar poured concentrated sugar syrup into the drains. The waste was organic and used up all the oxygen in the River Lark near Bury St. Edmunds in Suffolk and killed thousands of fish. The company was fined £2100.

The rapid growth of green algae in the River Bure in Norfolk is the result of nitrates and phosphates draining from the surrounding farmland. The algae lower the oxygen and light levels of the water resulting in the death of aquatic plants and fish. The algae can be toxic and can give drinking water an unpleasant taste. Filtering the water increases costs.

When you have written your answer read it carefully to check that you have an approximate balance between causes and effects and have included a number of examples with their causes and effects.

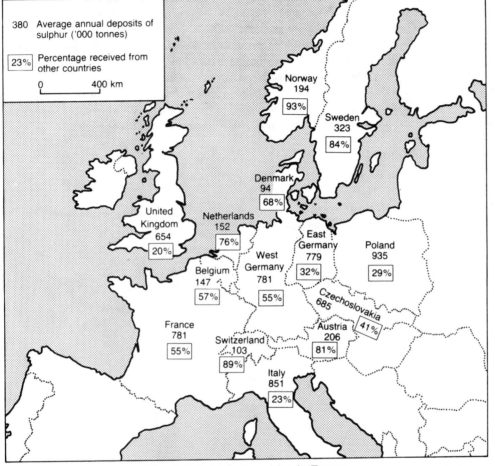

Fig. 30.5 Sulphur deposition in Europe

2 (a) Describe the main causes of acid rain and comment briefly on its effects.

(b) Study Fig. 30.5. Describe and account for the patterns of sulphur deposition shown on the map.

(c) Discuss the methods which can be adopted to reduce these problems and explain why they are harder to carry out in Europe than in North America.

(d) ' "Distant" environmental damage may also result from water-borne pollutants.' Discuss with reference to specific examples you have studied.

(Scottish Certificate of Education Higher Grade, May 1989)

Understanding the question Three of the four sections of this question require answers in two parts. For example, (a) asks for a description of the causes of acid rain and also for a brief comment on its effects. The fourth section of the question asks you to discuss, referring to actual examples. You should aim at providing at least three examples from different locations in your answer.

Answer plan (a) See 30.2.2 for a summary of the main causes of acid rain. Three main environments are particularly affected by acid rain: forests and woodland, stonework and building, and areas of water, particularly lakes where concentrations of the pollutant can build-up.

(b) A description of the sulphur deposition shown on the map should emphasize the countries which are heavily polluted by their own industries and those which receive most of their pollution from other countries. The UK, Italy, Poland and East Germany fall into the first category while Norway, Sweden, Denmark, Austria, Switzerland and the Netherlands are in the latter group. The pattern can be partly accounted for by industrialization, e.g. West Germany; the failure of some countries to check sulphur emissions, e.g. the UK; the prevalence of westerly and south-westerly winds which produce an easterly flow of pollution, e.g. from Britain to Norway; and also the use of out-of-date and inefficient power stations and factory power supplies, e.g. in East Germany (as it existed before unification with West Germany).

(c) Two methods should be described. (i) Those that attempt to cure existing problems such as the liming of lakes in Sweden. (ii) Those to prevent the emission of the gases at their source, particularly power stations burning coal and oil, by 'scrubbing' the gases emitted to remove sulphur dioxide. Changing from burning coal and lignite to using natural gas, the less polluting of the fossil fuels, can also reduce emissions.

In Europe emissions are more difficult to reduce than in North America because a larger number of countries are involved, with some, particularly those in Eastern Europe short of the necessary capital and with a legacy of old, out-of-date equipment. In North America the problems can be solved by co-operation between the USA and Canada and by using available capital to reduce emissions or switching to the large stocks of natural gas which are available.

(d) Water-borne pollutants can be carried long distances in rivers and by ocean currents. For example, pollution of the lower Rhine in the Netherlands may have come over 700 kilometres along the river from factories upstream. Rivers flowing into the North Sea carry high quantities of chemicals and toxic metals causing disease to seals and resulting in a ban on bathing on some of the beaches of Sweden's west coast. The oil slick which resulted from the holing of the tanker Exxon Valdez in Prince William Sound, Alaska, in March 1989 carried oil pollution for more than 2000 kilometres along the shoreline.

30.7 Further reading

Dix, H. M., *Environmental Pollution* (John Wiley, 1981)

Elsworth, S., *Acid Rain* (Pluto Press, 1984)

Gilpin, A., *Dictionary of Environmental Terms* (Routledge and Kegan Paul, 1976)

Mellanby, K., *The Biology of Pollution* (Edward Arnold, 1980)

Index